PRINT READING FOR THE MACHINE TRADES

Wilfred B. Pouler
Formerly
Technical Education Director
Society of Die Casting Engineers

DELMAR PUBLISHERS INC.®

NOTICE TO THE READER

Publisher and author do not warrant or guarantee any of the products described herein or perform any independent analysis in connection with any of the product information contained herein. Publisher and author do not assume, and expressly disclaim, any obligation to obtain and include information other than that provided to them by the manufacturer.

The reader is expressly warned to consider and adopt all safety precautions that might be indicated by the activities described herein and to avoid all potential hazards. By following the instructions contained herein, the reader willingly assumes all risks in connection with such instructions.

The publisher and author make no representations or warranties of any kind, including but not limited to, the warranties of fitness for particular purpose or merchantability, nor are any such representations implied with respect to the material set forth herein, and the publisher and author take no responsibility with respect to such material. The publisher and author shall not be liable for any special, consequential or exemplary damages resulting, in whole or in part, from the readers' use of, or reliance upon, this material.

For information, address Delmar Publishers Inc.
2 Computer Drive West, Box 15-015
Albany, New York 12212

Printed in the United States of America
Published simultaneously in Canada
By Nelson Canada
A Division of The Thomson Corporation

10 9 8 7 6 5 4 3 2

ISBN 0-538-33350-2
Library of Congress Catalog Card Number: 83-51093

CONTENTS

PREFACE

PRINT READING FOR THE MACHINE TRADES has been written for technical students, apprentices in the machine trades, and other personnel who must be skilled in reading industrial prints in their jobs.

The author has included actual industrial prints to familiarize the student with industry practices. New industry standards and current practices are represented. Prints are dimensioned in either fractional inches, decimal inches, or the metric measurement system.

The text develops a thorough explanation of the basic principles of pictorial communication including orthographic projection, pictorial sketching, arrangement of views, the alphabet of lines, and dimensioning. Industrial information is then presented in succeeding sections on tolerances and surface finishes, supplementary information, threads and fastening devices, and gears and splines. This material is drawn together in a comprehensive discussion of sectional views and title block information.

In accordance with industry practice, the book explains industrial materials, types of drawings found in industry, geometric dimensioning, and the impact of new drafting systems on machinists. All material presented follows current American Society of Mechanical Engineers standards.

Of special note is the inclusion of the American National Standards Institute *ASME Y14.5—1982* Standard "Dimensioning and Tolerancing" which primarily concerns subject material in Chapters 8, 15, and 16. Copies of this standard may be obtained from the American Society of Mechanical Engineers. Illustrations in this textbook conform to this standard; however, industrial prints include the drafting practices used at the time the prints were drawn.

Every effort has been made to assure the usefulness of this text as an aid to student learning. A functional second color has been included throughout and a spiral binding has been added to aid in convenient shop use. All chapters begin with learning objectives and conclude with *A* and *B* exercises to test and reinforce student comprehension. Additional *C* and *D* exercises are included in the appendix for optional study. All exercises are graded according to difficulty with *A* exercises being the least difficult and *D* exercises the most difficult.

The book is organized so that question pages may be answered and then torn out for correction without destroying the value of the text as a reference. All prints are book-size to avoid the cumbersome process of examining a fold-out print. The answers to summary questions and even-numbered chapter exercises are included in the appendix; answers to the odd-numbered chapter exercises and end-of-book problems are contained in a separate Instructor's Manual and Key.

CHAPTER 1

INTRODUCTION TO DRAWINGS AND PRINTS

OBJECTIVES

After studying this chapter, you will be able to:

- Discuss why drawings are so important to production planning.
- Describe the importance of industrial prints to manufacturing companies.
- Explain how different departments within a company use industrial prints.
- List the type of information displayed on industrial prints.

Humans have always produced utensils and other devices to make life easier. Advanced planning has been found to be an important part of producing these devices. Good planning can eliminate unnecessary production labor, thus reducing costs and saving time.

A drawing of a device is essential and basic to production planning. A drawing can describe a part more fully than numerous words. Even simple devices are hard to describe in words. More complicated devices may be impossible to describe in words.

Skilled craftspeople such as machinists and toolmakers must know the shape and size of each part, and how the various parts assemble, before they can produce the finished product. This information must be communicated from the part designer or engineer to the toolmaker or machinist. This can best be done by drawing a sufficient number of object views. The drawing must also contain dimensions and other necessary manufacturing information.

Original drawings must be stored or filed to protect them for record-keeping purposes. Therefore, duplicates, or prints, of the drawings must be made. These are called *industrial prints*, or simply *prints*.

The designer or engineer and drafter must be very thorough in drawing part plans. The part can then be expected to be produced properly, and should satisfy the design requirements. A poorly prepared drawing, however, can be almost worthless.

A completed drawing must allow the print reader to visualize exactly how the part will look. Dimensions and other manufacturing information on the drawing must also allow the craftspeople to understand how to produce the part.

The skills of accurately reading and interpreting industrial prints are developed by understanding drafting techniques. With the experience gained from practice, information can be quickly and easily taken from a print.

INDUSTRIAL PRINTS

The industrial prints of today do not resemble those of 25 or 50 years ago. At that time, the copy paper, when exposed to light, produced a blue background with white lines. Therefore, the prints of drawings were called *blueprints*. That term is still fre-

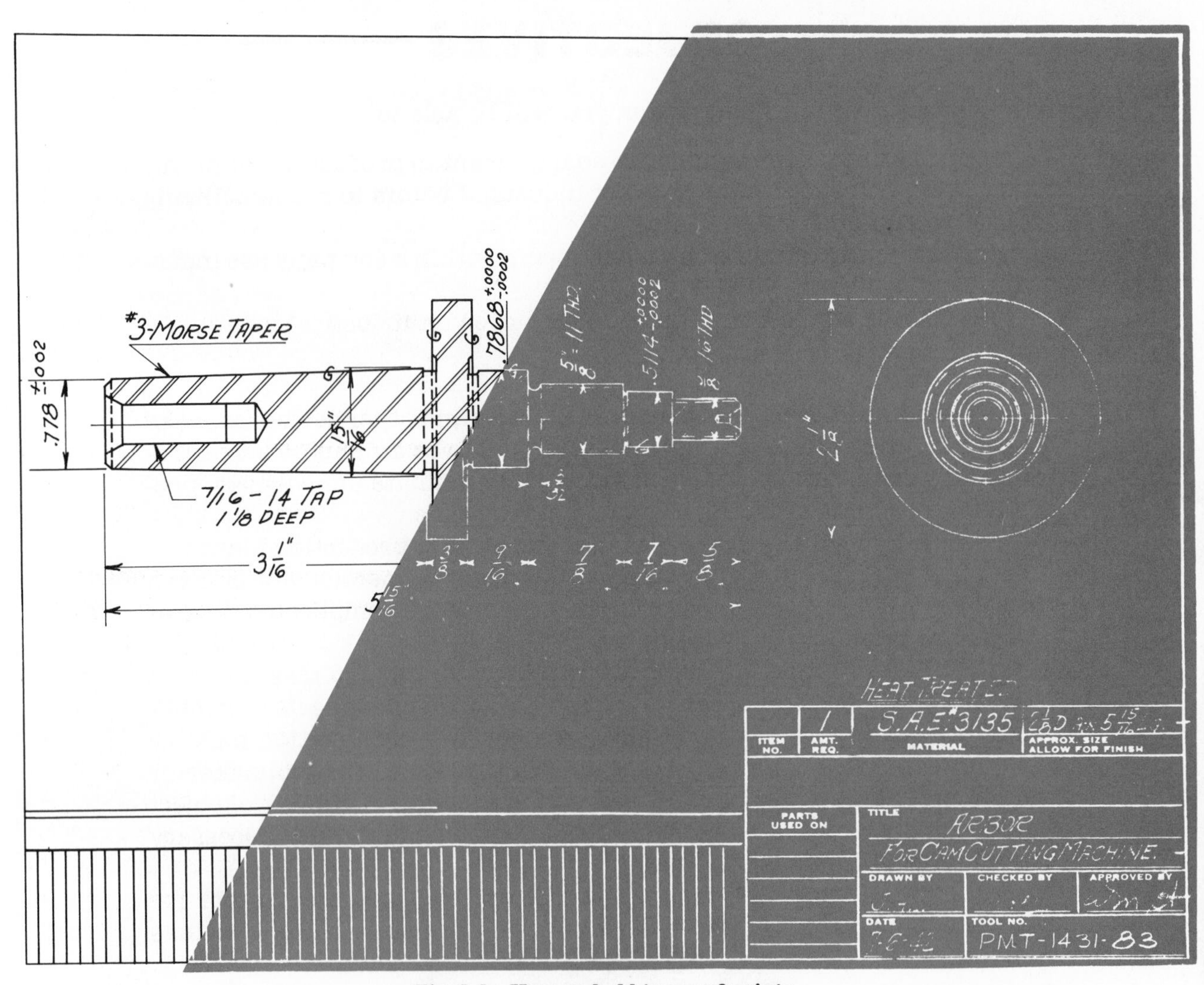

Fig. 1-1 New and old types of prints

quently used. However, the term *industrial print*, or *print*, is now preferred, since it is more accurate. The paper and chemical treatment used today produce a print with a white background and dark lines.

Engineering design changes are frequently required. These changes can be made easily on today's prints. Such changes were nearly impossible on the older blueprints with white lines on a blue background.

The older (blueprint) and newer types of prints are shown in Figure 1-1.

The technique of drawing the original on transparent paper or polyester film Mylar is still the same. However, changes have been made in the methods of making copies from an original drawing. One method is shown in Figure 1-2.

THE IMPORTANCE OF INDUSTRIAL PRINTS

The industrial print is the graphic language which conveys the information necessary to manufacture parts of an assembly. Each part is described in an industrial print. The print provides all the information necessary to produce the part.

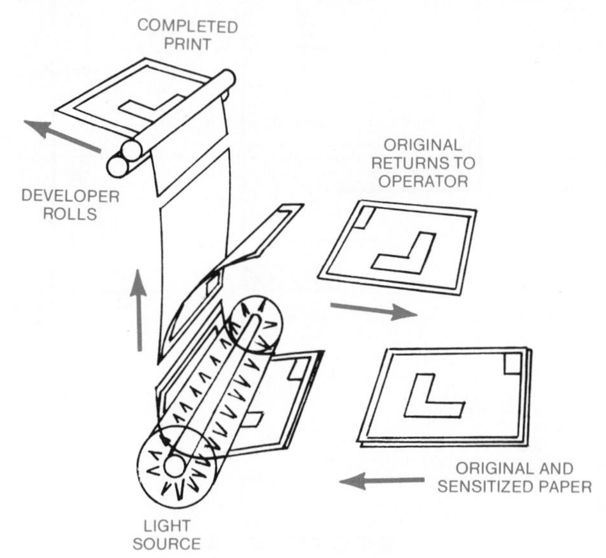

Fig. 1-2 Print copying process

Most industries depend upon other companies to produce or supply components for the finished product. In manufacturing automobiles and airplanes, for example, literally thousands of parts are used. The supplier companies may be far removed from the product company. Therefore, a good means of communicating manufacturing information must be available. The modern industrial print does this job so well that parts may be made many miles away from the assembly plant and still fit together with exact tolerance.

FROM IDEA TO PRODUCT

Engineering ideas develop into a finished product in different ways in different companies. One example of this process is described here.

1. The designer or engineer freehand-sketches a simple assembly such as that shown in Figure 1-3. Only the basic dimensions may be shown on this sketch.

2. Another designer or engineer may study this sketch to determine how it fits into the overall assembly, and if it can be produced economically.

3. After all design changes are made, the sketch is sent to the drafting department, where the finished drawing and its prints are made, Figure 1-4.

4. Prints are sent to the company's departments in which the part will be made.

Prints are used in the production process of some companies as follows. The prints will be studied by the *production engineering department* to determine if the part can be made by the company's production machines. The *purchasing department* will study the prints, and may send the prints to several outside supplier compa-

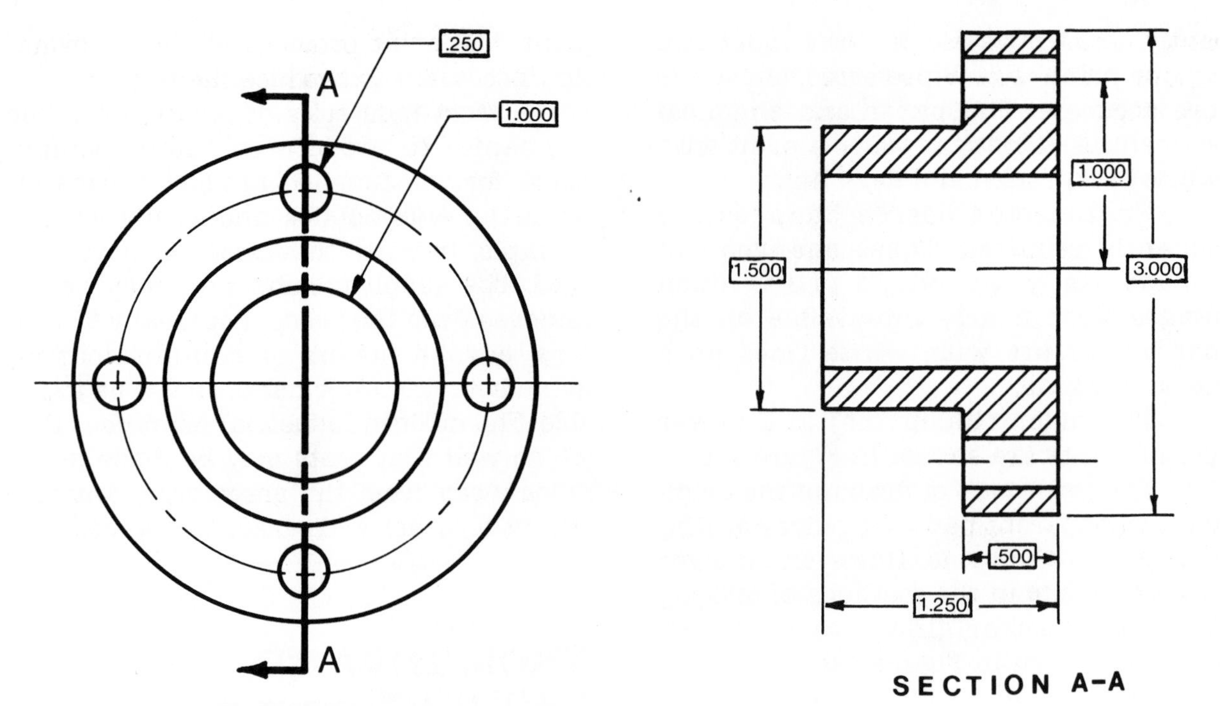

Fig. 1-3 Basic dimensions

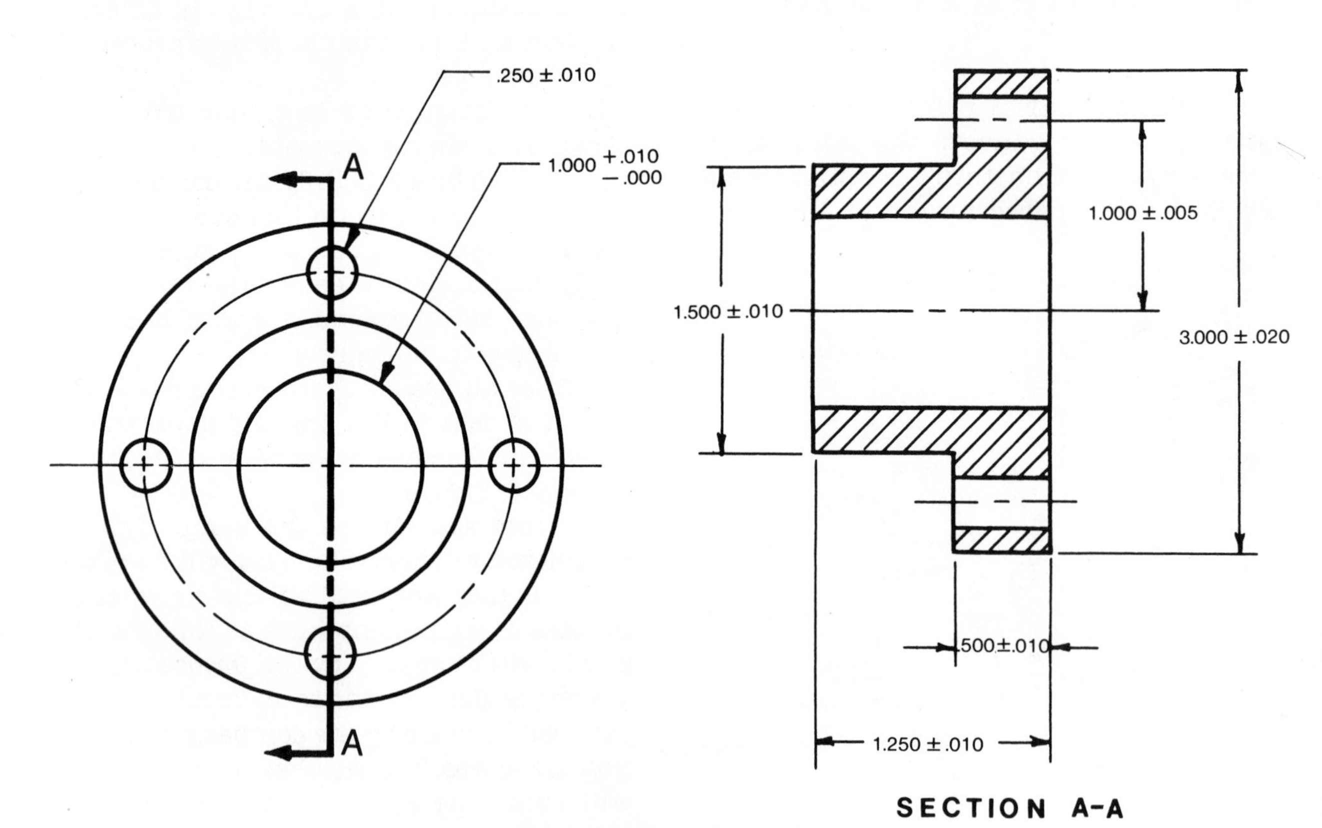

Fig. 1-4 Finish dimensions

nies for their cost estimates. After studying the outside cost figures, a decision may be made to produce the part within the plant.

The *production planning department* will study the prints to determine the machines involved in producing the part. In the *tool design department*, the methods engineers will design the tools to be used in production. Prints of the part are also necessary for the toolmaker to make the jigs and fixtures used to hold or position the part while it is being machined.

Prints are provided to the *diemaking* or *mould department*. If the part starts out as a casting, then dies must be produced well in advance of the production date. This will also require the creation of die design prints based on the original part prints.

The different *manufacturing process departments*, such as machining, plating, and heat treating, must be supplied prints in order to satisfy the print requirements for the part.

Prints are also supplied to the *inspection* or *quality-control department*. The mass production of products requires close inspection of each component in an assembly. This procedure ensures that each part conforms to the print requirements. For reasons of economics, the part may be inspected between every machining operation. This prevents further costly machining operations on defective parts. Inspections are performed with gages machined to the exact dimensions indicated on the prints. Inspectors perform a most important function. They verify that all dimensions, tolerances, and other design requirements have been satisfied.

Prints showing parts assembly must be provided to the *assembly department*. Prints may also be supplied to the *consumer*. In this way the buyer will know how the various parts fit together. This information is helpful in the repair and maintenance of the product.

THE PRINT IS A STORYBOOK

The print is a "storybook" which tells valuable information about an object.

- The print shows the shape of the part.
- It indicates the size of each feature of the part.
- The print indicates all additional information required to produce the part.
- The print indicates additional information which is not necessary to produce the part, but which is required to best utilize the part.

REVIEW

This review is provided to serve as reinforcement study material. Fill in the appropriate word(s) to complete the sentences below.

1. The older prints have ________________________ lines and (a)(an) ________________________ background.

2. The newer prints have ________________________ lines and (a)(an) ________________________ background.

3. Four departments of a company that may use industrial prints are:

a. ________________________ **c.** ________________________

b. ________________________ **d.** ________________________

4. The basic information provided on a print includes:

a. ________________________ **c.** ________________________

b. ________________________ **d.** ________________________

NAME ________________________

DATE ________________________

SCORE ________________________

CHAPTER 2

ORTHOGRAPHIC PROJECTION

OBJECTIVES

After studying this chapter, you will be able to:

- Explain the importance of obtaining a mental image of the object from the print.
- Recognize the viewing angles for the front, top, and side views.
- Discuss how the drafter selects views.
- Explain the relationships among surfaces, lines, and points.
- Outline the steps involved in visualizing an object.
- Identify the dimensions shown on the front, top, and side views.

When reading a print, it is important to visualize what the object will look like. A mental picture helps the worker to correctly produce the part.

It is helpful for the print reader to know the drafting methods used to draw the various views of an object. Different methods of viewing an object are discussed in this chapter to define the relationship of the views to the object.

VIEWING THE OBJECT

The *pictorial drawing,* Figure 2-1, shows the shape of an object from at least three different viewing angles. This type of drawing allows for easy visualization of the object.

The pictorial drawing may be used for simple objects. However, it is usually hard to draw pictorial drawings of more complex objects. Another disadvantage of the pictorial drawing is that features of the object, such as holes and recesses, must be drawn distorted because of the viewing angle. For example, holes that are actually round appear to

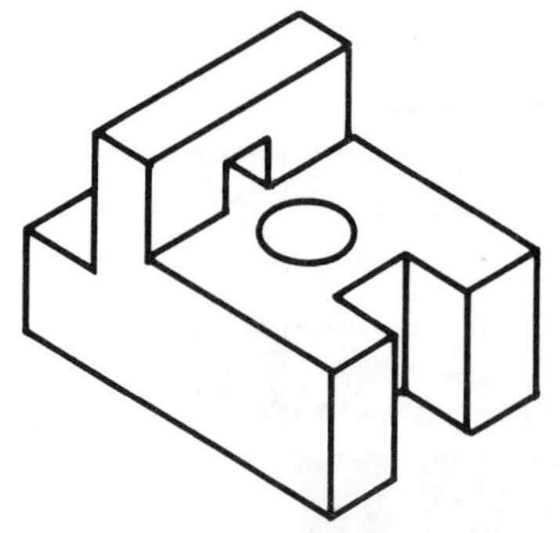

Fig. 2-1 Pictorial drawing

be oval in pictorial drawings. Surfaces that actually join at right angles appear to meet at a variety of angles.

A simple method of seeing the true shape of an object shown in a pictorial drawing is to draw one view at a time.

Front View

View the object in Figure 2-2A from angle *A*. Only the surface that is shaded can be seen from this angle. The outline of that surface is called the *front view.* The front view is shown in Figure 2-3.

Top View

Look straight down at the object in Figure 2-2B, from angle *B*. Only the

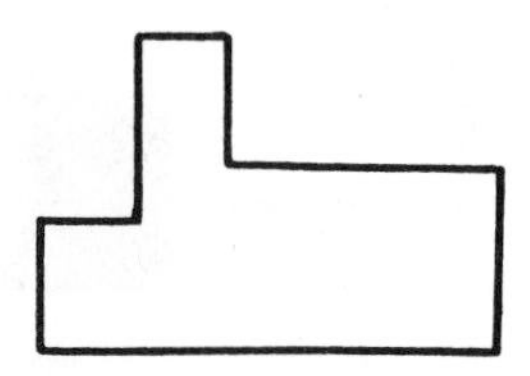

Fig. 2-3 Front view

shaded top surfaces can be seen from this angle. The drawn outline of these surfaces is called the *top view,* Figure 2-4.

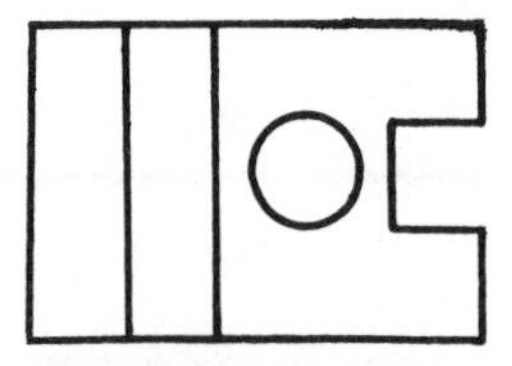

Fig. 2-4 Top view

Side View

View the object in Figure 2-2C from angle *C*. From this angle, the surfaces that are not shown in the top or side views can be seen. The drawn outline of these shaded surfaces is called the *side view,* Figure 2-5.

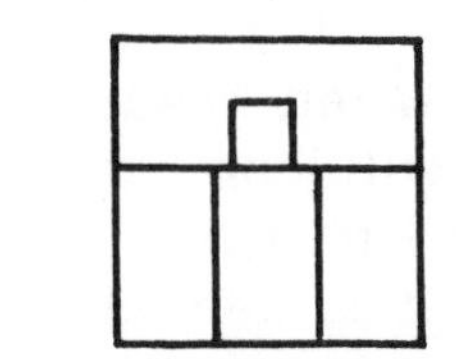

Fig. 2-5 Right side view

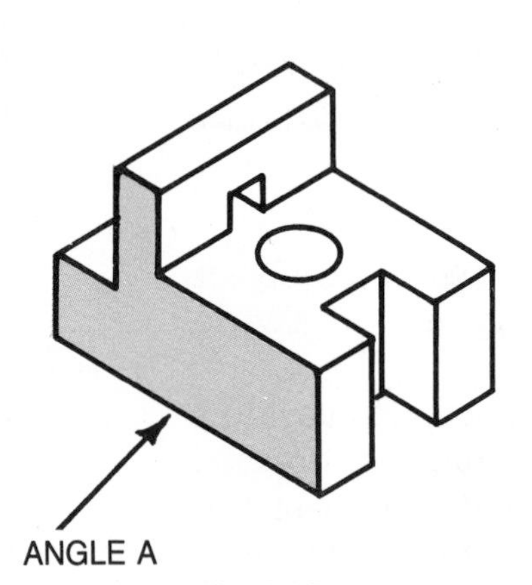

Fig. 2-2A

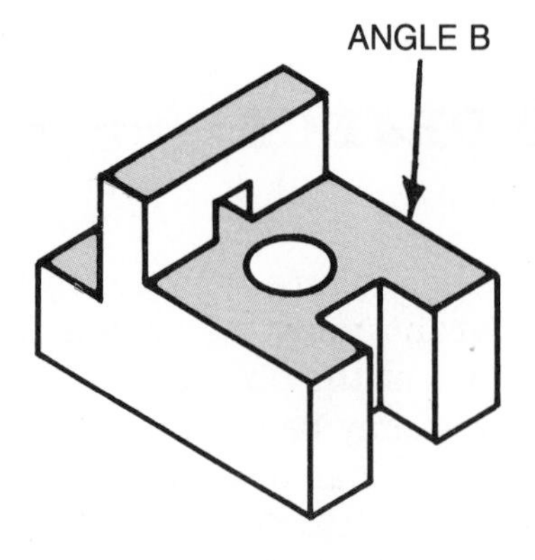

Fig. 2-2B

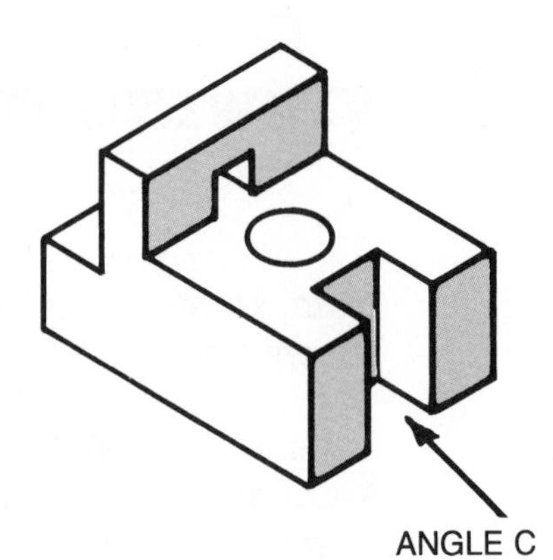

Fig. 2-2C

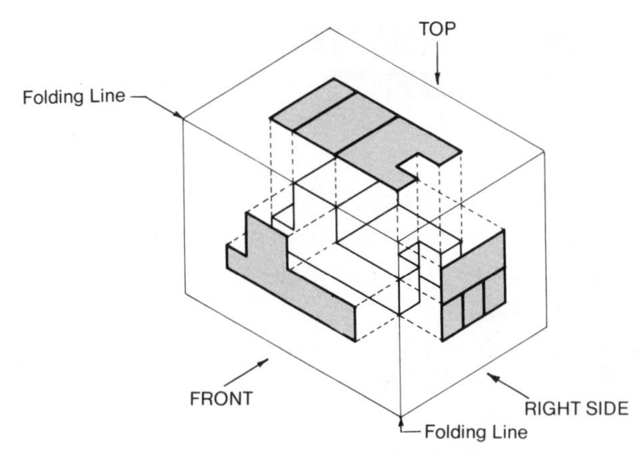

Fig. 2-6 Viewing directions

Orthographic Projection

Another method used to demonstrate the views of an object is shown in Figure 2-6. Imagine that the object is placed inside a transparent box. The outlines of the surfaces of the object, as seen from the direction of the arrows, are drawn on the sides of the box. The outlines show the same three views of the object (front, top, and side) shown in Figures 2-3 through 2-5.

In Figure 2-7, the three sides of the box are laid flat. The three views of the object are now in their proper positions relative to one another. This view arrangement is called an *orthographic projection.*

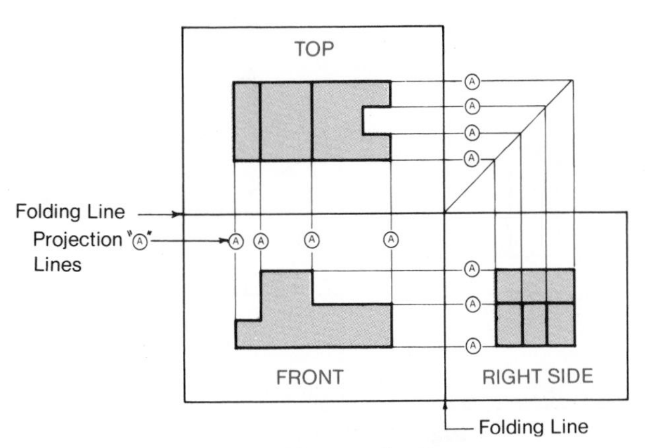

Fig. 2-7 View projections

SELECTION OF VIEWS

In the actual drafting method, the drafter usually decides which view best describes the object. The drafter calls this the *front view.* The object is usually viewed from its most natural position. The front view (the most descriptive view) is drawn first. Projection lines are then drawn to aid in the drawing of the top and side views.

An example of the results of this procedure is shown in Figure 2-7. The lengths of surfaces on the top view are projected up from the front view. The heights of surfaces on the side view are also projected from the front view. The widths of surfaces on the side view are transferred from the top view.

Surfaces, Lines, and Points

The drafter does not draw the actual surfaces of an object. Lines are drawn to indicate where one surface meets another surface. A surface that cannot be seen from a certain view is usually shown by a dashed (hidden) line. A line on one view becomes a point on another view.

The views of a simple object can be studied to learn about the relationships of surfaces, lines, and points. Such an object is shown in Figure 2-8. In this figure, the pictorial drawing is accompanied by the front, top, and side views.

SURFACES. The surfaces shown on the front, top, and side views are also shown on the pictorial view. In Figure 2-8:

Surface *A* (front view) is surface *A* (pictorial view).

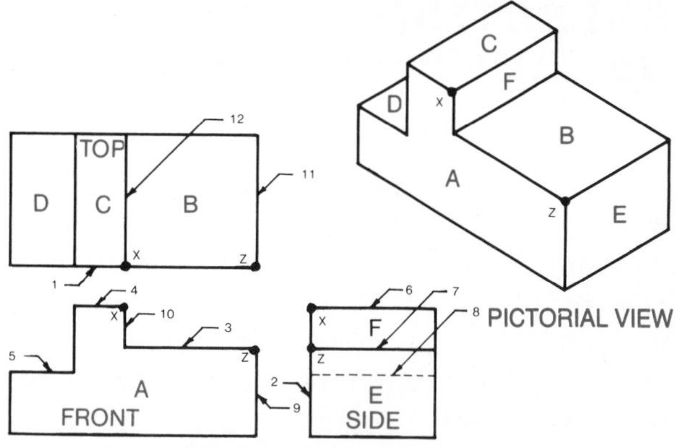

Fig. 2-8 Lines and points

- Surfaces *B*, *C*, *D* (top view) are surfaces *B*, *C*, *D* (pictorial view).
- Surfaces *E*, *F* (side view) are surfaces *E*, *F* (pictorial view).

LINES. A surface in one view becomes a line in another view. In Figure 2-8:

- Surface *A* (front view) becomes line *1* (top view).
- Surfaces *B*, *C*, *D* (top view) become lines *3*, *4*, *5* (front view) and lines *6*, *7*, *8* (side view). Line *8* is a dashed line (hidden line) which indicates surface *D*, which cannot be seen from the side view.
- Surfaces *E*, *F* (side view) are shown as lines *9*, *10* (front view) and lines *11*, *12* (top view).

POINTS. When lines meet, they form a point. In Figure 2-8:

- Lines *3*, *9* (front view) and line *7* (side view) form point *z* as shown in all three views and the pictorial view.
- Lines *4*, *10* (front view), lines *1*, *12* (top view), and line *6* (side view) form point *x*.

From these exercises it is apparent that surfaces from one view are represented by lines in adjacent views. Lines which represent the intersection of surfaces become points on other views.

VISUALIZING THE OBJECT

Some basic steps should be followed to visualize the object shown in an orthographic projection. The steps are outlined here.

1. Briefly study the entire print.

2. Choose the view that best shows the general shape of the object. This will be called the front view.

3. Look at the top and side views for lines that show the intersection of surfaces.

4. Study one shape at a time. Study the shape on all three views.

5. Sketch a pictorial view on paper to help visualize the object. Give careful attention to the location of each feature of the object.

It is very important to realize that no single view can completely describe an object. A detail in one view may be changed which does not alter the other two views. Study Figures 2-8 and 2-9. The top view is the same in both figures. The dashed line (hidden line) in the front view in Figure 2-9 should alert you to the angular surface *B*. Hidden lines are used to show the edges or features of an object that cannot be seen from that particular view.

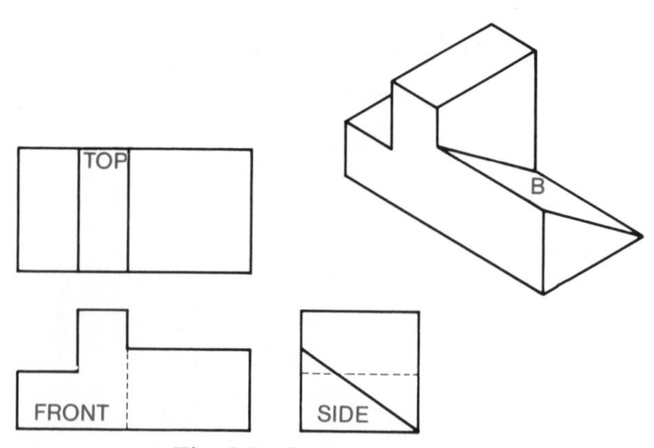

Fig. 2-9 Surface changes

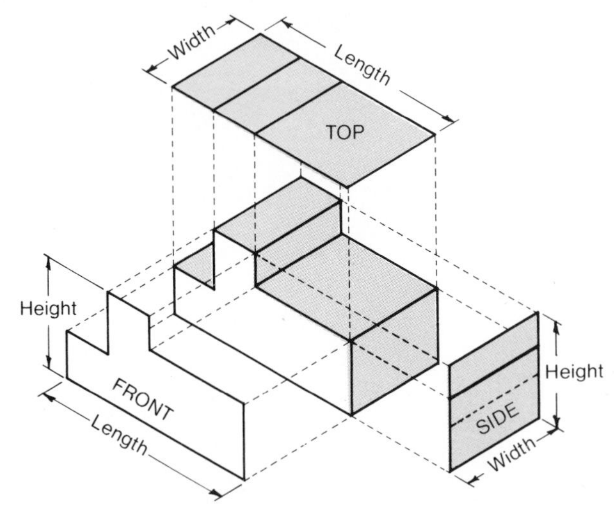

Fig. 2-10 Length, width, height

LENGTH, WIDTH, AND HEIGHT RELATIONSHIPS

Figure 2-10 indicates that the front view will not show the width of the object. The front view will only show the length and height of the object. Therefore, it is a waste of time to look for the width on the front view.

The top view will show the object's width and length. It will not show the object's height.

The length of the object cannot be found on the side view. The side view will show the width and height of the object.

On a three-view orthographic projection, features are shown at least twice. However, features are usually dimensioned only once. On newer types of prints (especially numerical control prints) the length, width, and height dimensions are referred to as the *X, Y, and Z coordinates.* See Figure 2-11.

X = Length
Y = Width
Z = Height

The X, Y, and Z coordinate system is discussed in more detail in a later chapter.

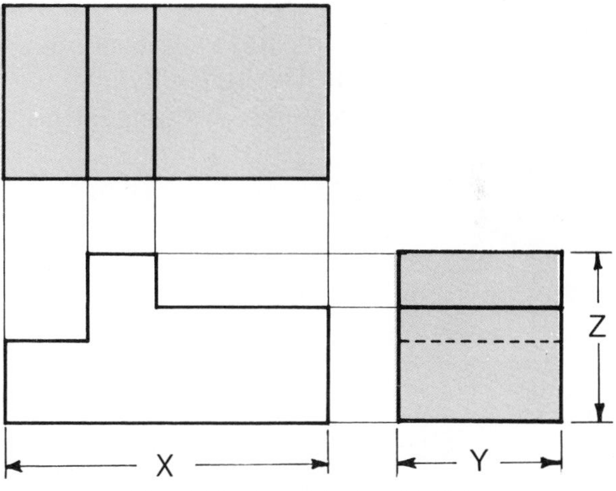

Fig. 2-11 Coordinates *X, Y, Z*

REVIEW

This review is provided to serve as reinforcement study material. Fill in the appropriate word(s) to complete the sentences below.

1. The views used most frequently on industrial prints are the ______, ______, and ______ views.

2. The most important view of an object is usually the ______ view.

3. The view arrangement shown in Figure 2-7 is called (a) (an) ______ projection.

4. Surfaces on one view become ______ on another view.

5. When lines meet, they form (a) (an) ______.

6. The ______ and ______ dimensions can be found in the front view.

7. The ______ and ______ dimensions can be found in the top view.

8. The ______ and ______ dimensions can be found in the side view.

9. On some of the newer prints, the length, width, and height dimensions are given as ______, ______, and ______ coordinates.

NAME ______

DATE ______

SCORE ______

EXERCISE A2-1

Surfaces, Lines, and Points

Match the numbers on the three-view drawing below with the letters on the pictorial drawing. Write your answers in the chart. The solutions to *A*, *F*, and *P* are given as examples.

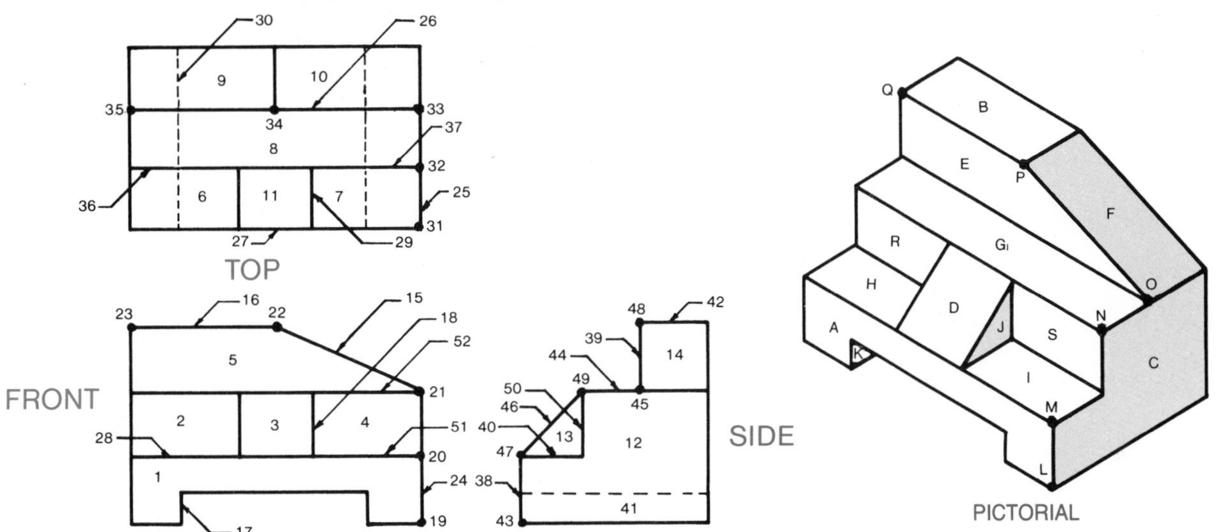

Letters on Pictorial	Surfaces, Lines or Points		
	Front View	Top View	Side View
A	1	27	38
B			
C			
D			
E			
F	15	10	14
G			
H			
I			
J			
K			
L			
M			
N			
O			
P	22	34	48
Q			
R			
S			

NAME ______________

DATE ______________

SCORE ______________

EXERCISE A2-2

Matching Pictorial with Orthographic

On the following two pages, match the three-view drawings with the pictorial drawings. Place your answers on the spaces provided in the pictorial drawings. The solution for number 1 is provided as an example.

NAME ________________

DATE ________________

SCORE ________________

(Exercise A2-2 continued)

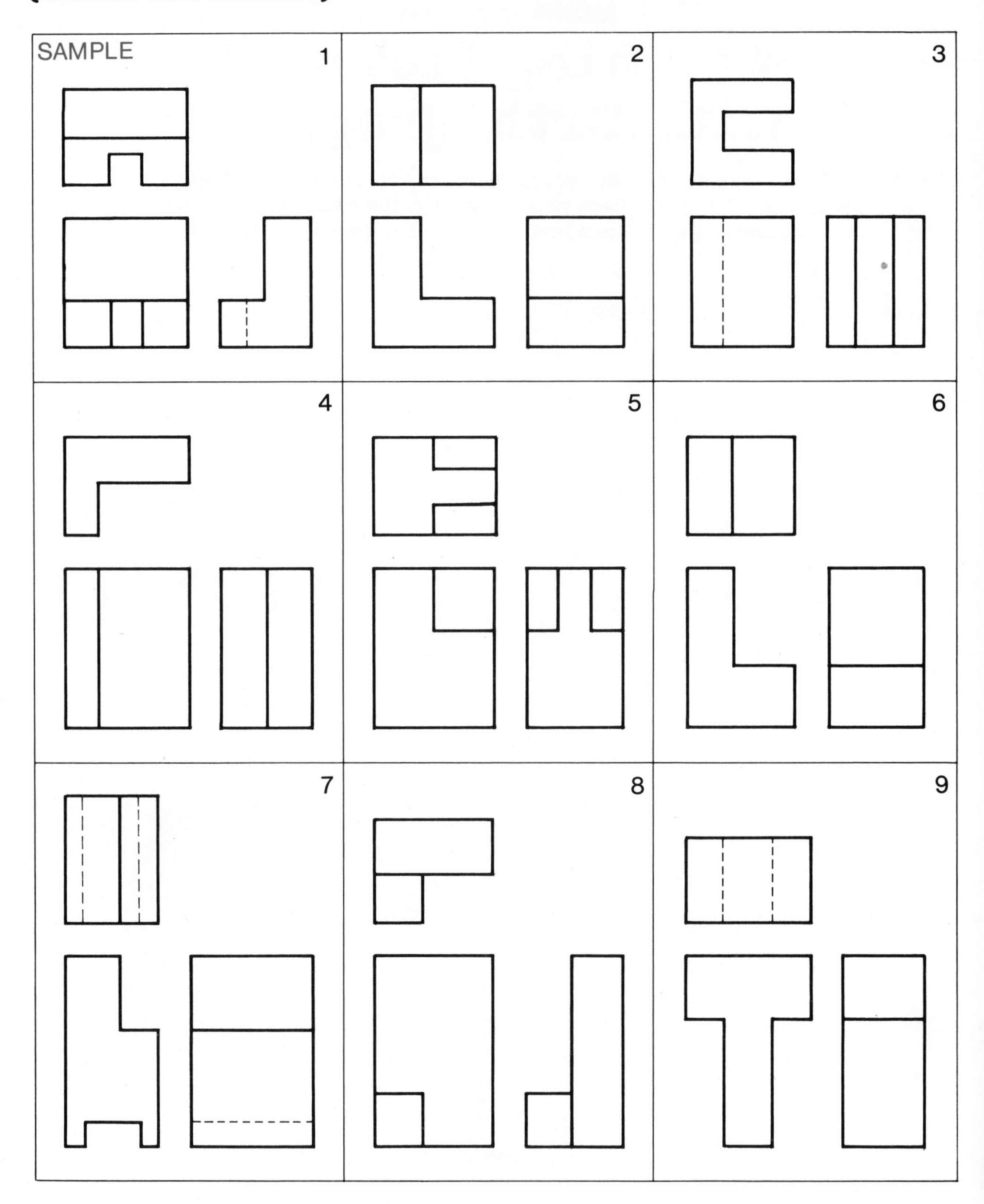

(Exercise A2-2 continued)

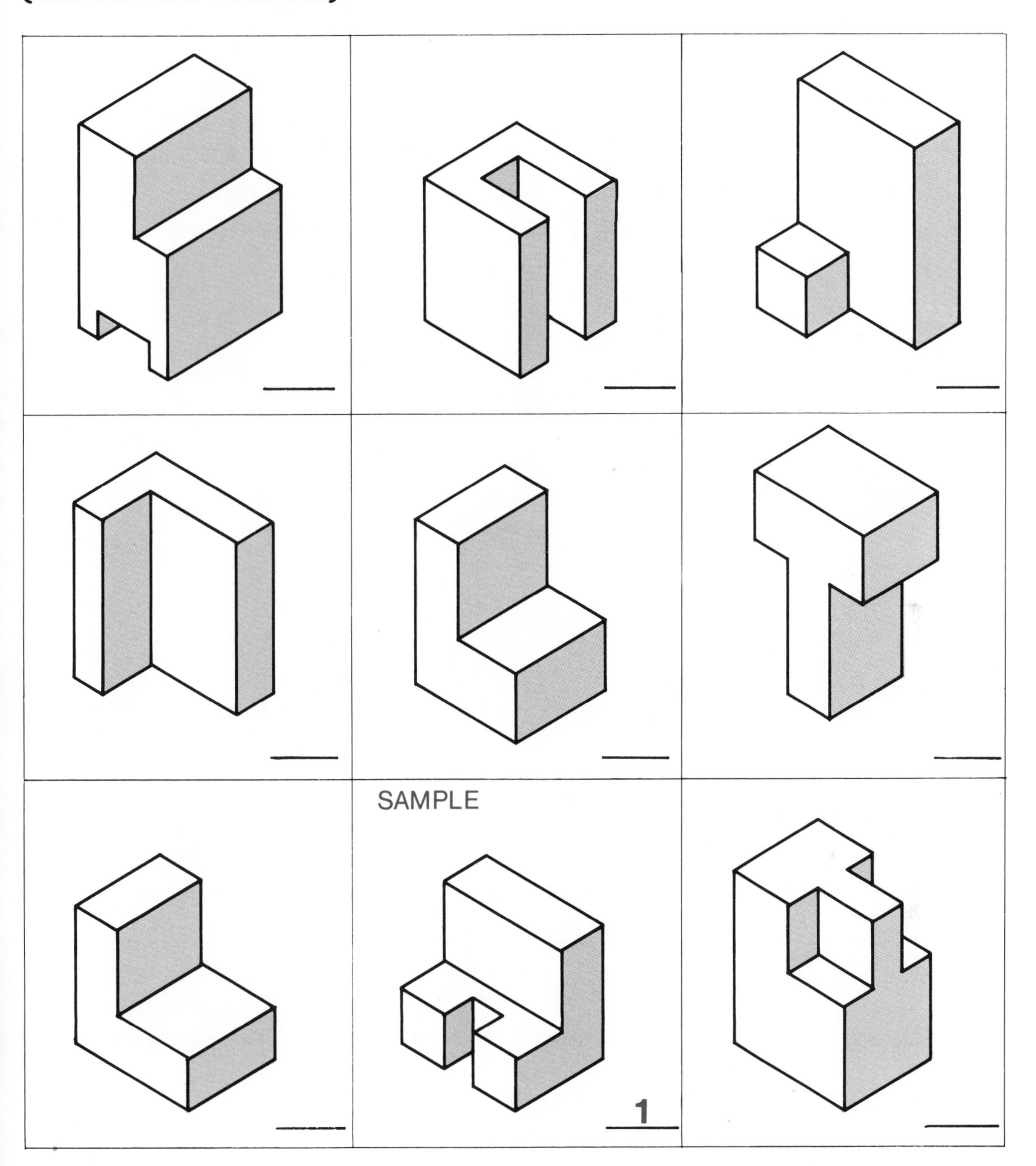

NAME ______

DATE ______

SCORE ______

CHAPTER 3

PICTORIAL SKETCHING

OBJECTIVES

After studying this chapter, you will be able to:

- Explain how pictorial sketching can help in the visualization of an object.
- Identify the most common type of pictorial sketch.
- Outline the steps used in isometric sketching.
- Sketch isometric circles.
- Outline the steps used in oblique sketching.
- Sketch oblique circles.

The views of an object are usually sketched by the designer or engineer. Later, the sketch is redrawn by the drafter. The drafter adds the dimensions and other details needed by the production departments.

The designer or engineer may also sketch a pictorial view to better illustrate the details of the object. The drafter or production supervisor may draw a pictorial view from the print to clarify certain details for people who are not trained to read orthographic (multiview) drawings. Those who have trouble visualizing an object from an orthographic projection may find it helpful to draw a pictorial sketch. Very little practice is required to learn the techniques of drawing pictorial sketches.

The *isometric* and the *oblique* are the two most frequently used types of pictorial sketches. Isometric projection is used most often because it represents how the eye would actually see the object. However, this method creates sketching problems with cylindrical shapes. An oblique sketch is much easier to make, but it also creates distortion of some features.

ISOMETRIC SKETCHING

A pictorial view can be sketched of nearly any object by following a few basic rules. Most often, the object will be visualized before the sketch is completed. The steps for sketching a pictorial view are shown in Figure 3-1.

STEP 1. Three lines are drawn to represent the three main axes: length, width, and height. The height axis is sketched vertically. The two other axes are drawn approximately 30 degrees (30°) from the horizontal.

STEP 2. An isometric box is sketched, with dimensions equaling the object to be illustrated. (*Note:* All vertical lines are parallel, and all horizontal lines are drawn parallel to the 30° axis.)

STEP 3. The features are sketched in, as shown on each view of the print.

STEP 4. All unnecessary construction lines are erased, and the remaining lines are darkened.

Isometric Circles

More complicated objects such as isometric circles can be "blocked in." This technique is shown in Figure 3-2. Isometric circles are not drawn round. They are drawn as ovals, as they would be seen by the eye. An easy method of drawing an oval is to construct an isometric square and sketch the oval within that square.

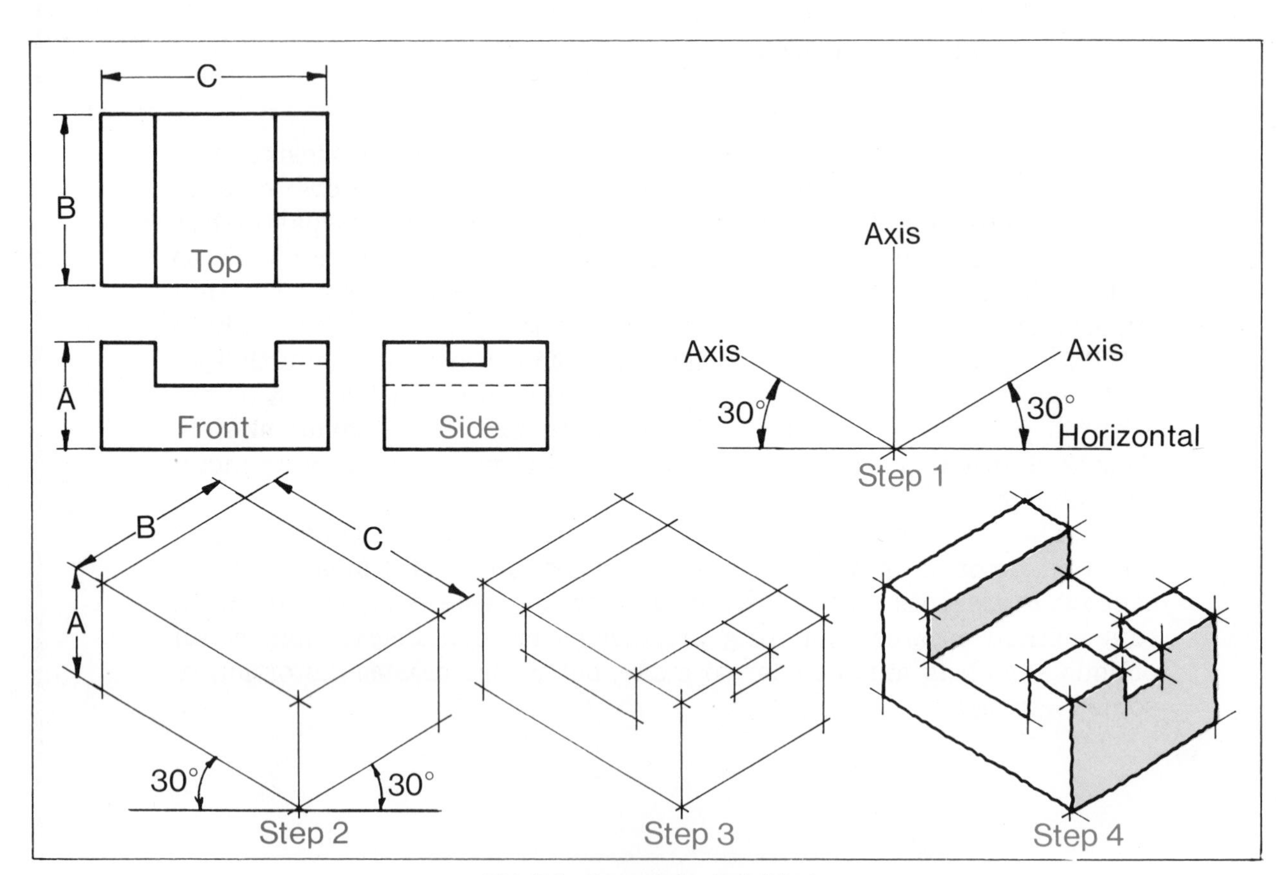

Fig. 3-1 Isometric sketching

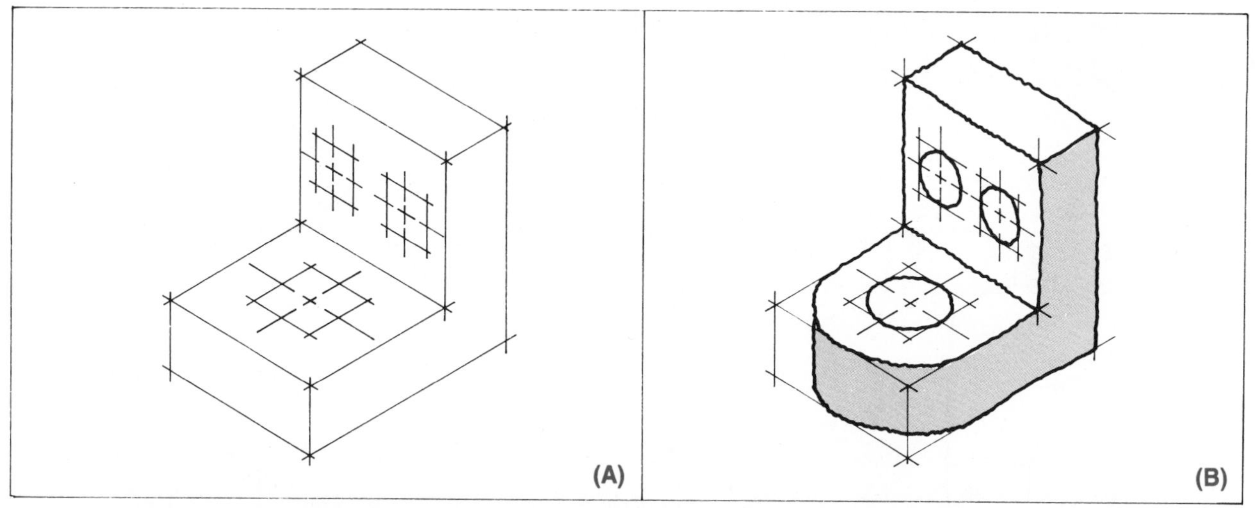

Fig. 3-2 Blocking in an isometric sketch. (A) Rough overall block view used as the boundary of the finished sketch. (B) Finished sketch. The heavy lines show the actual edges of the object.

The steps used to construct isometric circles are illustrated in Figure 3-3. First, an isometric square is constructed around the locating center line of the circle. The midpoints of each side of the isometric square are connected by lines. The midpoint of each triangle formed by these lines is located. In the second step, the short arcs are sketched through the midpoints of smaller triangles. In the third and final step, the long arcs are sketched through the midpoints of larger triangles.

An isometric circle will change in shape, depending on whether it is in a horizontal or vertical plane of projection. The three positions for an isometric circle are shown in Figure 3-4.

OBLIQUE SKETCHING

The advantage of an oblique sketch is that it shows the front view in true shape. This is the same as it is shown in an orthographic projection (refer to Figure 2-7).

The steps involved in sketching an oblique drawing are similar to those used for an isometric drawing. However, cylindrical shapes on the front view are drawn round rather than oval. In addition, horizontal lines in the front view are not drawn at an angle to the horizontal. Figure 3-5 illustrates the steps used in sketching oblique drawings.

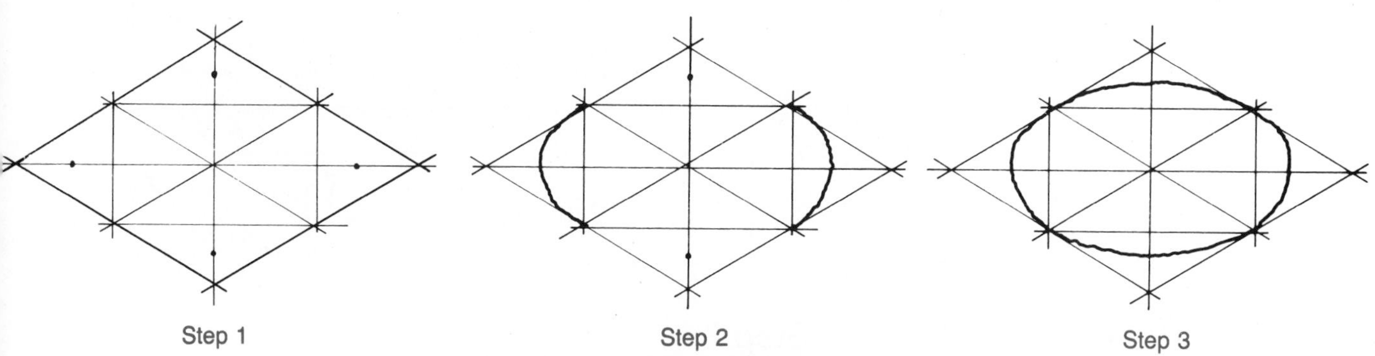

Fig. 3-3 Sketching an isometric circle

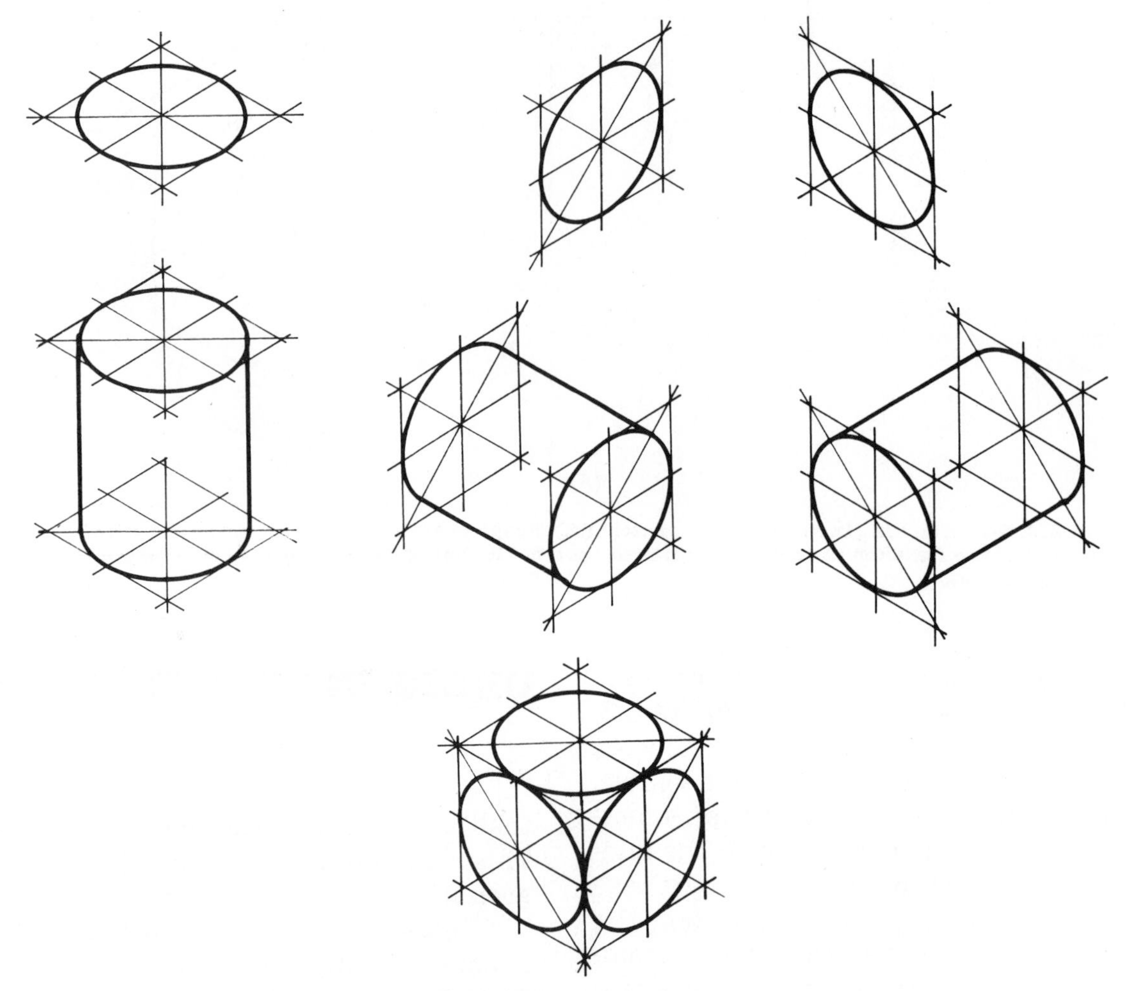

Fig. 3-4 Isometric circles

Oblique projection is not recommended for objects that have circular or irregular shapes on views other than the front view. The side view can be sketched at any angle from the horizontal. However, a 45° angle is used most often because it is most pleasing to the eye.

The features of more complicated objects can be blocked in as shown in Figure 3-6.

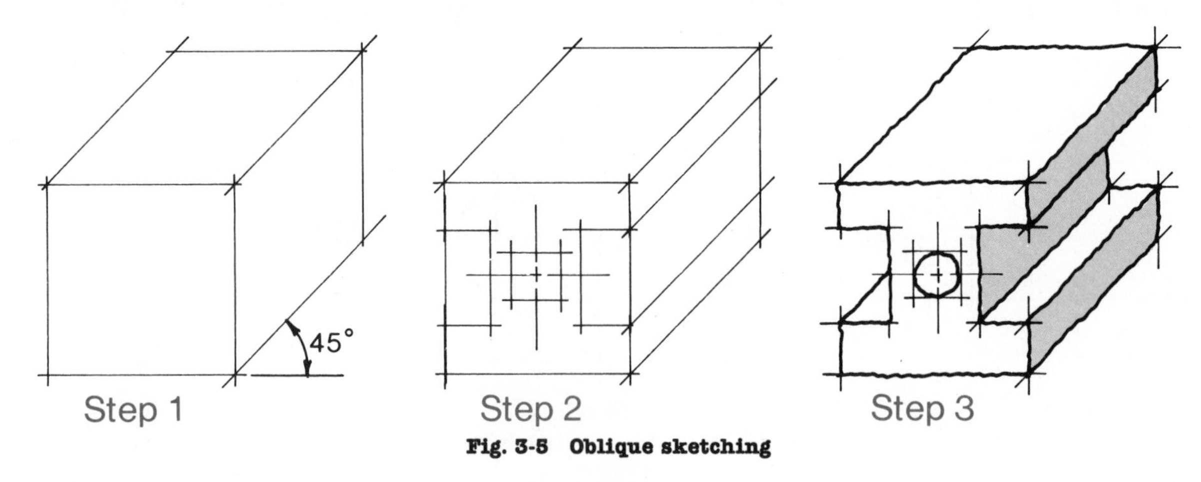

Fig. 3-5 Oblique sketching

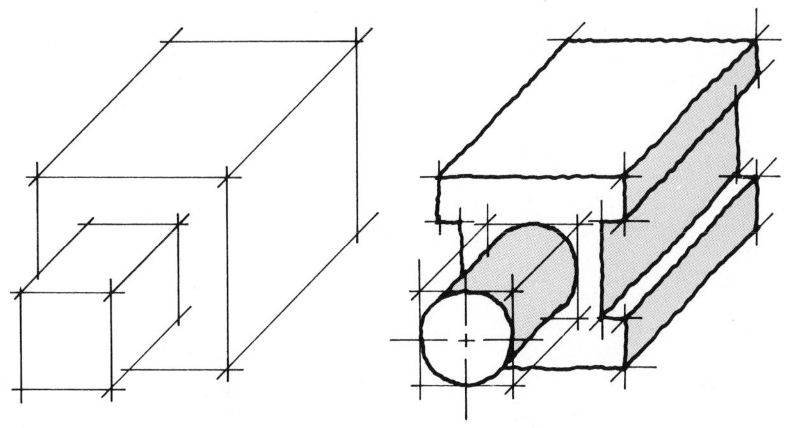

Fig. 3-6 Blocking in an oblique sketch

Oblique Circles

As we have mentioned, circles on the front plane of an oblique projection are drawn as true circles. Circles on planes which are at an angle to the front plane are drawn as ovals. This is accomplished by using the same steps illustrated for isometric sketching (Figure 3-3).

REVIEW

This review is provided to serve as reinforcement study material. Fill in the appropriate word(s) to complete the sentences below.

1. The ______________________ usually provides the drafter with a part sketch.

2. The two most common types of pictorial sketches are the ______________________ and ______________________ sketches.

3. On an isometric sketch, the length and width axes are drawn at a ______________________ angle to the horizontal base line.

4. On all views of an isometric pictorial sketch, a circle will have (a) (an) ______________________ shape.

5. On an oblique sketch, the length and width axes are usually drawn at a ______________________ angle to the horizontal base line.

6. On an oblique pictorial sketch, a cylindrical feature will have (a) (an) ______________________ shape on the front view.

NAME ______________________

DATE ______________________

SCORE ______________________

EXERCISE A3-1

Isometric Sketching

Study the sample sketches below. In Sample A, the front isometric view is given on the isometric grid. The dashed lines show the rough overall block view of the object. The purpose is to draw in the object lines of the isometric sketch. In Sample B, the finished isometric sketch of the object is shown. All object lines are drawn heavy. Dashed lines are shown for construction purposes only.

After studying the sample sketches, complete the sketches on the following page.

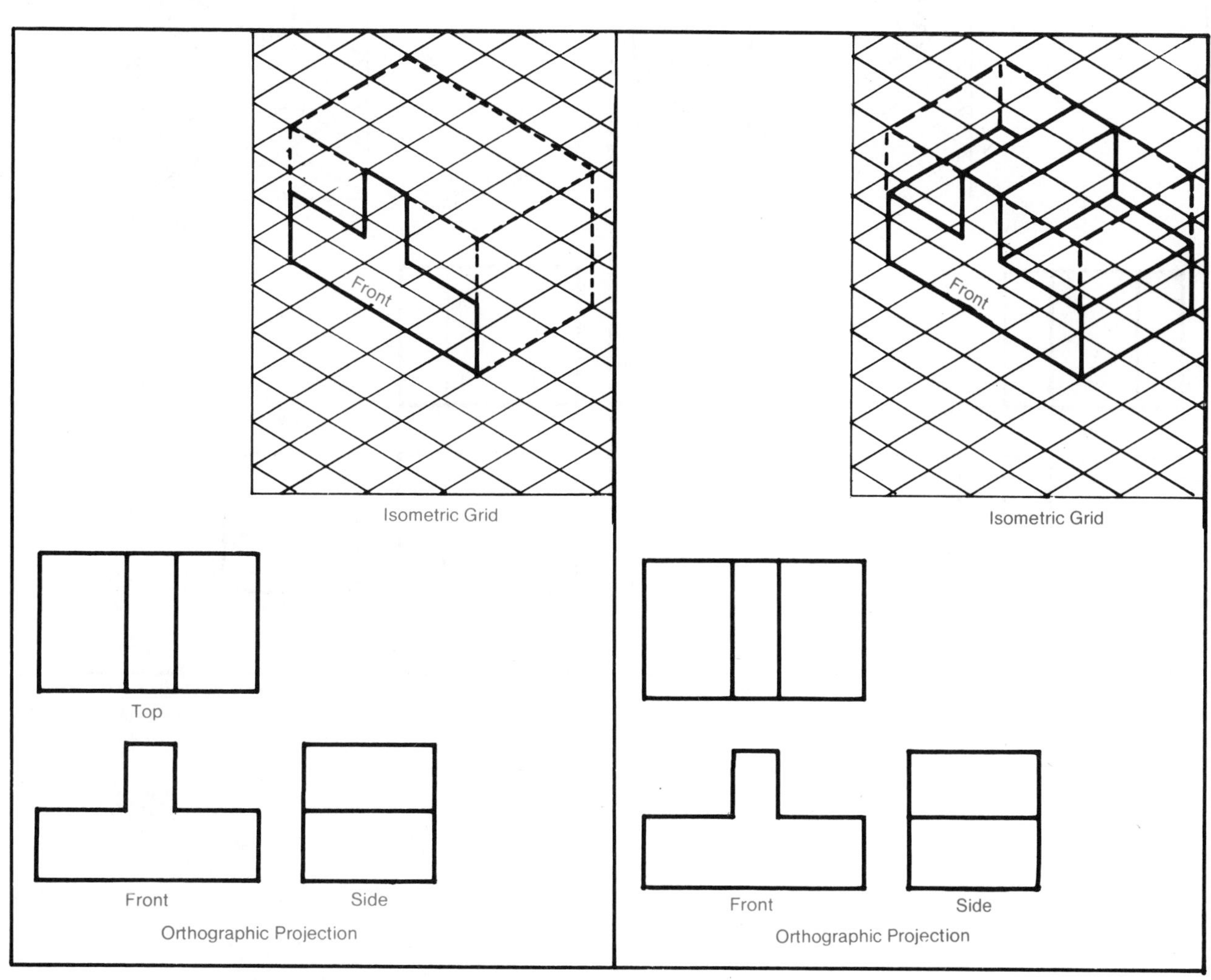

Sample A **Sample B**

NAME________________

DATE________________

SCORE________________

(Exercise A3-1 continued)

Draw isometric sketches of the objects shown in the orthographic projections.

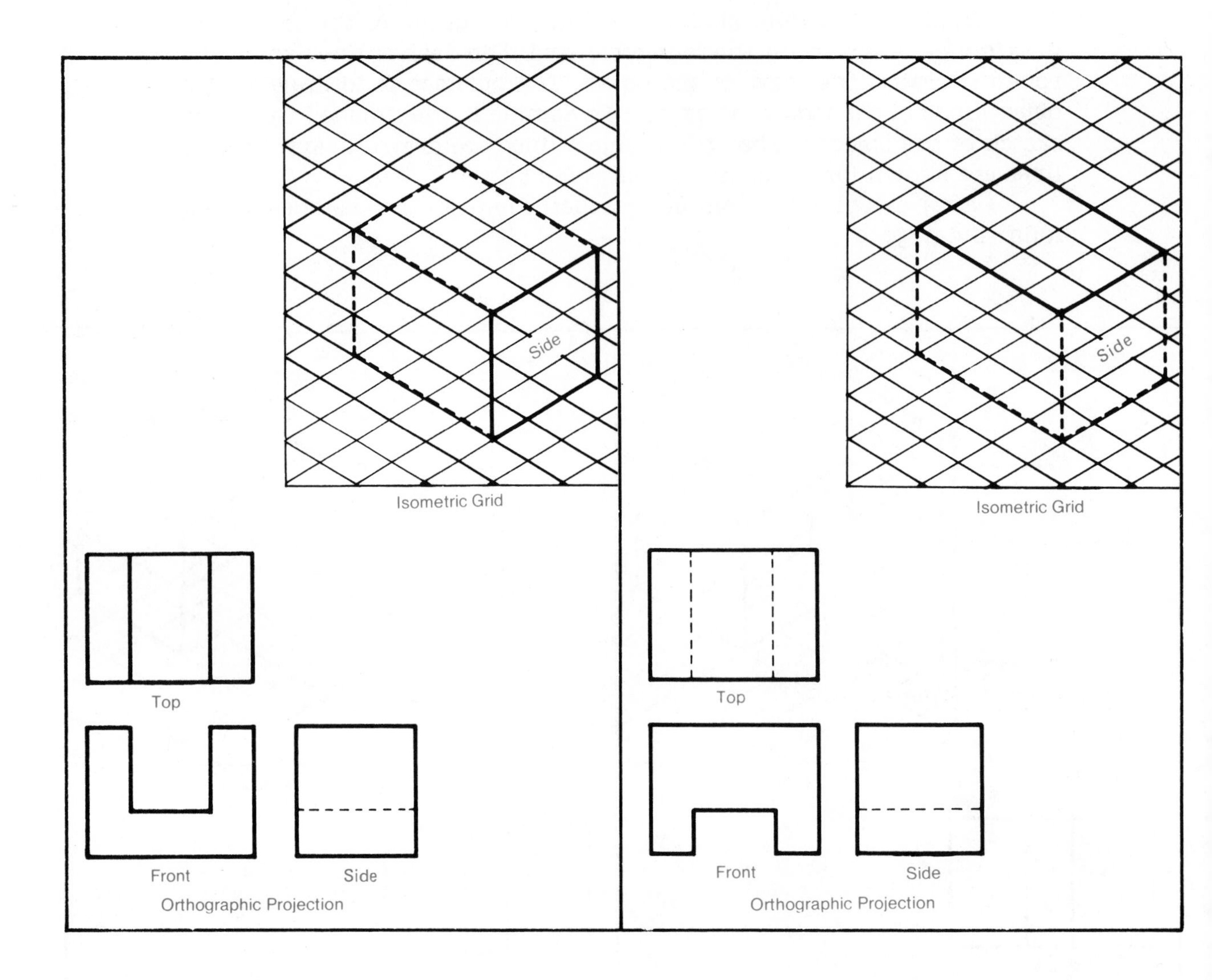

EXERCISE A3-2

Isometric Sketching

Complete the isometric sketches to conform with the three-view orthographic projections.

NAME____________________

DATE____________________

SCORE____________________

EXERCISE A3-3

Isometric Sketching

Complete the isometric sketches to conform with the three-view orthographic projections.

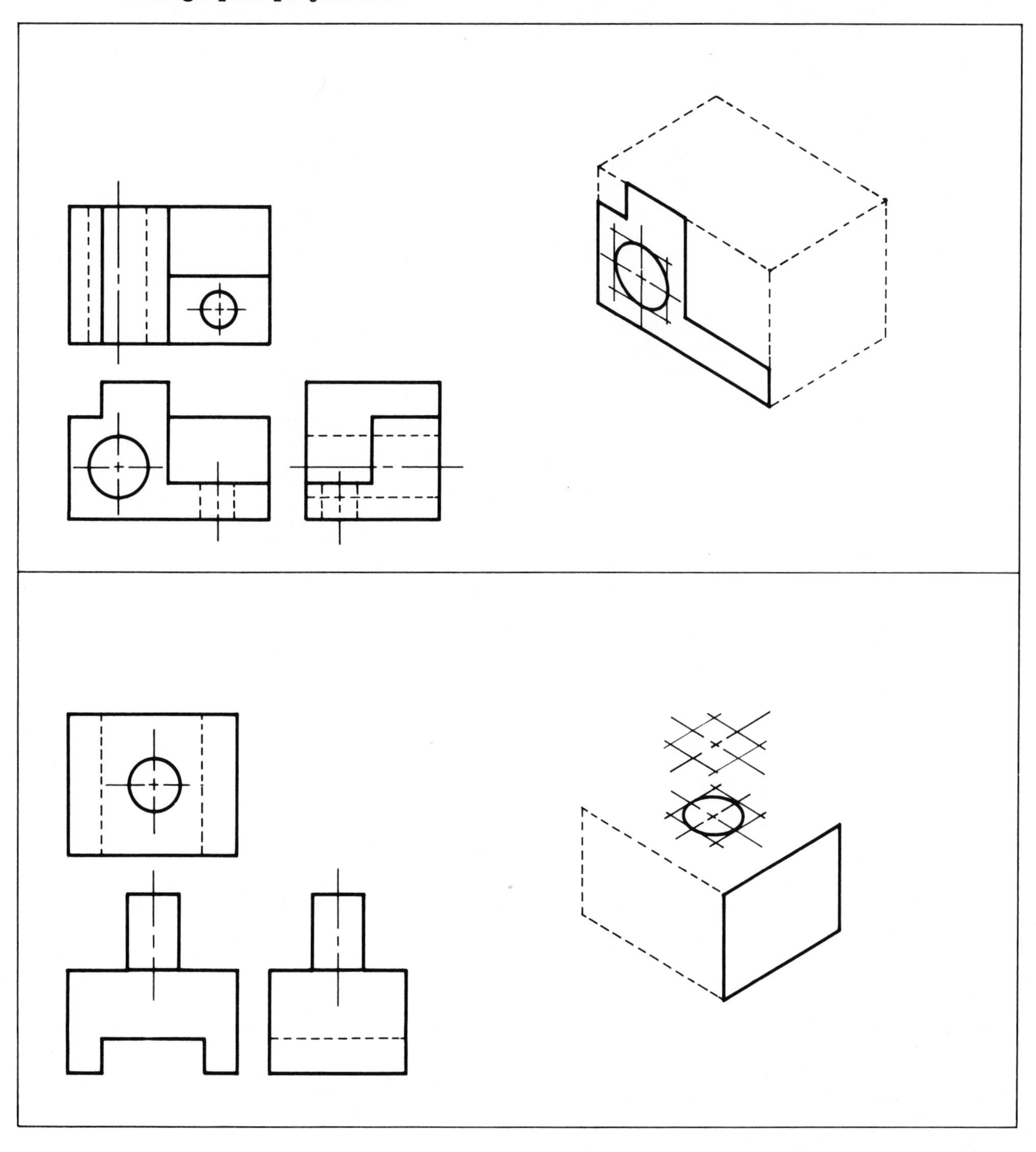

NAME ____________

DATE ____________

SCORE ____________

CHAPTER 4

ARRANGEMENT OF VIEWS

OBJECTIVES

After studying this chapter, you will be able to:

- Explain the importance of the relative positions of views on an industrial print.
- Discuss why certain views are selected for drawing.
- List the factors that determine which view is selected as the front view.
- Explain why the drafter selects the top view or the bottom view.
- Explain why the drafter selects the right side view or the left side view.

The illustrations in Chapters 2 and 3 follow a definite system of arranging the various views of an object. Proper view arrangement is very important.

An improper view arrangement can confuse the print reader. Such an arrangement is shown in Figure 4-1. It is hard to visualize the object from that arrangement of the three views. Sketching a pictorial view of the object would be a very frustrating job.

RELATIVE POSITIONS OF VIEWS

The drafter must first decide which view best describes the object. This view is called the *front view.* The front view is usually drawn in the lower left corner of the page. The top view is usually drawn above the front view. The right side view is usually drawn to the right of the front view. This arrangement is known as an *orthographic projection* or *multiview drawing.* A proper rearrangement of the views in Figure 4-1 is shown in Figure 4-2. The object is easier to visualize

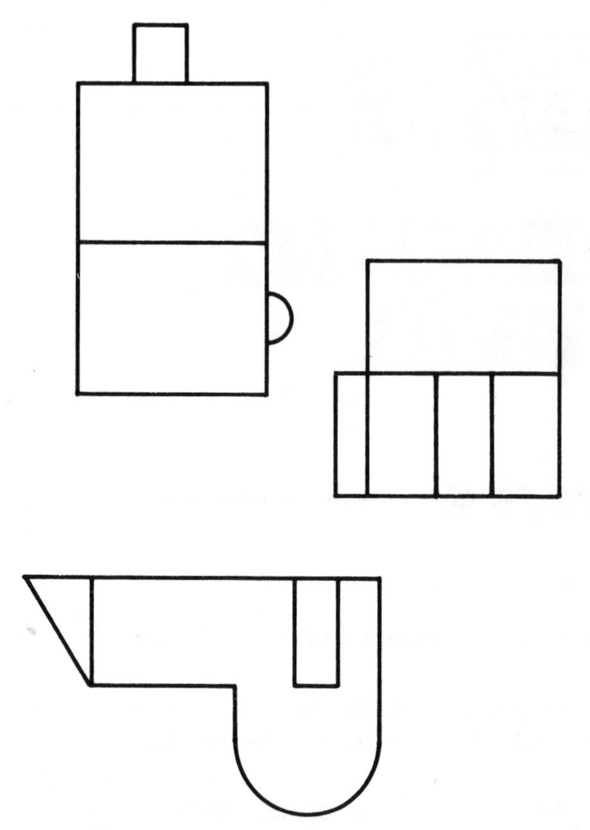

Fig. 4-1 Improper view arrangement

from the orthographic projection in Figure 4-2.

Other views of the object may also be drawn. For example, a rectangular box,

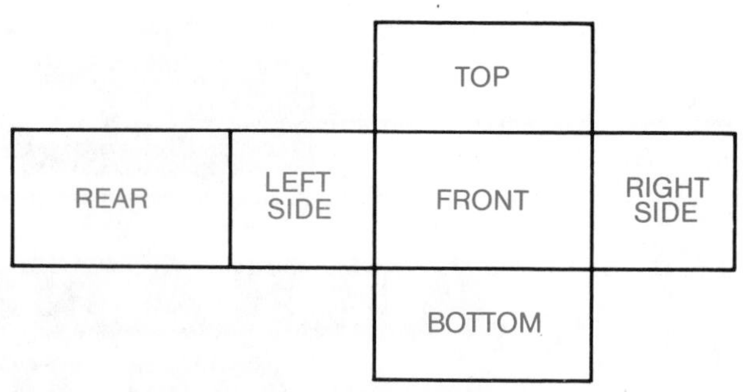

Fig. 4-3 Six views

cut apart, will have a total of six sides, as shown in Figure 4-3. If the six sides of the box are to be shown, then the views should be separated from one another. The drawing would now look like Figure 4-4. The spaces between the views are used to insert the necessary dimensions to indicate the actual size of the object.

This method of arranging views is standard in the mechanical industries.

PRIMARY VIEWS

Refer again to Figure 4-2. The views shown are the front, top, and right side views. The possible additional views of

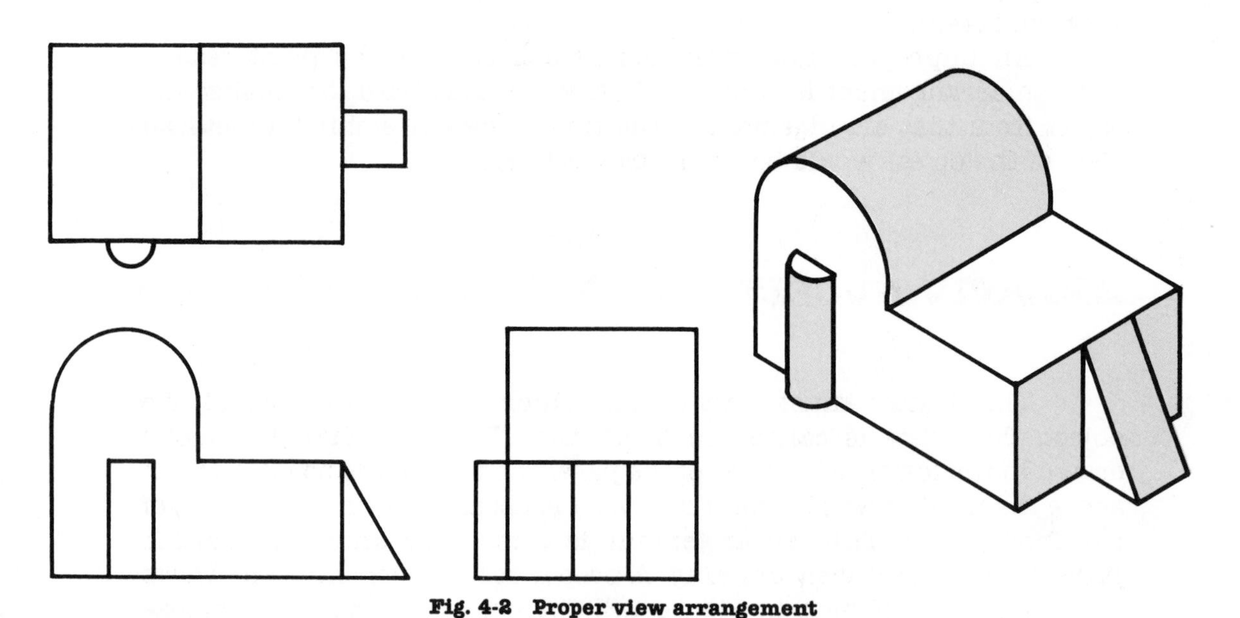

Fig. 4-2 Proper view arrangement

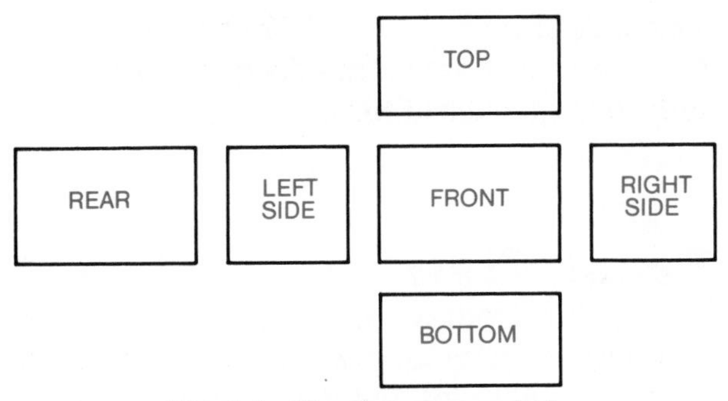

Fig. 4-4 Six views separated

the object are the left side, bottom, and rear views.

All six of the possible primary views of the object are shown in Figure 4-5. The following observations can be made from Figure 4-5.

- The shape of the rear view is the same as the front view, in reverse.
- The shape of the bottom view is the same as the top view, in reverse.
- The shape of the left side view is the same as the right side view, in reverse.

The main difference in the three additional views (rear, bottom, and left side) is the use of dashed (hidden) lines. Hidden lines are used in the place of some of the solid (object) lines in the front, top, and right side views. These illustrate two important principles in print reading.

1. The intersection of surfaces that *can* be seen in the view are represented by object lines.
2. The intersections that *cannot* be seen in the view are shown by hidden lines.

A cylindrical projection is shown on the front view of the object in Figure 4-5. This feature is shown in the same location on the rear view by hidden lines. Other features of the object are also shown as hidden lines in the rear, bottom, and left side views. These hidden lines, as shown, duplicate the object lines on the front, top, and right side

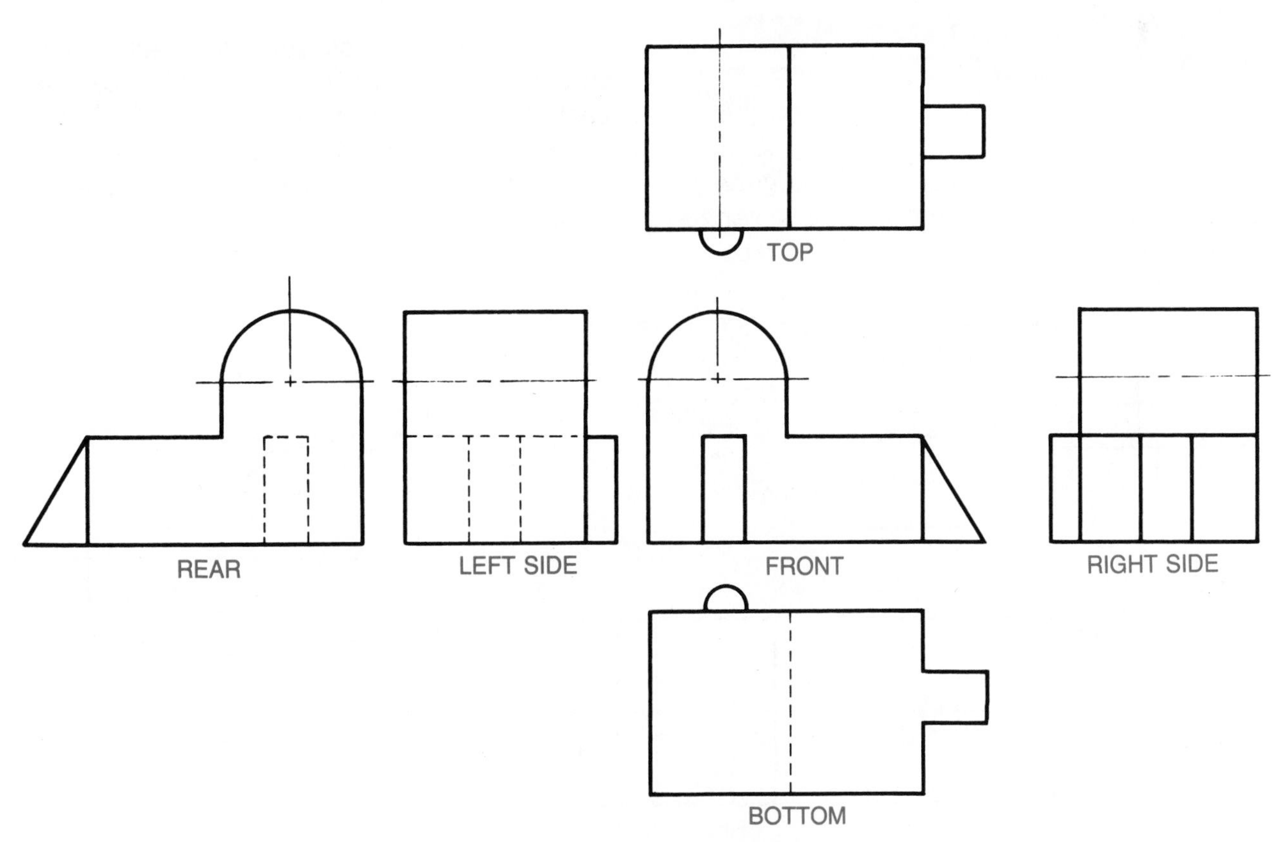

Fig. 4-5 Six object views

views. They provide no additional information about the object.

As is now apparent, the three additional views (rear, bottom, and left side) do not add to the visualization of the object. These additional views, then, are unnecessary. They would just add to the drafting costs. The front, top, and right side views completely describe the object.

A general rule in drafting is to draw only the views necessary to completely describe the features of the object. The validity of this rule is clear in the case just described. Indeed, most objects can be accurately represented by three views. The most common views shown are the front, top, and right side views. The relative positions of the views as shown in Figure 4-2 are also preferred. This is the arrangement used most often in industry.

VIEW SELECTION AND ARRANGEMENT

This section presents some general guidelines for view selection and arrangement. It should be noted that many situations warrant departure from the standard methods discussed here.

Front View

The front view is usually the most important view of the object. The selection of the front view is determined by the following guidelines.

1. The front view of the object should be the view that best describes its shape.
2. The front view should be the view that reveals the most detail or information about the surface features of the object.
3. The longest dimension of an object is usually important; therefore, the front view should show this dimension.
4. When an object has a definite functional position, the front view should show that position, if possible.

These are not "hard and fast" rules. Quite often, not all of the rules can be satisfied. Usually, though, the view that shows the greatest amount of detail and best reveals the object's shape is selected as the front view.

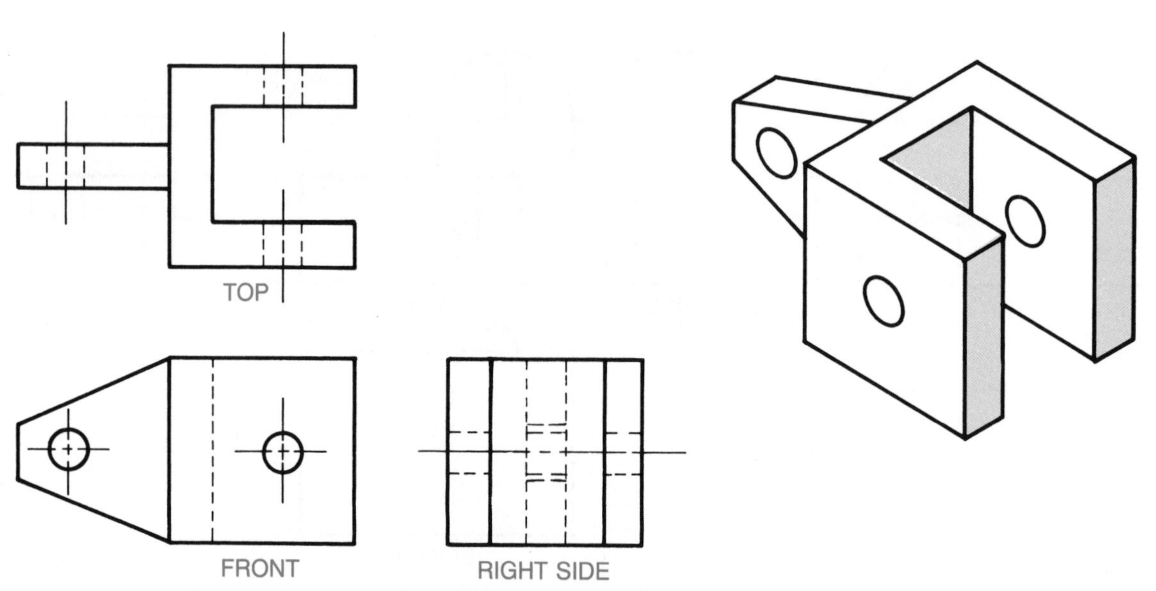

Fig. 4-6 The right side view is positioned to the right of the front view.

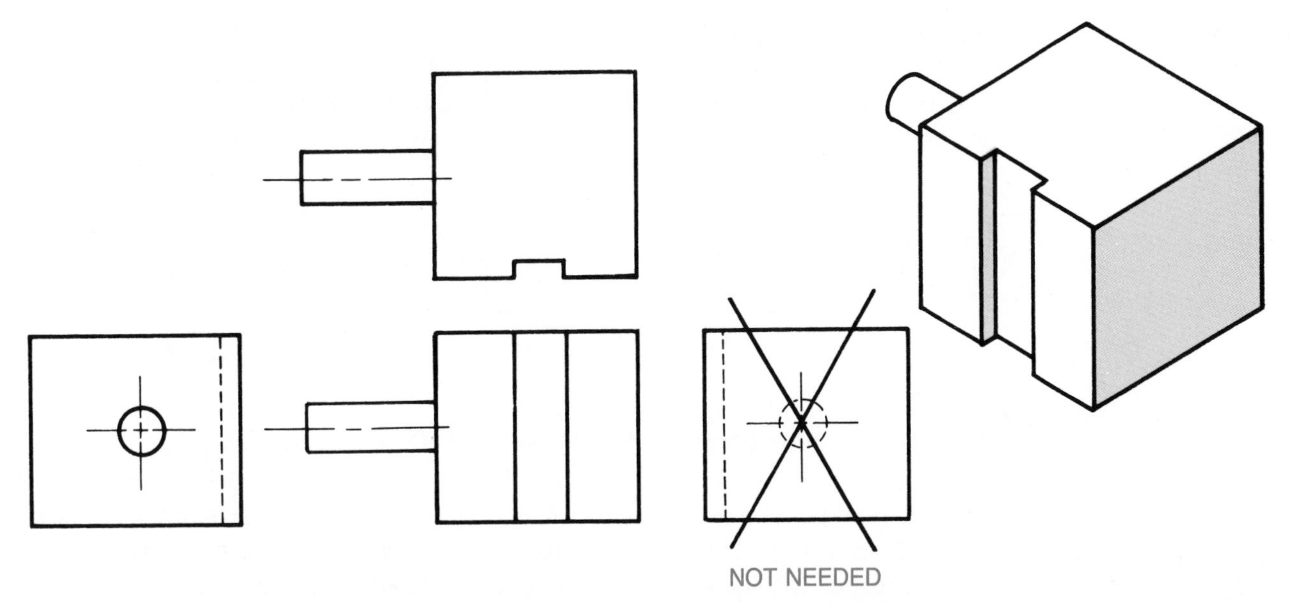

Fig. 4-7 The right side view is unnecessary in this case.

Top or Bottom View

After the front view has been chosen, the top or bottom view is selected. When the object is not very complicated, usually only one of these views is drawn.

The front view must be positioned on the drawing in a way that allows room for the top or bottom view. If the top view is to be shown, then the front view will be positioned toward the bottom of the drawing. This will allow the top view to be drawn above the front view. An opposite view arrangement is required when the bottom view is selected.

Side View

Regardless of whether the top or bottom view is selected, the side view is positioned to one side of the front view. The left side view is positioned to the left of the front view. The right side view is positioned to the right of the front view. A right side view is shown in Figure 4-6.

Figure 4-7 shows that the left side view contains fewer hidden lines. Therefore, in this case the left side view of the object is the better choice. It is good practice to select the view that shows the shape of the object's features and contains the least number of hidden lines.

Occasionally, it is necessary to show both side views to clearly show the shape of each side of the object. Such a case is illustrated in Figure 4-8.

In Figure 4-9, the left side view is the same as the right side view. The top view is the same as the front view. Such

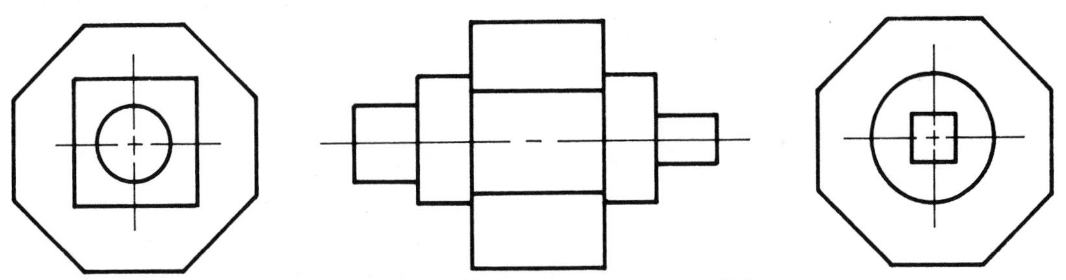

Fig. 4-8 Two side views are necessary in this case.

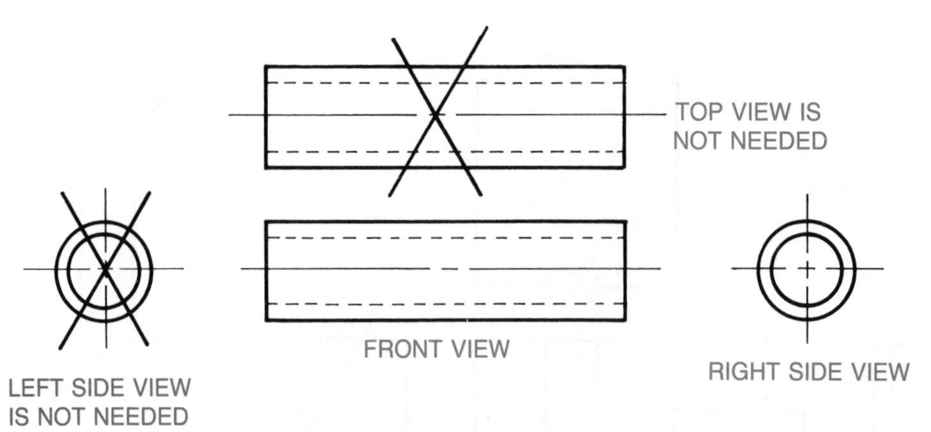

Fig. 4-9 Unnecessary views should not be drawn.

duplicative views should not be shown. This is in accordance with the rule stating that the minimum number of views that completely describe the object should be shown. This makes it easier for the print reader to visualize the object.

REVIEW

This review is provided to serve as reinforcement study material. Fill in the appropriate word(s) to complete the sentences below.

1. The first decision of the drafter is to determine which view should be the ____________________ view.

2. If all possible views of a rectangular object were drawn, there would be ____________________ views.

3. Most objects can be represented accurately by ____________ views.

4. The most important view on a print is usually the ____________ view.

5. The guidelines that should be followed to determine the front view are:

   a. ____________________

   b. ____________________

   c. ____________________

   d. ____________________

6. A general rule in drafting is to show the ____________________ number of views that will completely describe the object.

NAME ____________________

DATE ____________________

SCORE ____________________

EXERCISE A4-1

Matching Views

Match the six pictorial views with the three-view drawings. Place your answers in the spaces provided in the three-view drawings. A sample is provided.

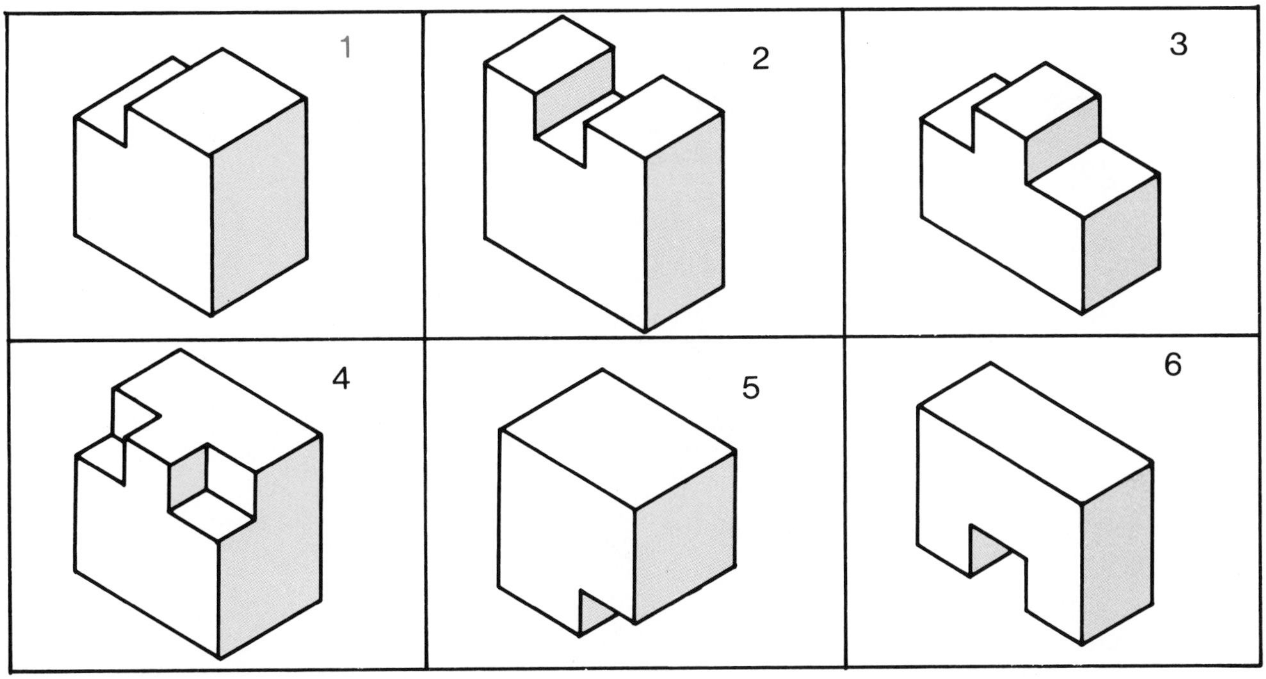

1

2

3

4

5

6

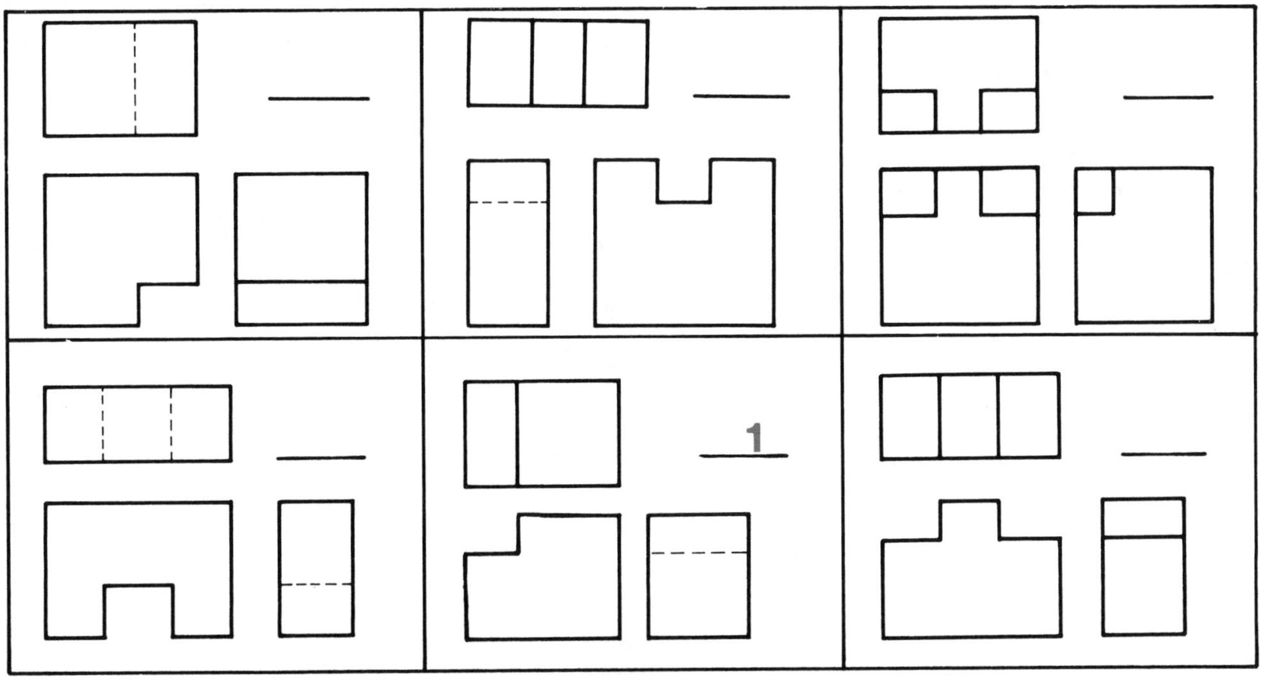

____

____

____

____

1

____

NAME __________

DATE __________

SCORE __________

EXERCISE A4-2

Sketching

Study the sample sketches. Then, on the bottom two sketches, sketch in the missing lines.

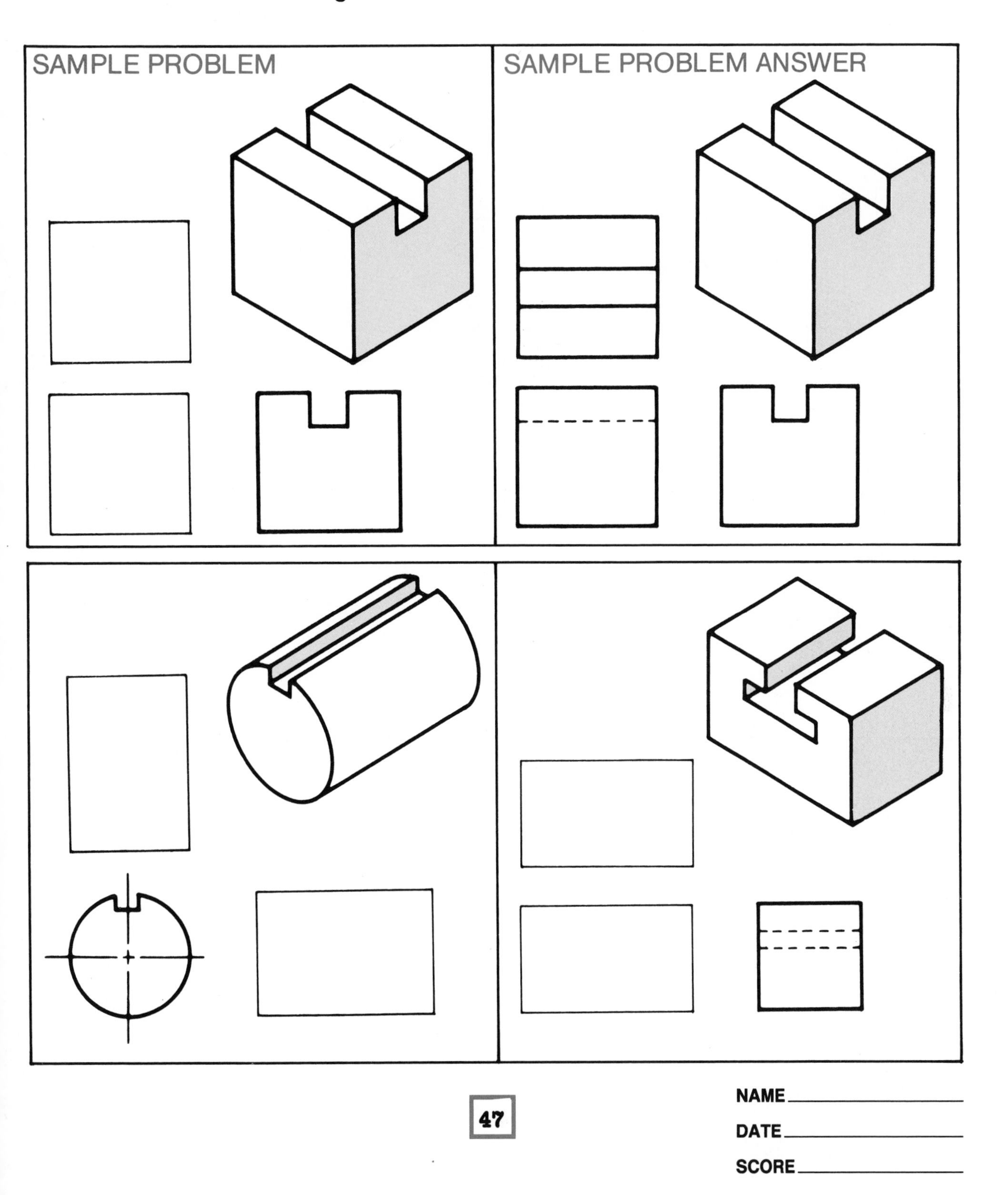

NAME ______________________

DATE ______________________

SCORE ______________________

EXERCISE A4-3

Sketching

Sketch in the missing lines and complete the isometric views.

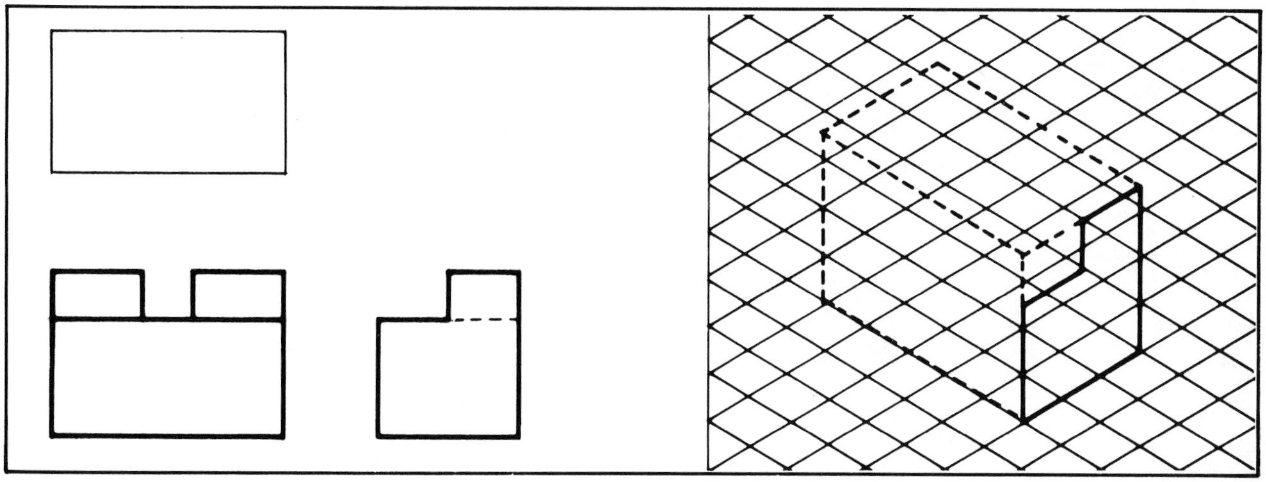

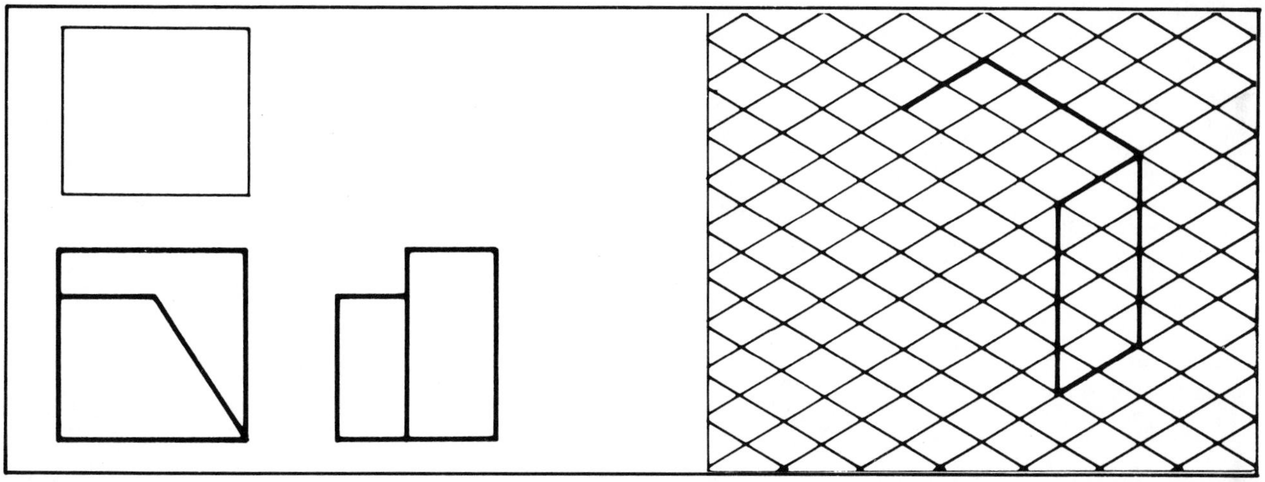

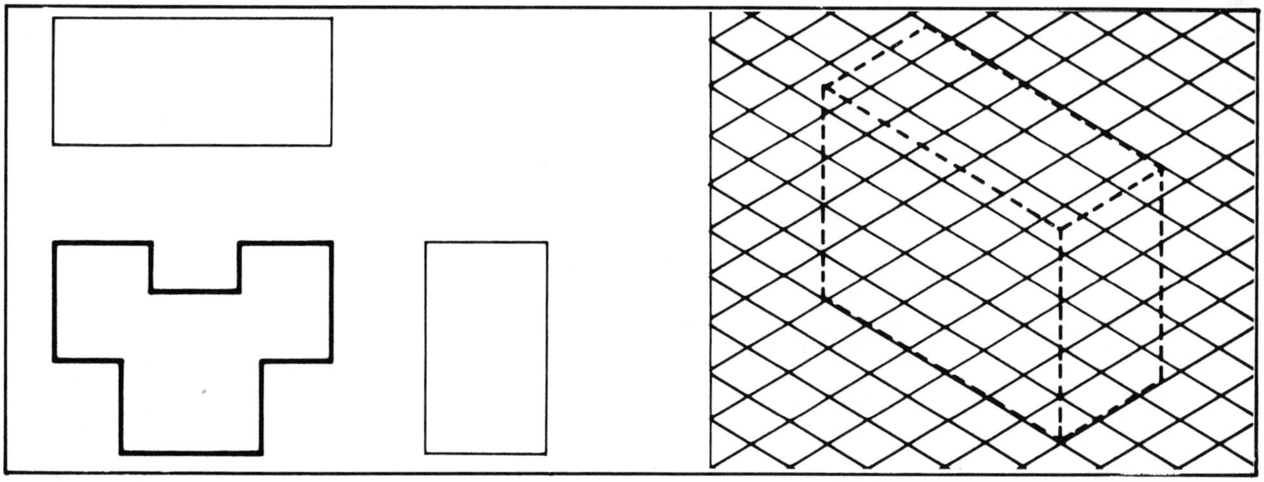

NAME ____________

DATE ____________

SCORE ____________

EXERCISE A4-4

Sketching

Sketch the front, top, and side views for each isometric drawing.

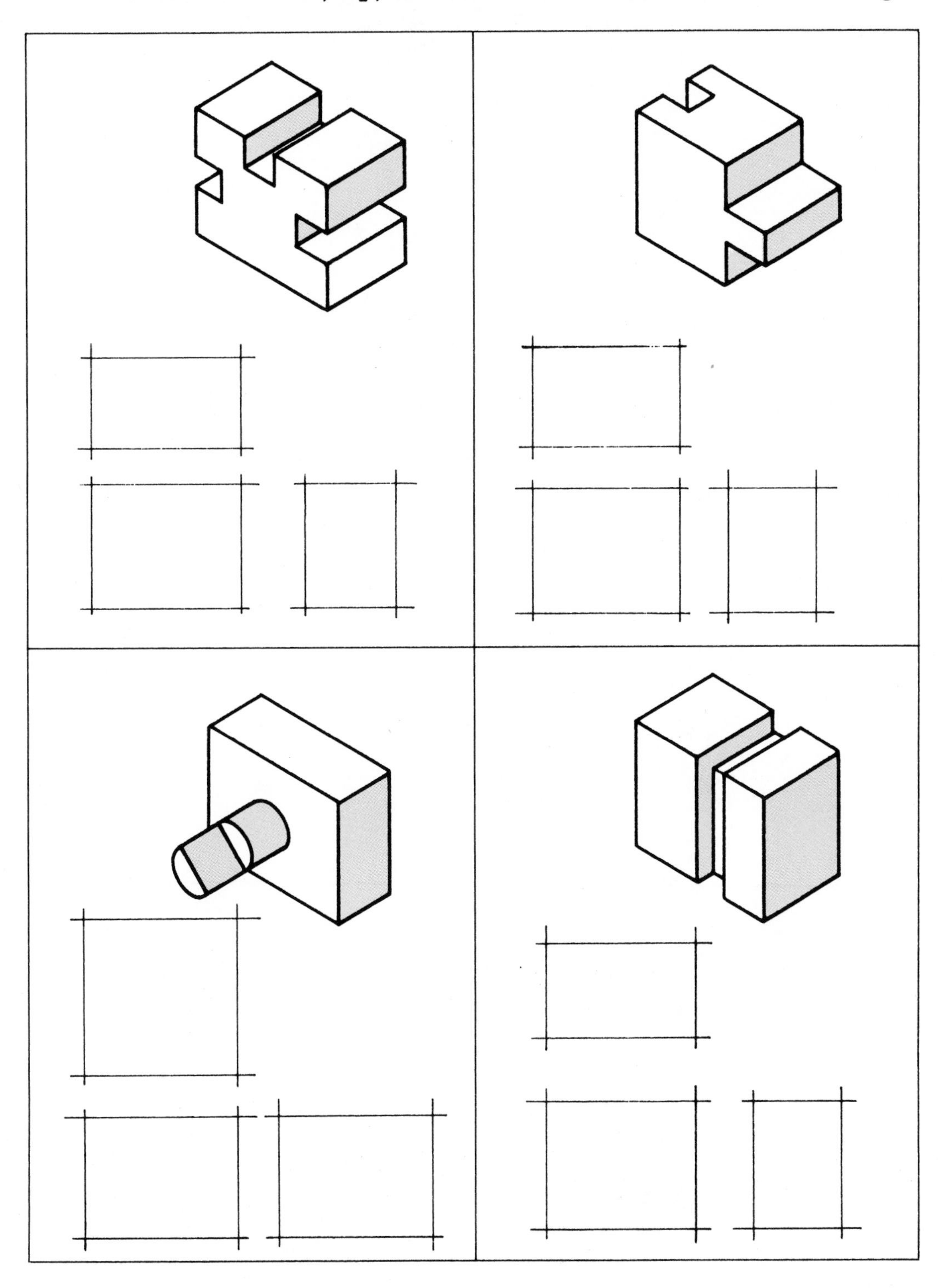

NAME ____________________

DATE ____________________

SCORE ____________________

EXERCISE A4-5

Matching Views

View number 1 is identified as the top view. Below the six views are six view areas arranged in the standard positions. In the circle in each view area, write the view number that satisfies the view arrangement. A sample is provided.

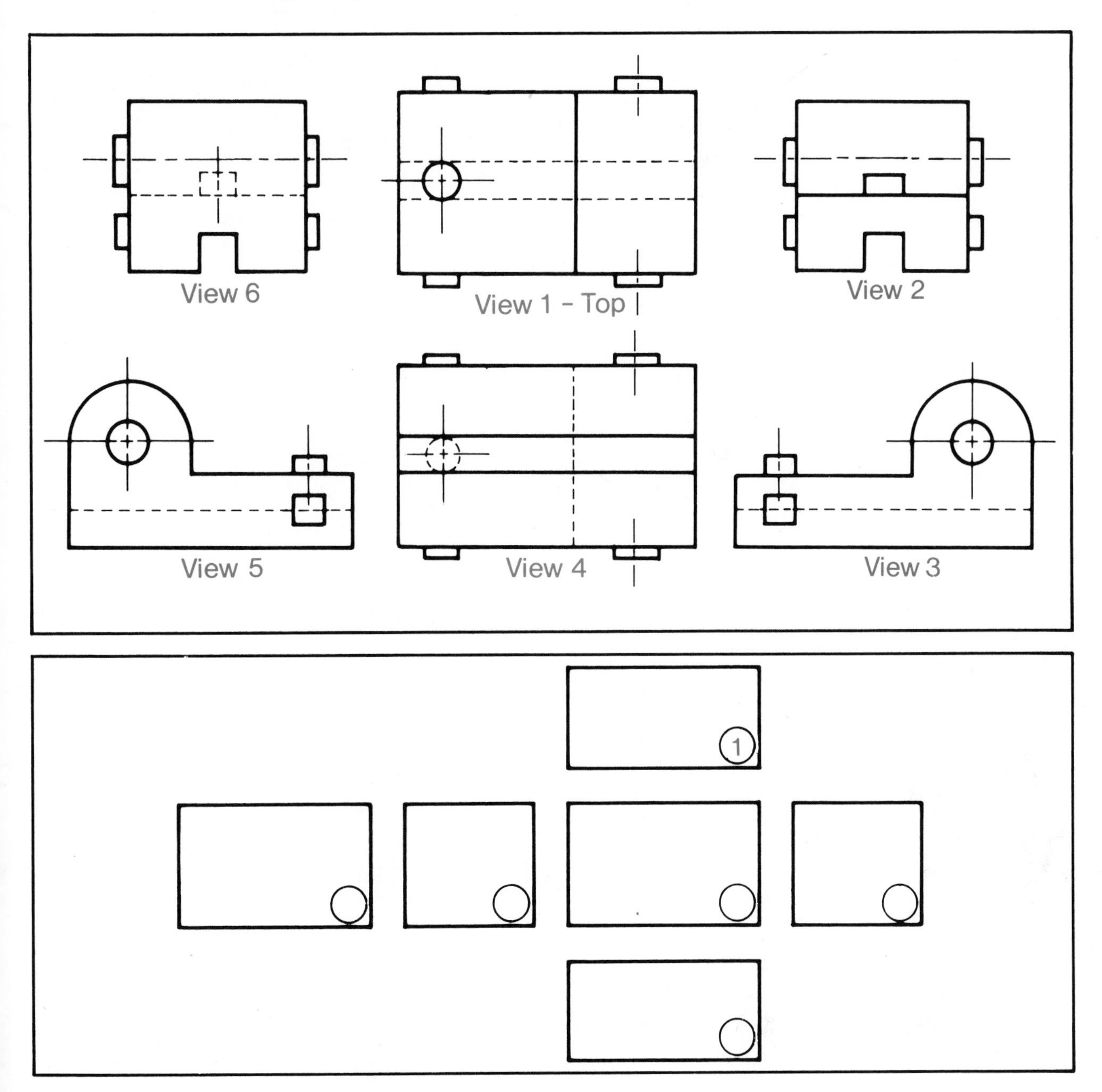

NAME

DATE

SCORE

NAME ____________

DATE ____________

SCORE ____________

EXERCISE A4-6

Front Views

Match these two sets of drawings. In the circles provided in the lower set, indicate the drawing number that belongs with each front view.

Drawing #1

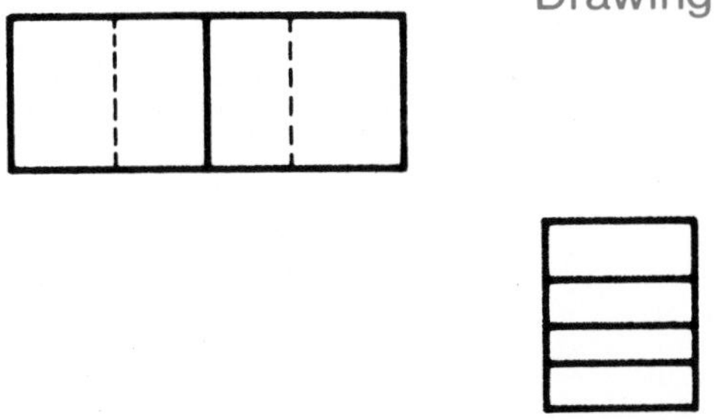

Drawing #2

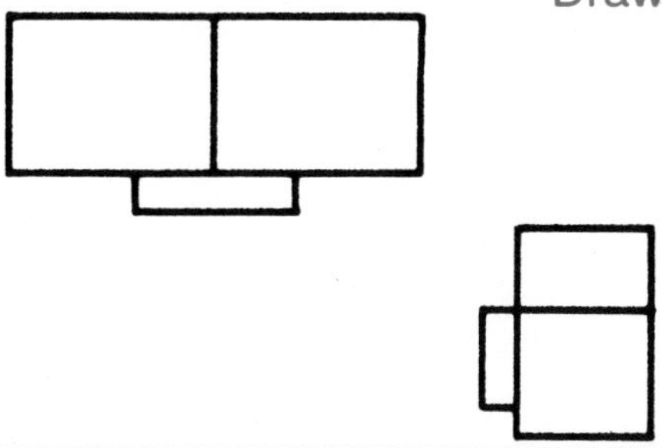

Drawing #3

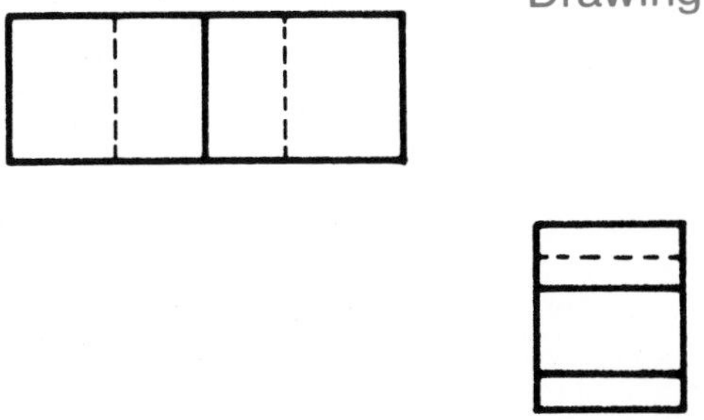

Drawing #4

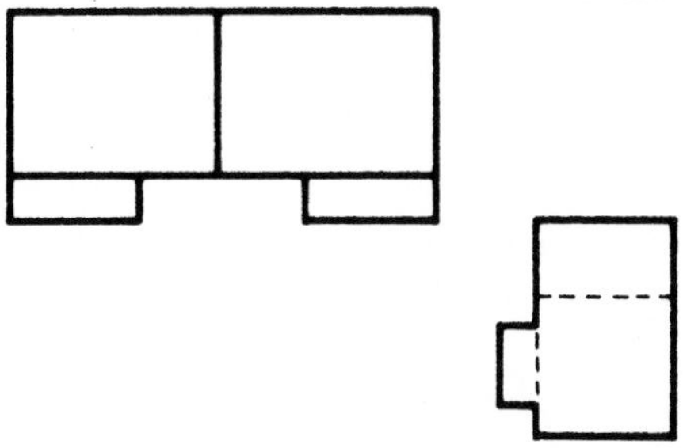

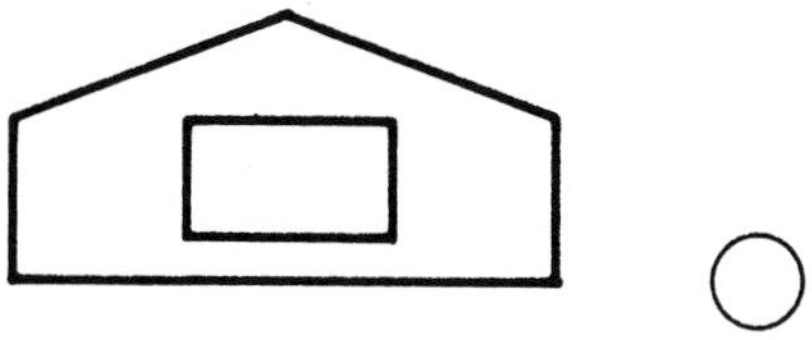

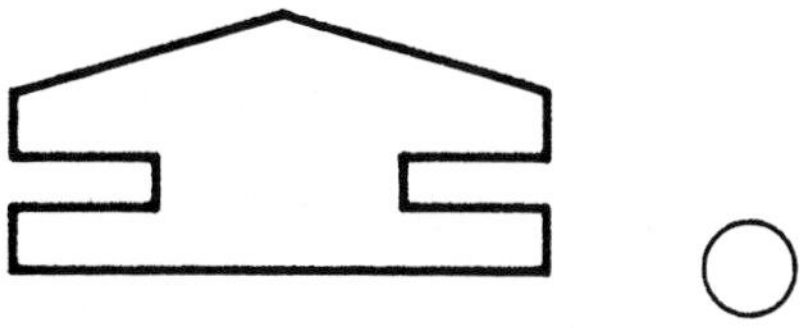

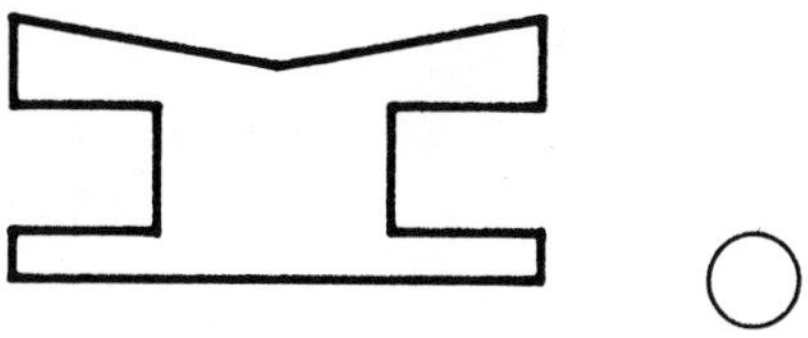

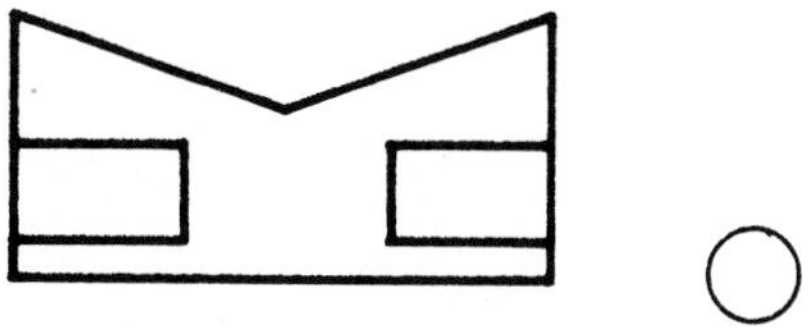

NAME ____________

DATE ____________

SCORE ____________

EXERCISE A4-7

Top Views

Match these two sets of drawings. In the circles provided in the lower set, indicate the drawing number that belongs with each top view.

Drawing #1

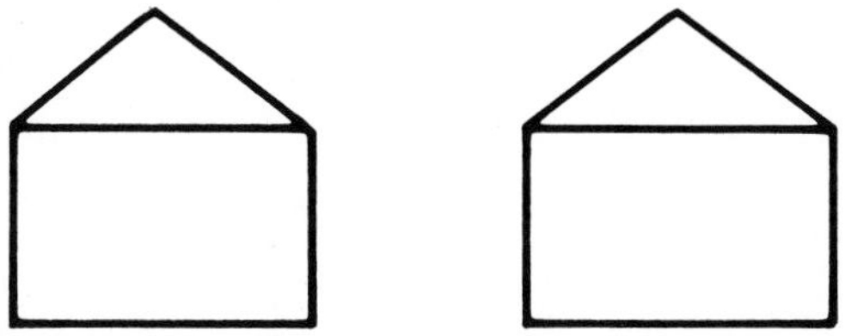

Drawing #2

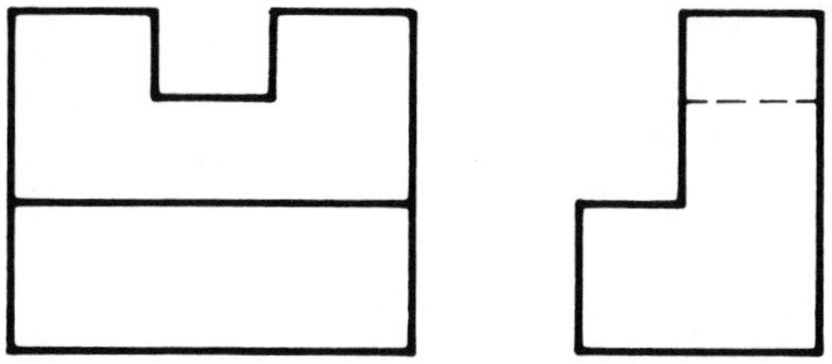

Drawing #3

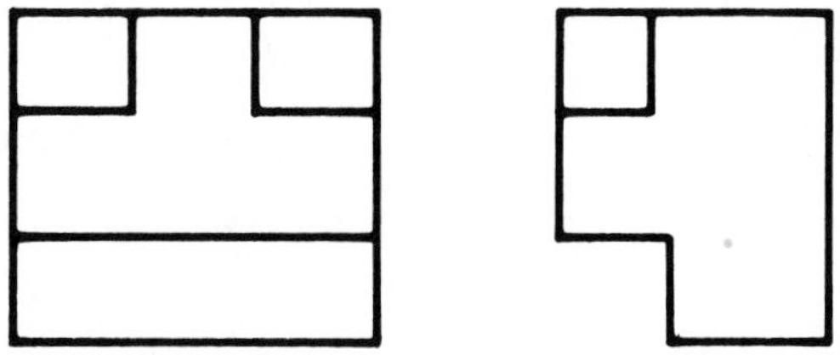

Drawing #4

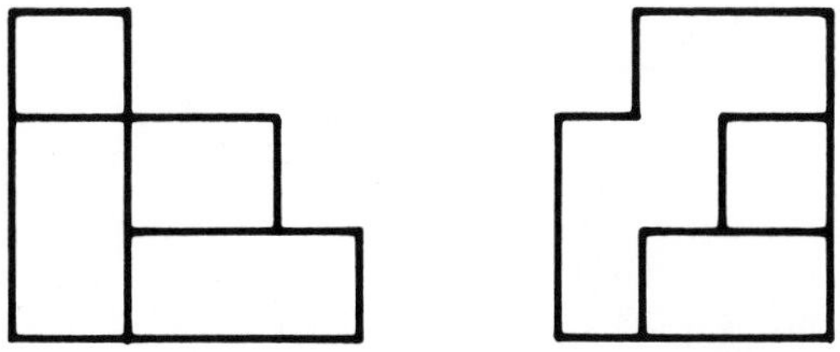

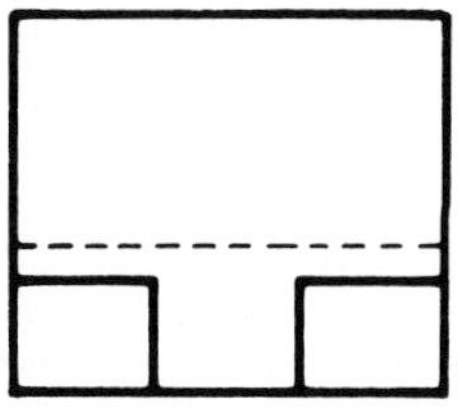

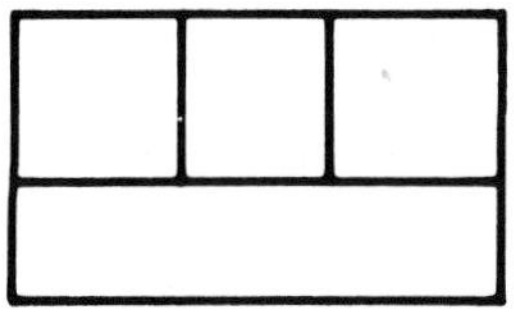

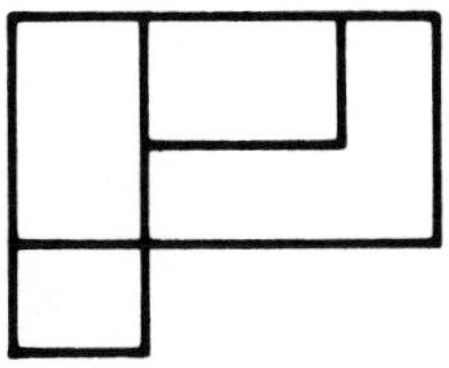

NAME ____________

DATE ____________

SCORE ____________

EXERCISE A4-8

Side Views

Match these two sets of drawings. In the circles provided in the lower set, indicate the drawing number that belongs with each side view.

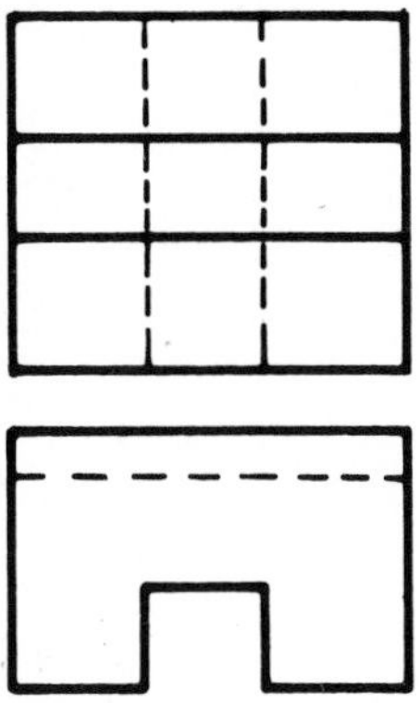

Drawing #1

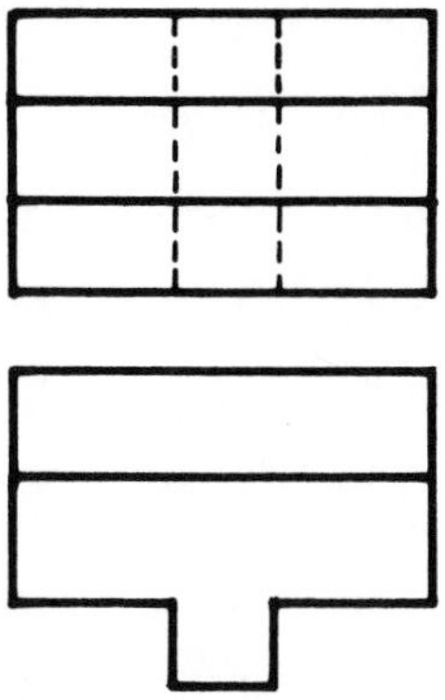

Drawing #2

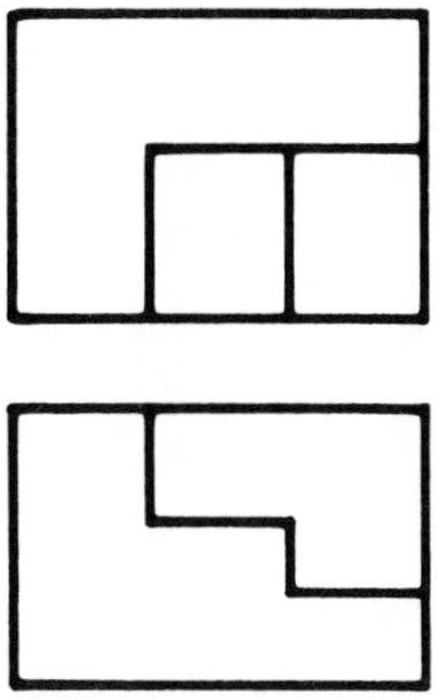

Drawing #3

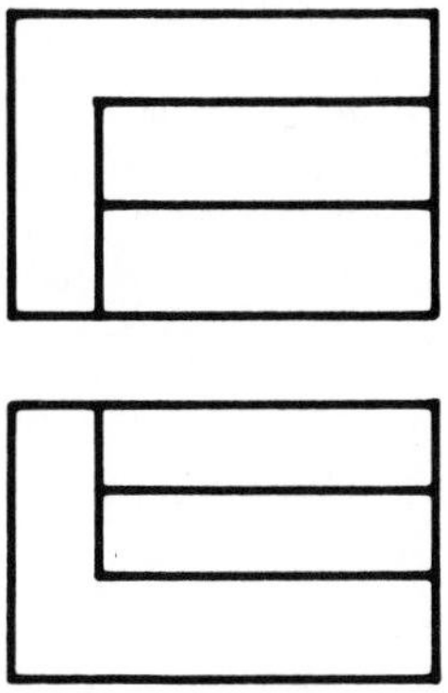

Drawing #4

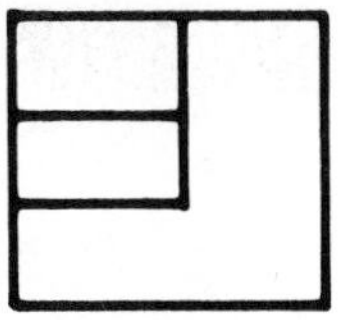

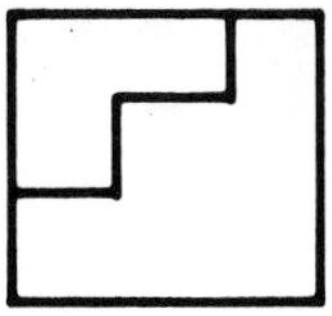

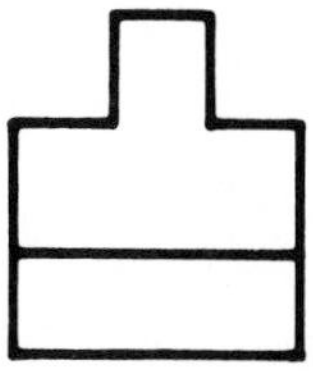

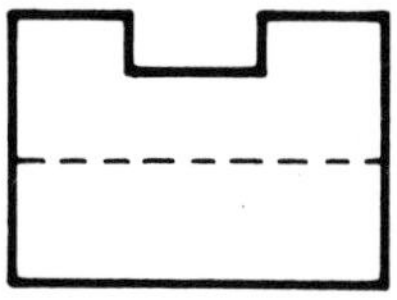

NAME ______________

DATE ______________

SCORE ______________

EXERCISE A4-9

Surfaces, Lines, and Points

Study the print. Then complete the statements following the print.

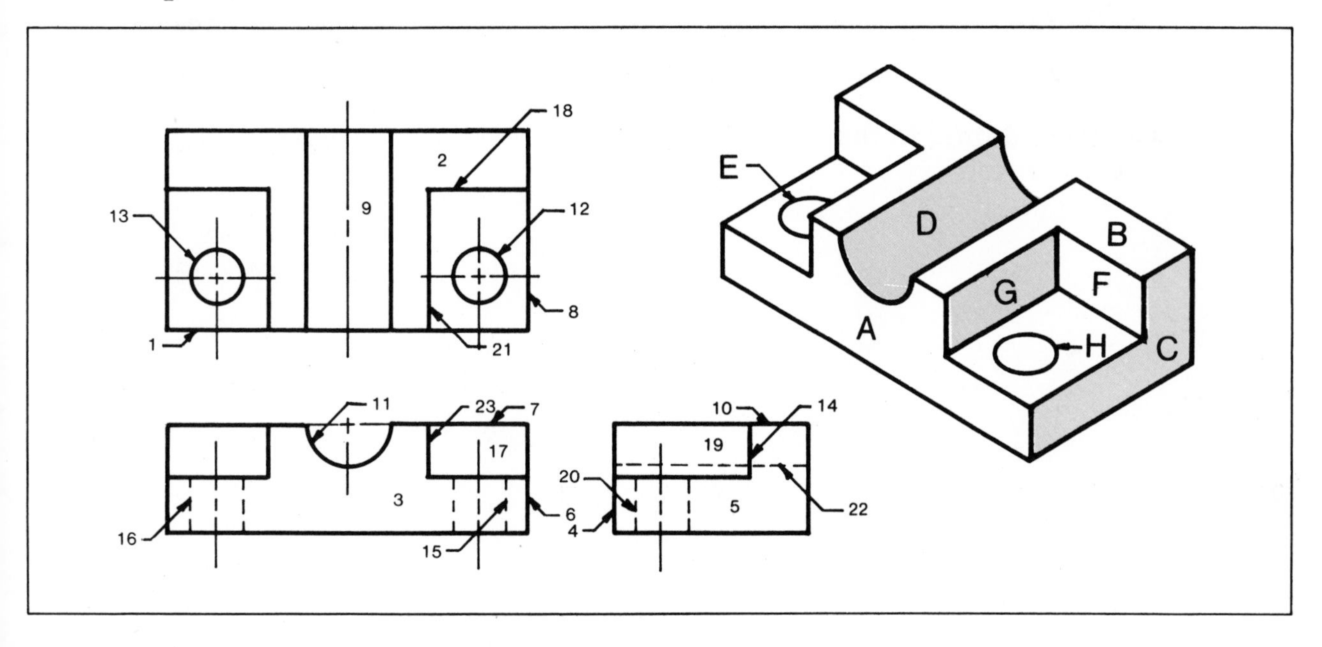

1. Surface *A* in the pictorial view is represented by surface ______ in the front view, line ______ in the top view, and line ______ in the side view.

2. Surface *B* in the pictorial view is represented by surface ______ in the top view, line ______ in the front view, and line ______ in the side view.

3. Surface *C* in the pictorial view is represented by surface ______ in the side view, line ______ in the front view, and line ______ in the top view.

4. Surface *D* in the pictorial view is represented by surface ______ in the top view and line ______ in the front view.

5. Hole *E* in the pictorial view is represented by line ______ in the top view and line ______ in the front view.

NAME ______________

DATE ______________

SCORE ______________

6. Surface *F* in the pictorial view is represented by surface ______, line ______, and line ______ in the orthographic projection.

7. Surface *G* in the pictorial view is represented by surface ______, line ______, and line ______ in the three-view print.

8. Surface *H* in the pictorial view is represented by line ______ in the front view.

9. Holes ______ and ______ in the top view are represented by lines ______ and ______ in the front view, and by line ______ in the side view.

10. Dashed lines in the side view represent the depth of surface ______ in the top view.

CHAPTER 5

THE ALPHABET OF LINES

OBJECTIVES

After studying this chapter, you will be able to:

- Identify the different types of drafting lines.
- Explain what each type of line represents and how it is used.
- List the basic steps followed by a drafter in making a drawing.
- Describe the location of auxiliary lines, such as dimension, extension, leader, and center lines, which are not part of the object.

The ability to visualize an object by studying a working drawing is essential. It is equally important to be familiar with other parts of the drawing. Dimensions provide information about the size of the object. Notes and symbols provide the additional information necessary to manufacture the part.

Drawings are made up of lines. The different types of lines and their uses are discussed in this chapter.

THE ALPHABET OF LINES

In writing, different letters of the alphabet are used to form words. By a similar concept, in drafting, different types of lines are used in making a drawing. The *alphabet of lines* includes many different types of lines, Figure 5-1. These lines may be thick, medium, or thin, and they may be continuous, broken, or irregular. Each type of line conveys different kinds of information to the print reader.

The manufacturing industries have established a commonly accepted standard for line drawing. This standard is known as the American Standards Association Conventional Line Symbols.

Mechanical drawings consist of ten commonly used types of lines.

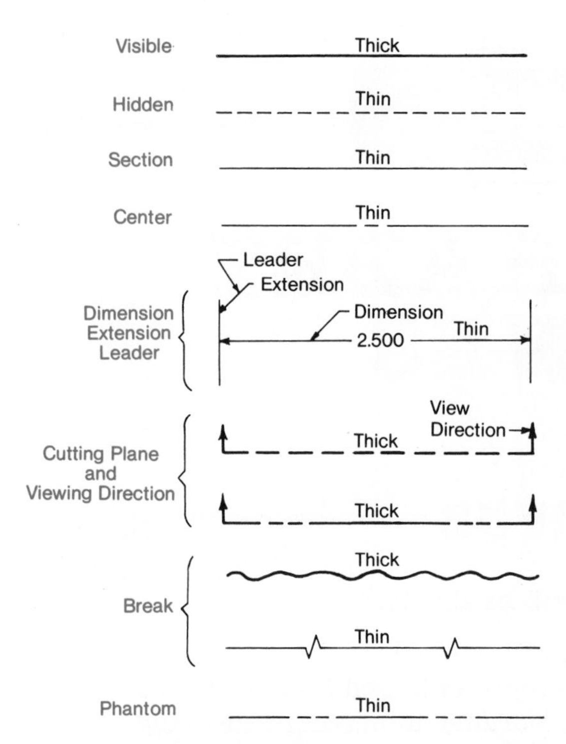

Fig. 5-1 Alphabet of lines and line weights

Each type provides different information about the object.

By following the basic steps of making a drawing, it is easy to understand how each type of line is used. Most drawings are started by locating the center of the object and the object's features. The center lines should be located to allow equal spaces between the views and the border lines of the drawing. This will ensure room for dimensions and other necessary information. The drafter also tries to arrange the views so that the drawing is pleasing to the eye.

Visible Object Lines

Visible object lines are thick continuous lines. Object lines outline the object and represent all edges that are visible in the view. See Figure 5-2.

Hidden Lines

Hidden lines are short dashes of medium weight, Figure 5-2. They are used to show edges and other features of the object that cannot be seen from the particular view. Hidden lines should alert the print reader to study other views for more information.

Hidden lines are used to clarify hidden details. Sometimes hidden lines are omitted if they would confuse the print reader or complicate the drawing.

Section Lines

Section lines are thin, continuous, and closely spaced lines usually drawn at a 45° angle. Section lines represent the cut surface in a sectional view, Figure 5-3. Some industries use other line forms to represent different materials. (Refer to Chapter 11, Sectional Views.)

Center Lines

Center lines are thin broken lines of long and short dashes, Figure 5-2. They are used to center objects that are symmetrical, especially circular parts. Center lines are also used as reference lines for dimensions. The drafter usually draws these lines first.

Dimension, Extension, and Leader Lines

The dimensions of the features of an object are indicated by *dimension lines* and *extension lines.* These lines are thin and continuous, and limit the extent and direction of dimensions, Figure 5-2.

Extension lines are normally continuations of the object lines, locating the features to be dimensioned.

Dimension lines are drawn between the extension lines, and terminate with arrowheads.

Holes and other features may be dimensioned or described by notes with a *leader line.* The leader line is usually

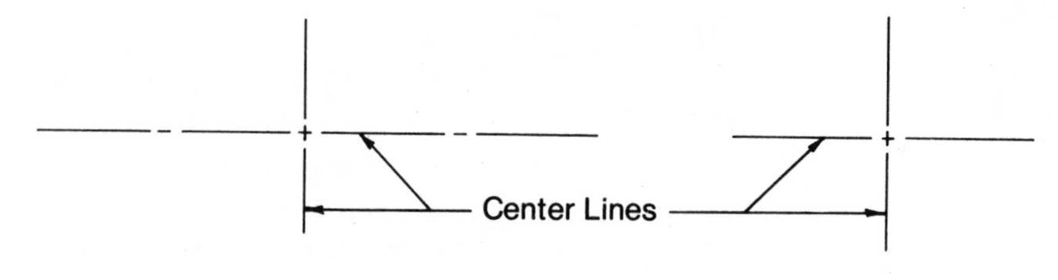

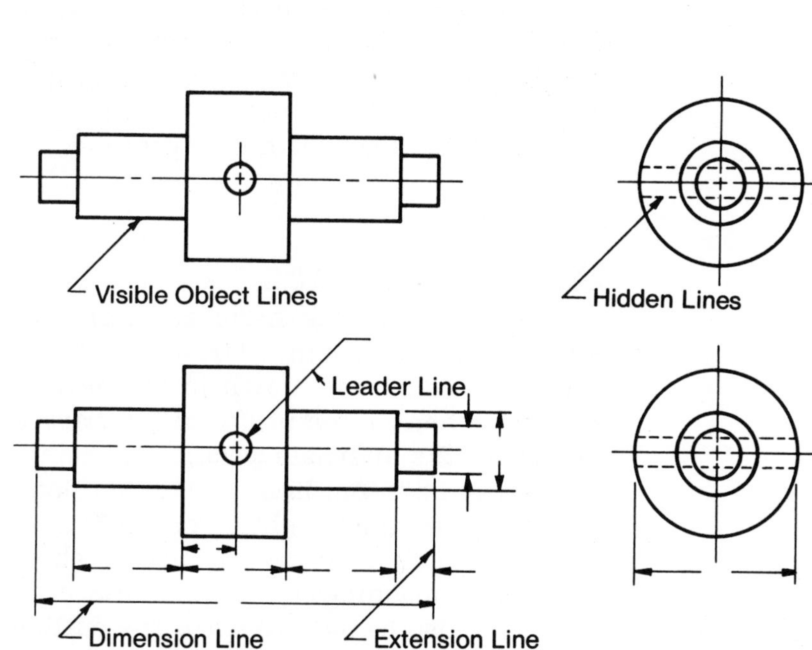

Fig. 5-2 Common types of lines

drawn at an angle from the note to the feature being described, Figure 5-2.

Cutting Plane Lines

The regular views of complicated objects cannot always show the inside details clearly. Therefore, a *cutting plane line* is drawn to represent the location of the cutting plane. The cutting plane line is a heavy line consisting of alternating long dashes and pairs of short dashes. Arrowheads are drawn at the end of the line to indicate the direction from which the view is to be observed, Figure 5-3.

Capital letters are used at the end of the arrowheads to identify the section

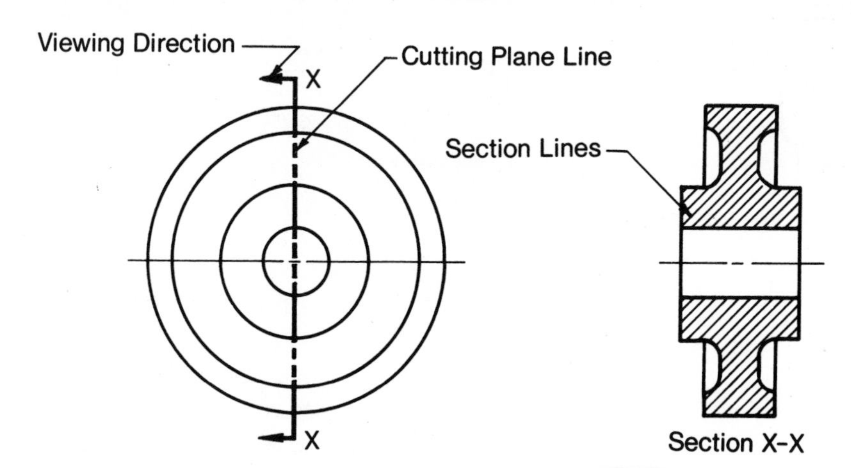

Fig. 5-3 Section view lines

view. (See Figure 5-3.) This is most useful when more than one section view are drawn.

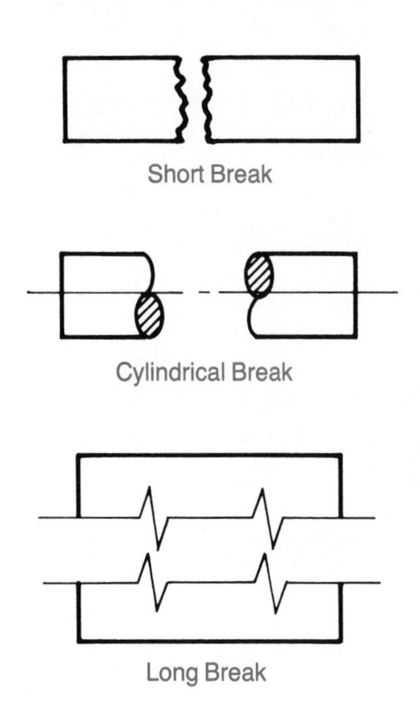

Fig. 5-4 Break lines

Break Lines

Long objects that have a uniform shape can be drawn shorter than their actual length to reduce space on the drawing. This is accomplished by *break lines.* If the cross-sectional area is small, a heavy wavy line is used to represent a short break. When the section is long, a light line with a zigzag is used. A special "S" break line can be used for cylindrical objects. These are shown in Figure 5-4.

Phantom Lines

Many objects have components which may be moved into various positions. The movable part is drawn in the normal manner when in one position. The part is shown in the alternate position by *phantom lines.* Phantom lines are broken long dashes of medium weight, Figure 5-5.

Phantom lines are also used to show finished or machined surfaces on castings. They can also be used to show the outline of the rough casting on a finished part.

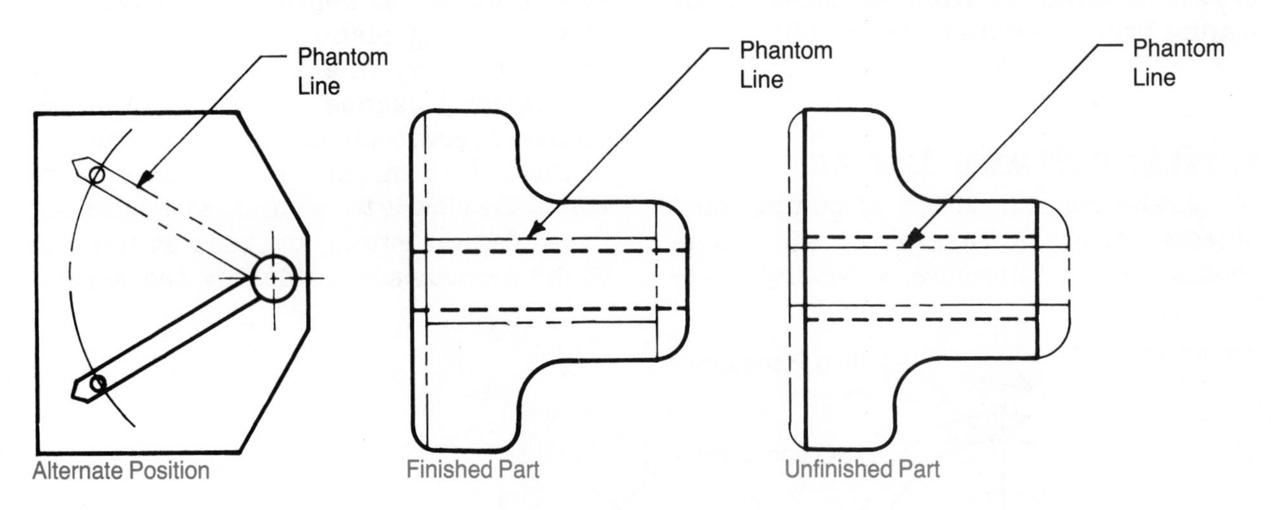

Fig. 5-5 Phantom lines

REVIEW

This review is provided to serve as reinforcement study material. Fill in the appropriate word(s) to complete the sentences below.

1. Heavy continuous lines are called ______________________.
2. Heavy dashed lines are called ______________________.
3. Thin continuous lines are called ______________________,

______________________, and ______________________.

4. Thin dashed lines are called ______________________.
5. Heavy wavy lines are called ______________________.
6. A combination of thin, continuous, and dashed lines is called

______________________.

NAME ______________________

DATE ______________________

SCORE ______________________

EXERCISE A5-1

Line Identification

Identify the lines (*A* through *J*) noted on the drawing. Write your answers in the spaces provided below the drawing.

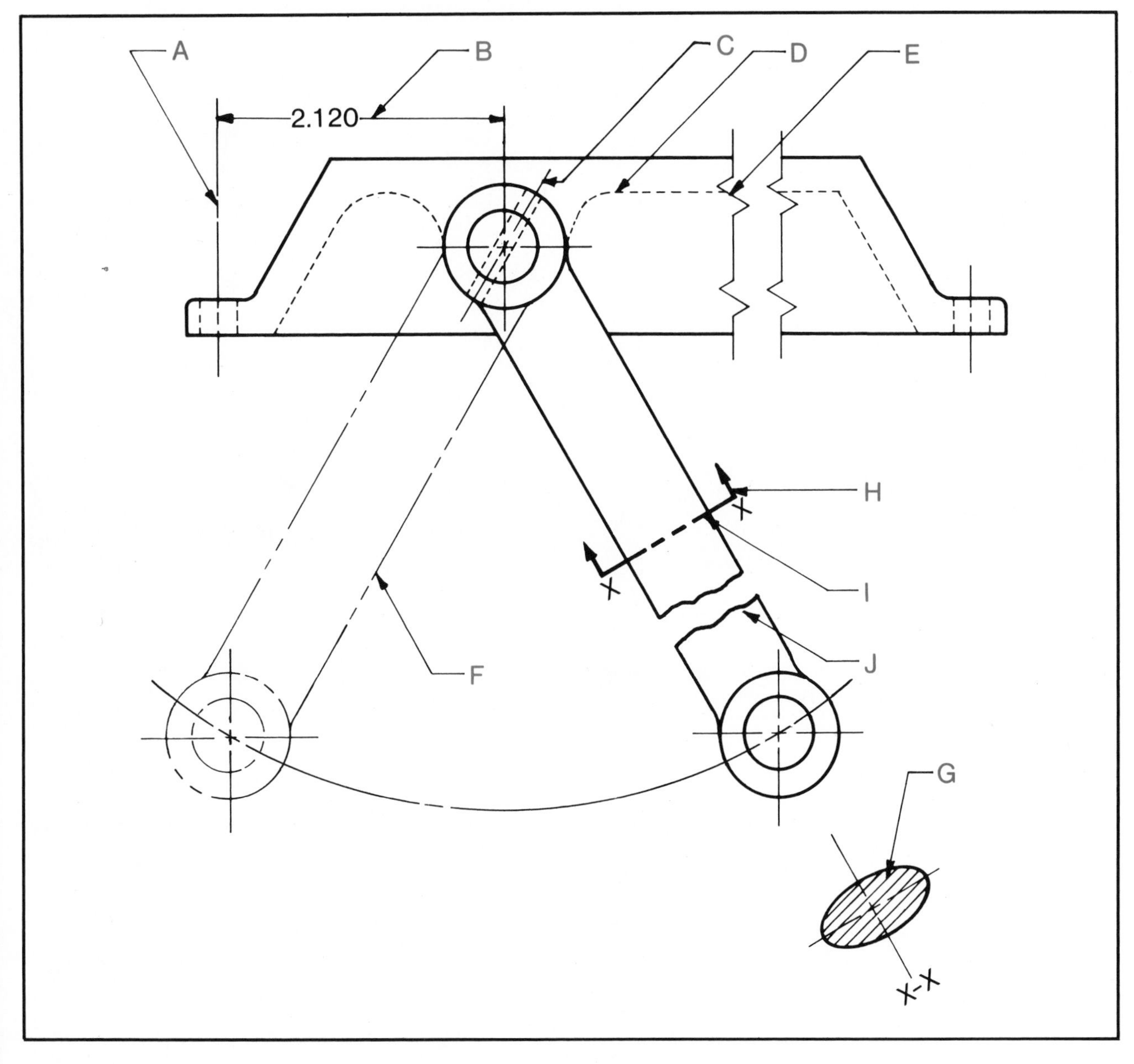

A ______________________ E ______________________ H ______________________

B ______________________ F ______________________ I ______________________

C ______________________ G ______________________ J ______________________

D ______________________

NAME ______________________

DATE ______________________

SCORE ______________________

EXERCISE A5-2

Missing Lines

Draw in the missing line or lines on the orthographic views. A sample is provided.

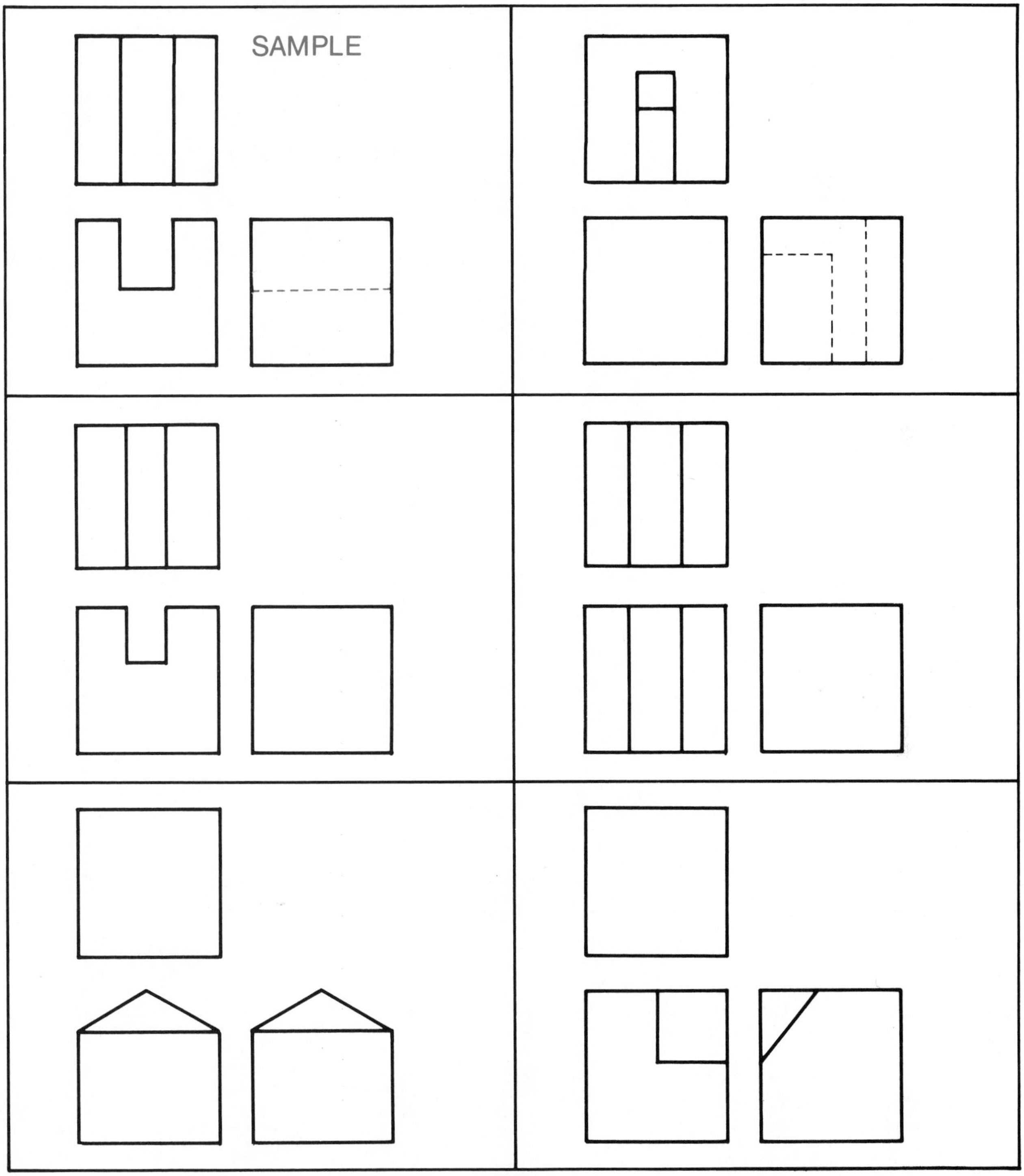

SAMPLE

NAME ____________________

DATE ____________________

SCORE ____________________

EXERCISE A5-3

Isometric Sketching

Draw isometric sketches on the grids of these three-view drawings.

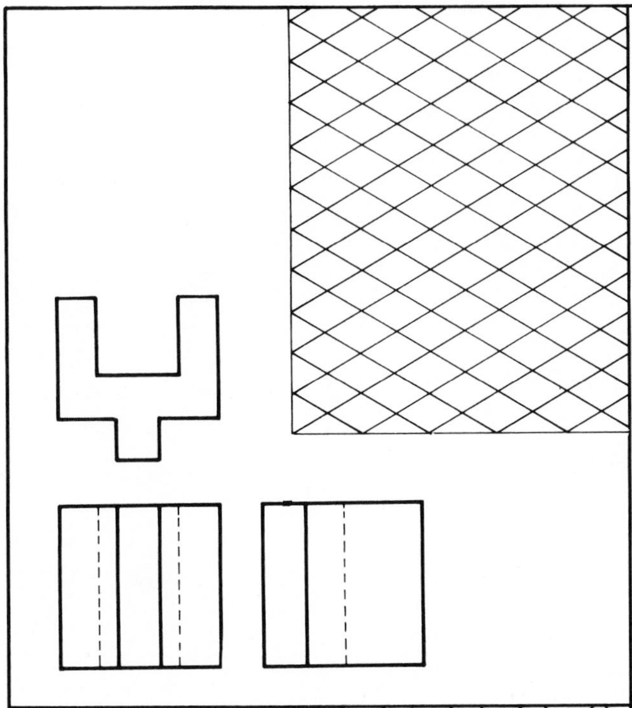

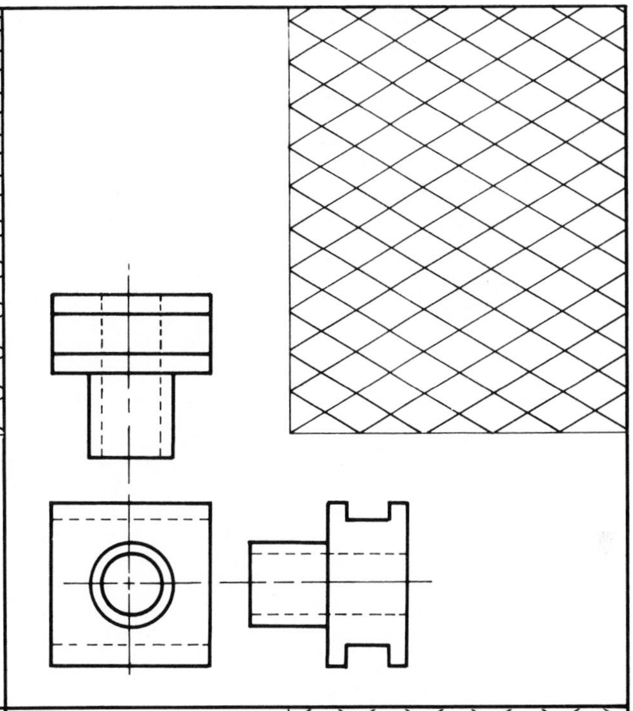

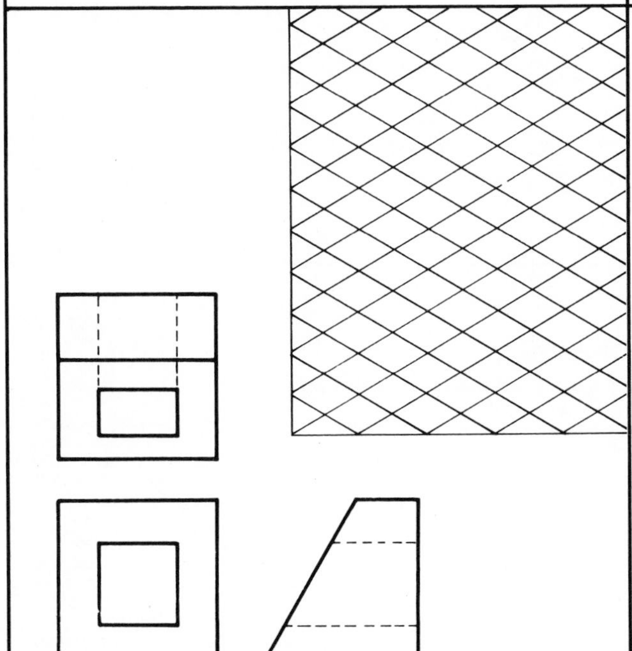

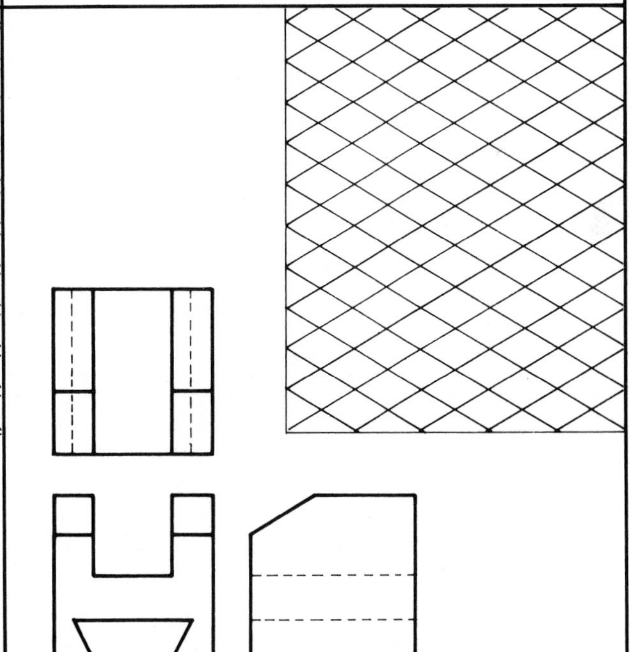

NAME ____________

DATE ____________

SCORE ____________

EXERCISE A5-4

Three-View Sketching

Sketch the three views of each isometric drawing.

NAME ____________

DATE ____________

SCORE ____________

EXERCISE A5-5

Line Identification

Study Print #315409 below. Identify lines *A* through *H* shown on the print. Place your answers in the spaces provided.

Note that Print #315409 uses the aligned dimensioning method.

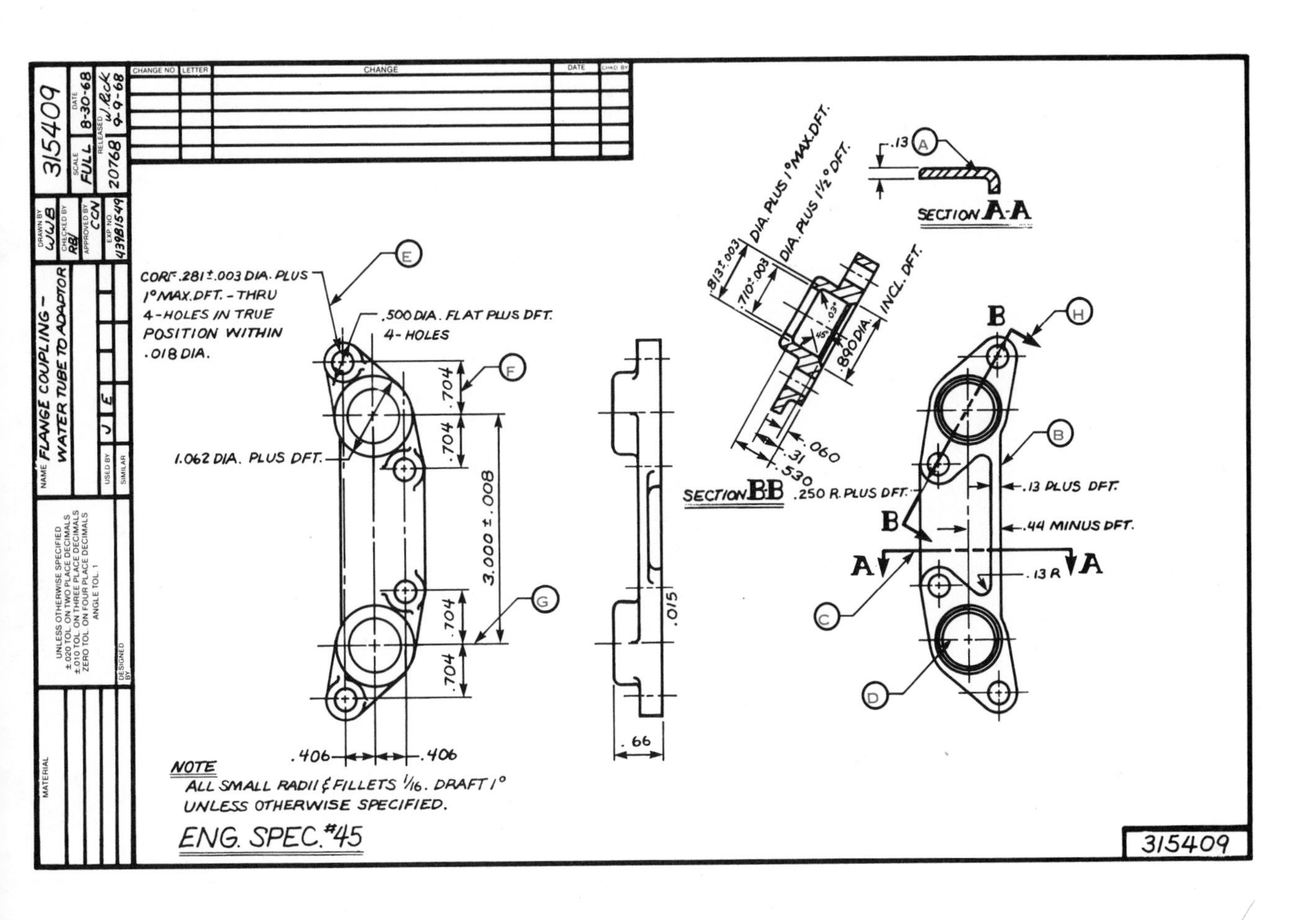

Line A ____________________

Line B ____________________

Line C ____________________

Line D ____________________

Line E ____________________

Line F ____________________

Line G ____________________

Line H ____________________

NAME ____________________

DATE ____________________

SCORE ____________________

CHAPTER 6

DIMENSIONING

OBJECTIVES

After studying this chapter, you will be able to:

- Explain the difference between size and location dimensions.
- List the rules of finding the size and location dimensions.
- Explain how to find the object and feature dimensions.
- Describe the types of dimensioning systems commonly used on industrial prints.
- Describe the ways of displaying fractional, decimal, and angular dimensions.

The worker making a part must know the size and location of all features of the object. This information is indicated on a print by dimensions placed between the extension lines of surfaces or center lines. A leader line with a dimensional number is often used for the sizing of holes. See Figure 6-1.

The production procedures will usually determine which dimensions are to be included on a print. The dimensions included should be those that will be useful to the machinist or toolmaker, rather than to the drafter. The machinist should not have to add or subtract to determine required machining information.

TYPES OF DIMENSIONS

Two types of dimensions are provided on prints. These types are *size dimensions* and *location dimensions.*

Size Dimensions

Size dimensions tell the size of the various features of the part, such as holes, slots, flat surfaces, arcs, and circles. The extension lines

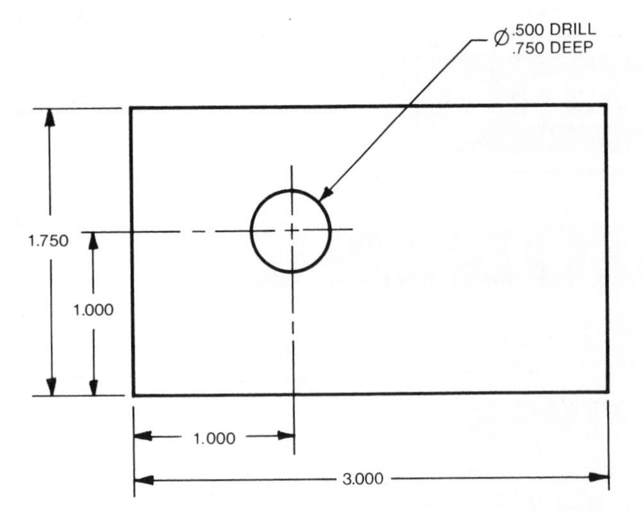

Fig. 6-1 Dimension, extension, and leader lines

(extensions of a surface) limit the extent of the dimensions. See Figures 6-1 and 6-2.

Location Dimensions

Location dimensions determine the position of one feature relative to another. If a hole is located at an indicated position from a surface, two dimensions are usually required. These should be the same dimensions used by the machinist in the production operation. Center lines are frequently used as references for location dimensions. See Figures 6-1 and 6-2.

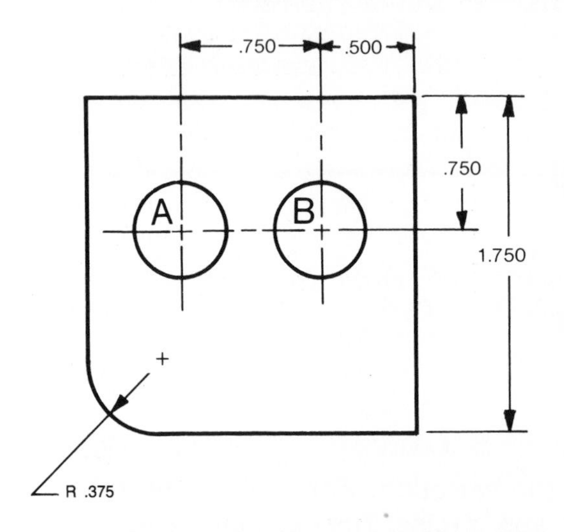

Fig. 6-2 Location dimensions for holes A and B

DIMENSIONING RULES

Understanding the basic rules of dimensioning will simplify the procedure of finding information on a print. It will also save time. Twelve common dimensioning rules are listed here. These rules cover most dimensioning situations.

1. Dimensions of the smaller part features are shown closest to the object.

2. The next-larger features are then dimensioned.

3. The longest or overall size is shown furthest from the object.

4. Dimensions are indicated on the view which shows the shape of the part feature.

5. Two dimensions are usually required, for feature size as well as location (see the hole in Figure 6-1).

6. Dimensions are placed off the object if possible.

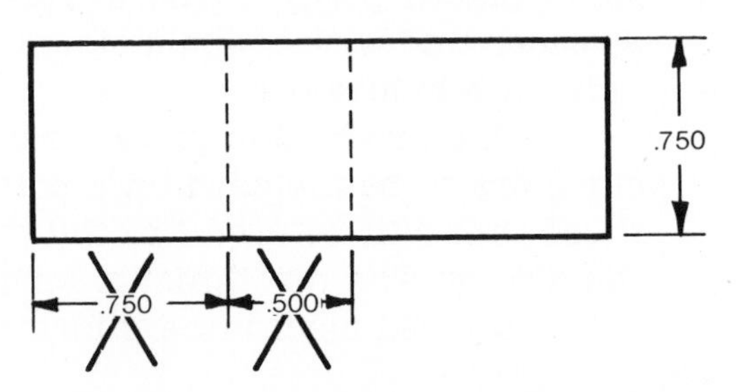

(A) Incorrect: Dimensions taken from invisible lines.

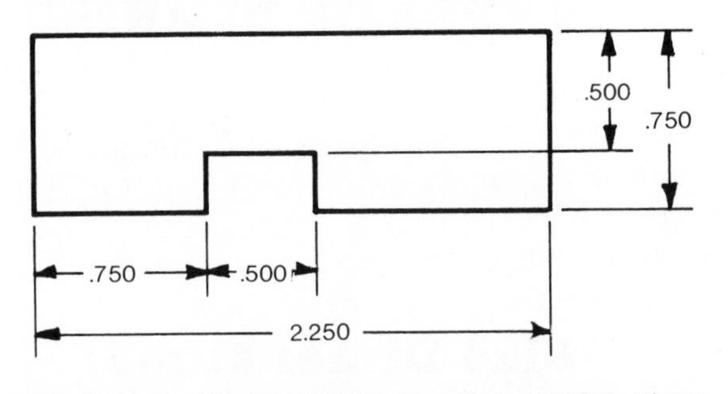

(B) Correct: Dimensions taken from visible lines.

Fig. 6-3 Dimensions from object lines

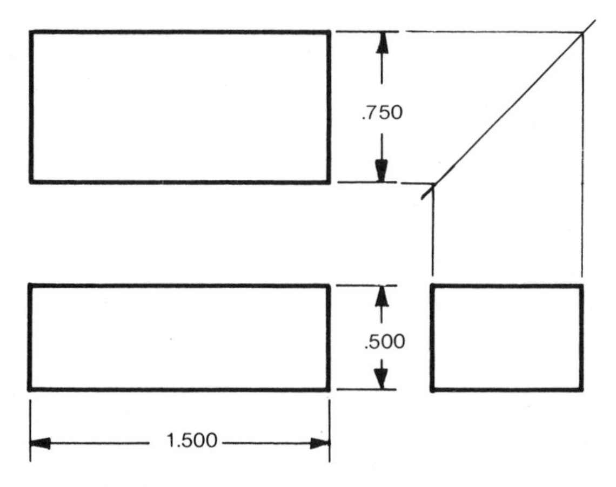

Fig. 6-4 Unidirectional dimensions

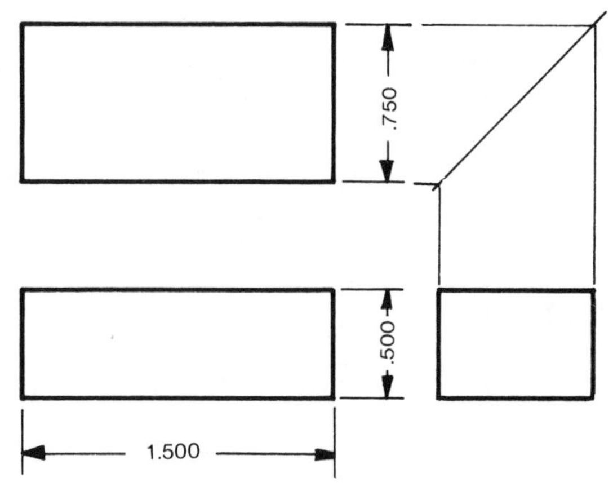

Fig. 6-5 Aligned dimensions

7. Most dimensions are placed in the spaces between the views.
8. Dimensions are shown from object lines, not from hidden lines (Figure 6-3).
9. Dimensions are normally located from a finished surface, center line, or base line.
10. Dimensions of circles or holes are given as diameters.
11. All circular features are located to their center.
12. All arcs (parts of a circle) are dimensioned by radii (one-half of a circle's diameter). Arcs are located to their centers (Figure 6-2).

PLACEMENT OF DIMENSIONS

Prints with *unidirectional dimensions* are most common today. On these prints, dimensions are placed to read only from the bottom (unidirectional), Figure 6-4. Unidirectional dimensions are the easiest to read.

The older practice of *aligned dimensions* is seldom used. However, aligned dimensions can still be found on older prints. In this system, dimensions are placed to read both from the bottom and the right side, Figure 6-5.

DIMENSIONING SYSTEMS

Rectangular Coordinate Dimensioning

In the *rectangular coordinate dimensioning system,* all dimensions originate from two or three surfaces that are perpendicular to one another.

The dimensioning systems used on many industrial prints do not follow any standard technique. Dimensions are usually given from finished surfaces or center lines. However, the location of these dimensions seems to be arbitrary. Dimensions are located from the front, back, or top side, or from any other surface that can be identified easily, Figure 6-6.

Base Line Dimensioning

Dimensions are given from base lines on precision-work prints. *Base lines* are usually finished surfaces at

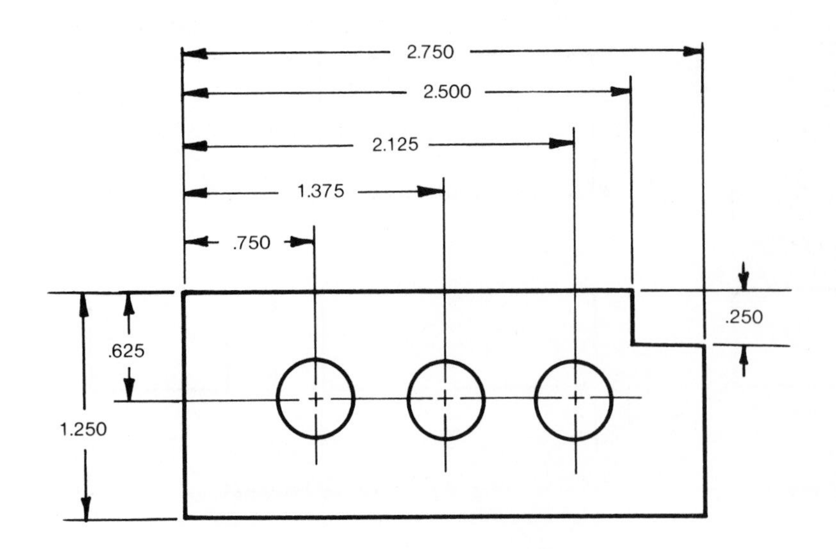

Fig. 6-6 Rectangular coordinate dimensioning

right angles, or important center lines, Figure 6-7. These base lines are often referred to as *datum planes. Datums* are points, lines, planes, cylinders, and so on, assumed to be exact for purposes of computation. From datums, the locations or geometric relationships (form) of features of a part may be established. Datums are usually identified by boxed-in letters, Figure 6-8.

Features which are selected to serve as datums must be easily recognized. These features should also be based primarily on the functional use of the design. To be useful for measuring, a datum feature on an actual piece must be accessible. This will allow measurements from the datum to be made during manufacture and inspection.

Joining features on mating parts may also be used as datums to ensure assembly and to aid in tool and fixture design.

Ordinate Dimensioning

Ordinate dimensioning is a newer type of base-line dimensioning. In this system, all dimensions are measured from two to three perpendicular datum planes, Figure 6-9. The datum planes are indicated as zero coordinates. Dimensions from these datum planes are shown on extension lines

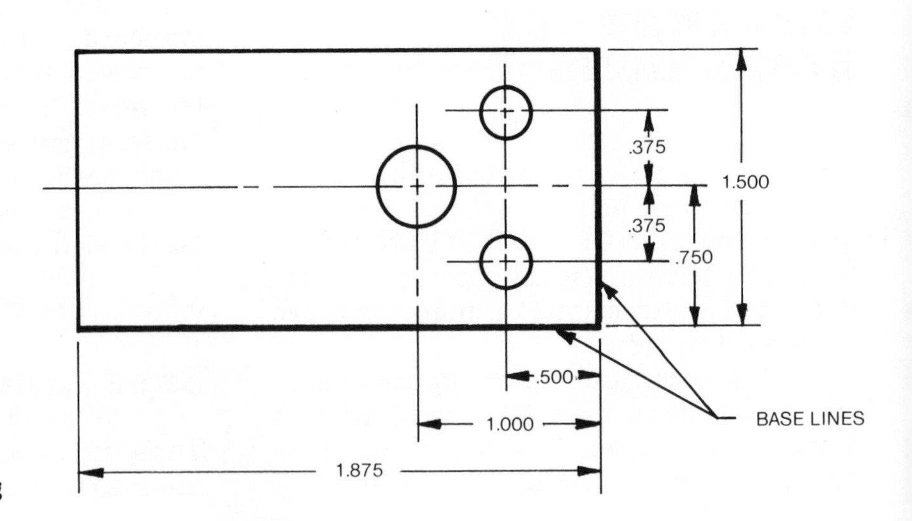

Fig. 6-7 Base line dimensioning

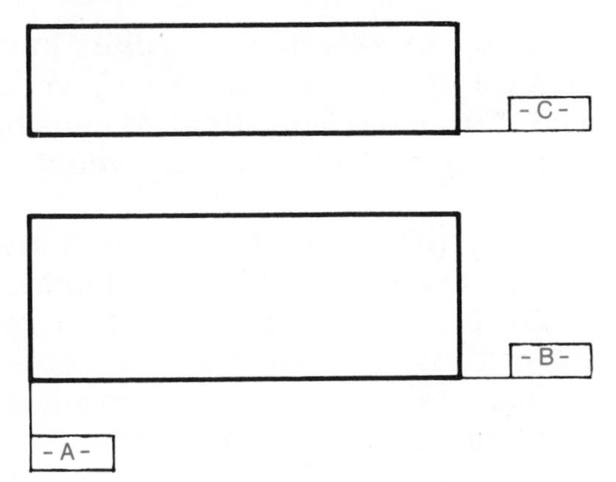

Fig. 6-8 Datum surfaces

without the use of dimension lines or arrowheads. Feature sizes such as hole diameters are indicated in a separate table.

Tabular Dimensioning

Tabular dimensioning is another type of base-line dimensioning. In this system, dimensions from perpendicular datum planes are listed in a table rather than on the pictorial portion of the drawing. See Figure 6-10. The *X*, *Y*, or *Z* coordinate dimensions from the indicated datums are listed in the table. This method is used for drawings which require a large number of similarly shaped features. Both the ordinate and tabular dimensioning systems are used extensively for numerical control prints.

DIMENSIONAL NOTATIONS

The main purpose of any industrial print is to guide the machinist or production department in producing the object illustrated on the print. The object lines on the print indicate the shape of the object. The dimensions indicate its size. Three main types of dimensions are commonly used in industrial prints: *fractional, decimal,* and *angular dimensions.*

Fractional Dimensions

Before the time of precision machining, all dimensions on prints were in fractional units. The inch was the standard

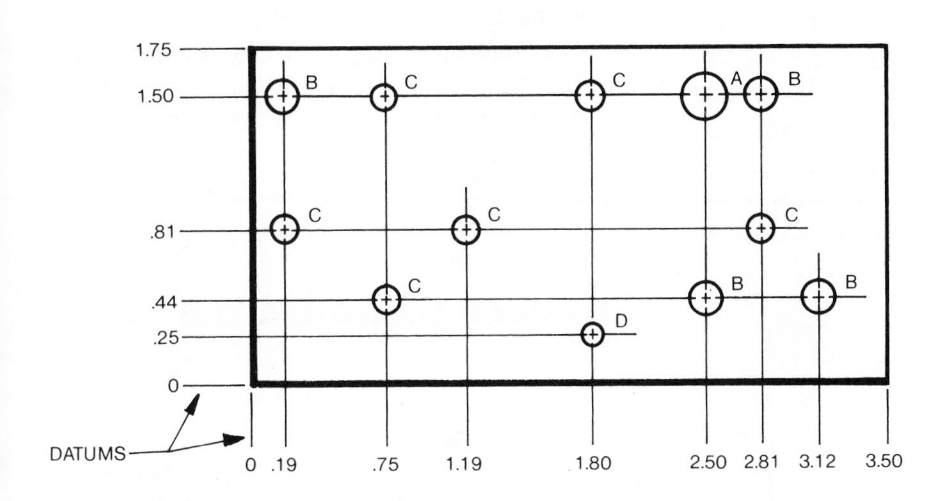

SIZE SYMBOL	A	B	C	D
HOLE DIAMETER	.350	.188	.156	.125

Fig. 6-9 Ordinate dimensioning

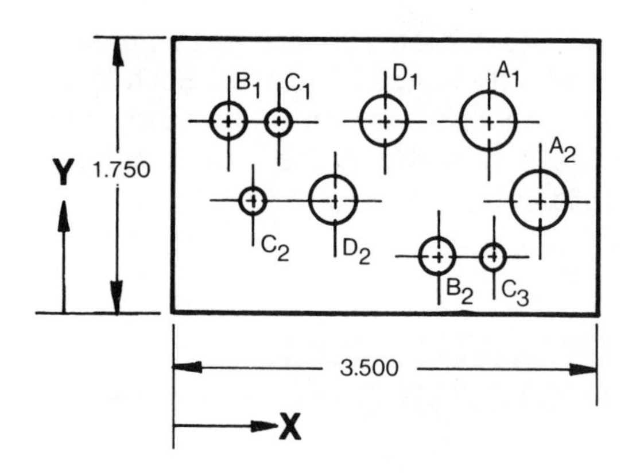

	REQD.	2	2	3	2
	HOLE DIA.	.250	.190	.150	.200
POSITION		HOLE SYMBOL			
X →	Y ↑	A	B	C	D
3.000	1.500	A_1			
3.250	.780	A_2			
.500	1.500		B_1		
2.750	.500		B_2		
.780	1.500			C_1	
.680	.780			C_2	
3.100	.500			C_3	
2.250	1.500				D_1
1.500	.780				D_2

Fig. 6-10 Tabular dimensions

unit of measurement. Fractional divisions of the inch were also used, such as $\frac{1}{2}$, $\frac{1}{4}$, $\frac{1}{8}$, $\frac{1}{16}$, $\frac{1}{32}$, and $\frac{1}{64}$. Most older measuring tools included these units. The *tolerance* is the amount of allowable variance from a desired dimension. Tolerance was also specified in fractional units. For example, a dimension of $4\frac{5}{16}$ inches might have a tolerance of plus or minus (±) $\frac{1}{16}$ of an inch. Therefore, any final dimension between $4\frac{4}{16}$ and $4\frac{6}{16}$ inches would be acceptable.

Some prints contain a combination of both fractional and decimal dimensions. The fractional dimensions are used for less critical part features. The decimal dimensions are used for features requiring more accurate machining or location.

±.020 TOL. ON TWO PLACE DECIMALS
±.010 TOL. ON THREE PLACE DECIMALS
ZERO TOL. ON FOUR PLACE DECIMALS
ANGLE TOL. ± 1°
DO NOT SCALE

DESIGNED
FOR

Fig. 6-11 Tolerance block

Decimal Dimensions

Decimal dimensions are used on prints when the design requires a very close fit. The base unit used is the inch. The divisions of the inch used are hundredths (.01), thousandths (.001), and ten-thousandths (.0001). Sometimes hundred-thousandths (.00001) are used.

Most industries have standardized on the decimal inch system. All dimensions on their prints are made in decimals. The only change from this standard would be for components that are still produced with large tolerance, such as rope, cable, and some fastening devices.

Tolerances for decimal dimensions are listed in the tolerance block, Figure 6-11, or are indicated as part of the dimensions, Figure 6-12.

Angular Dimensions

The size of an angle is given in degrees. A circle contains 360 degrees (360°). One degree equals 60 minutes (60′). One minute equals 60 seconds (60″). A notation of *46° 10′ 30″* means 46 degrees, 10 minutes, and 30 seconds.

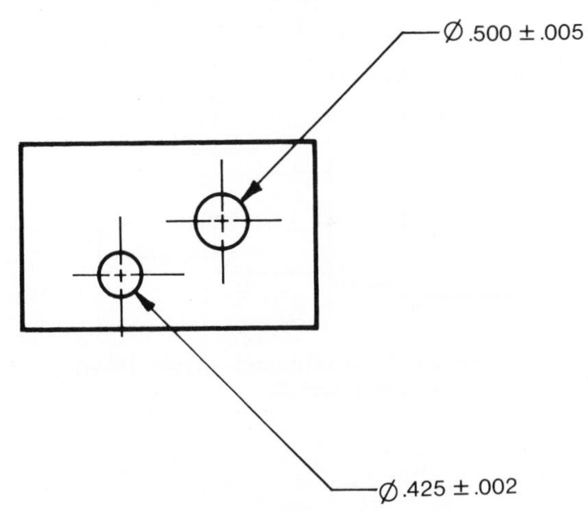

Fig. 6-12 Decimal tolerance

The angular tolerance standard may also be specified in the tolerance section of the title block or indicated on the print along with the dimension. See Figure 6-13.

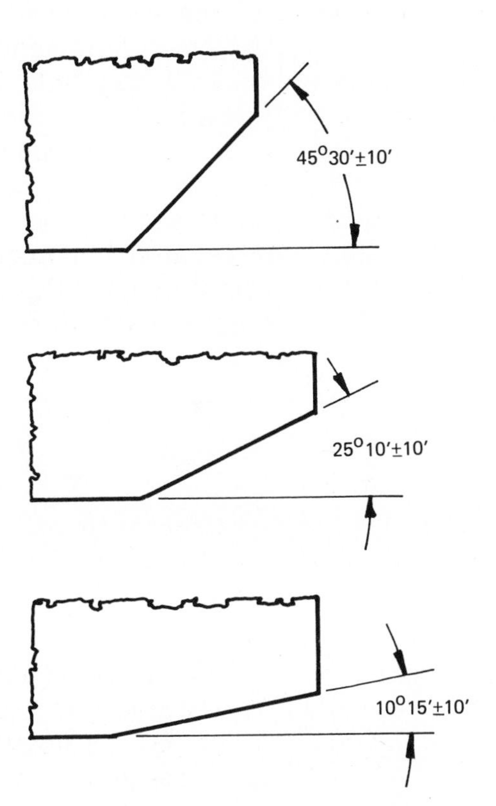

Fig. 6-13 Angle dimensions

METRIC DIMENSIONS

Most countries have standardized on the *metric dimensions system.* This system employs the millimeter rather than the inch as the standard unit of length.

The metric system also uses different values for weight and volume. However, these are of little consequence to the print reader, since most dimensions on a print are units of length.

TOLERANCE ACCUMULATION

How the dimensions are placed on the print may influence the actual dimensions between features on the finished part. There are three basic methods of dimensioning that control tolerance accumulation. Additional information on dimension tolerance is given in Chapter 7.

Chain Dimensioning

The maximum variations between two features is equal to the sum of the tolerances between these two features. In Figure 6-14(a) the tolerance accumulation between X and Y is ±0.15.

Base-Line Dimensioning

The maximum variation between two features is equal to the sum of the tolerance of the two dimensions from the base line to the features. In Figure 16-14(b), the tolerance accumulation between X and Y is ±0.10.

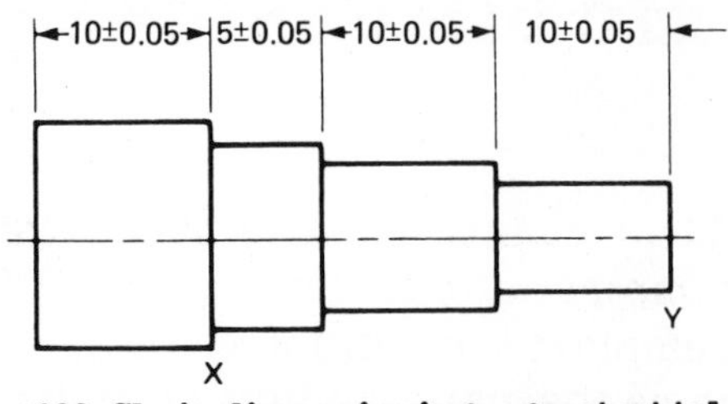

(A) Chain dimensioning—greatest tolerance accumulation between X and Y.

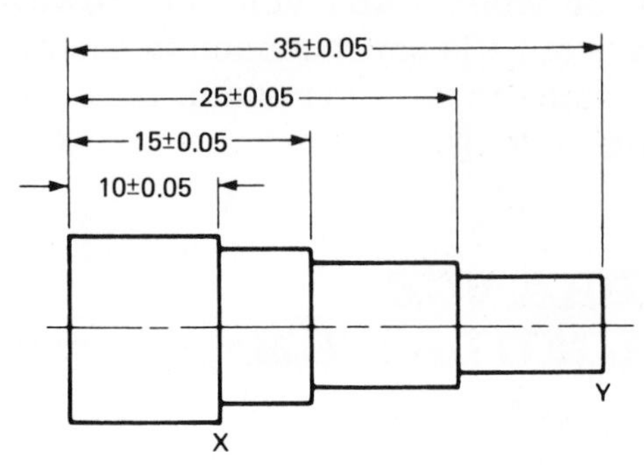

(B) Base-line dimensioning—lesser tolerance accumulation between X and Y.

Fig. 6-14 Tolerance accumulation (*ASME, from ANSI Y14.5M-1982*)

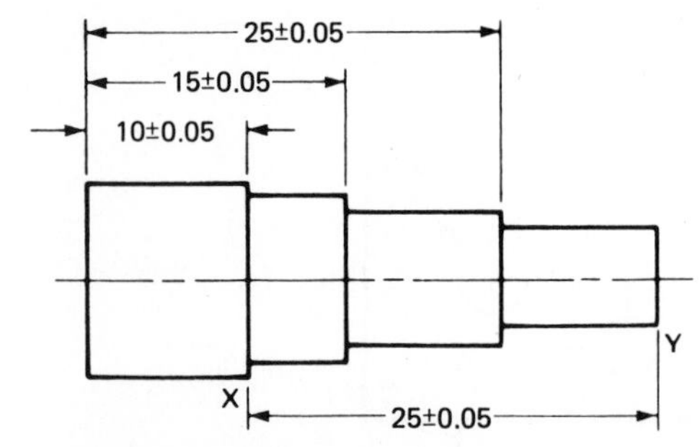

(C) Direct dimensioning—least tolerance between X and Y.

Direct Dimensioning

The maximum variation between two features is equal to the dimension between these features. In Figure 16-14(c) the tolerance between *X* and *Y* is the least amount or ±0.05.

REVIEW

This review is provided to serve as reinforcement study material. Fill in the appropriate word(s) to complete the sentences below.

1. Dimensions are used to indicate the ______________________ and ______________________ of features.

2. Dimensions of smaller features are located ______________________ to the object.

3. Overall dimensions are located ______________________ from the object.

4. Each feature size or location requires ______________________ dimensions.

5. Most dimensions are located ______________________ in the views.

6. Circles and holes are sized by ______________________ dimensions.

7. Arcs are sized by ______________________ dimensions.

8. Dimensions are located from ______________________ lines.

9. The four basic types of dimensioning systems are:

   a. ______________________ **c.** ______________________

   b. ______________________ **d.** ______________________

NAME ______________________

DATE ______________________

SCORE ______________________

EXERCISE A6-1

Missing Lines

Study the isometric views and draw the missing lines on the three-view drawings.

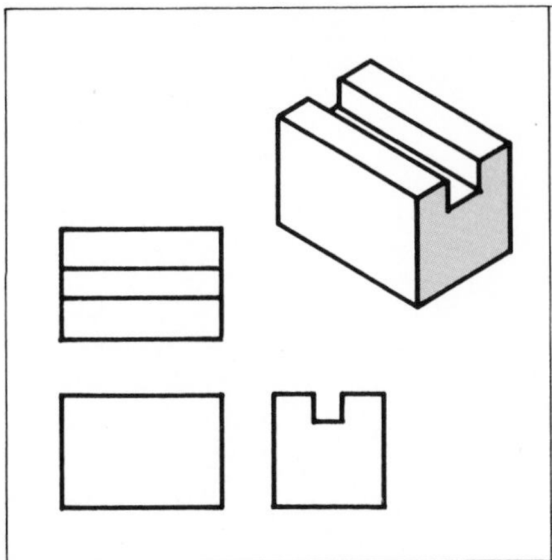

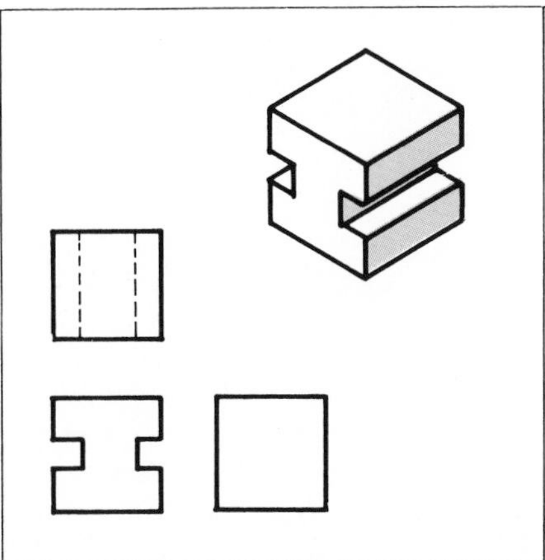

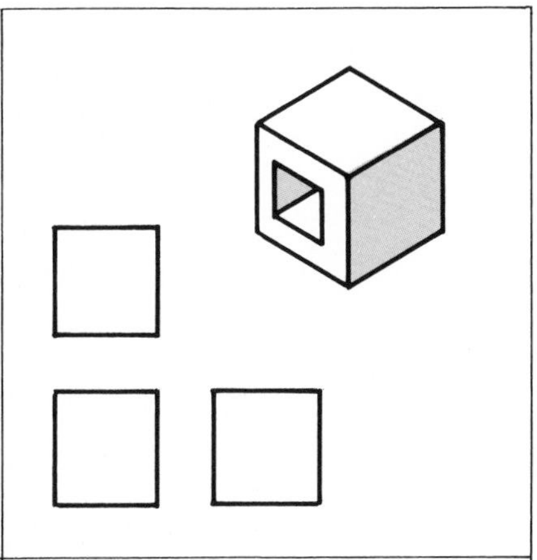

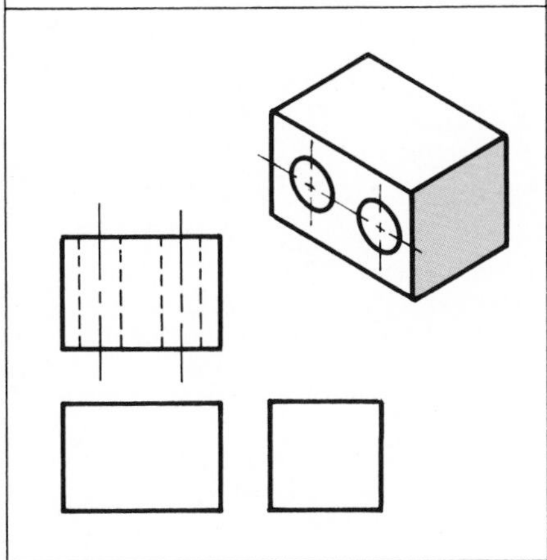

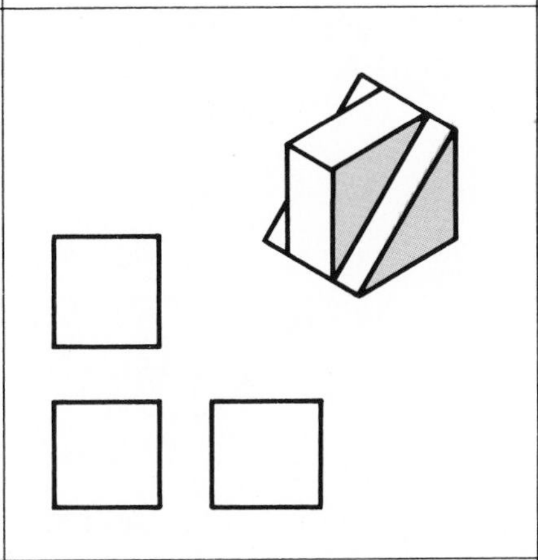

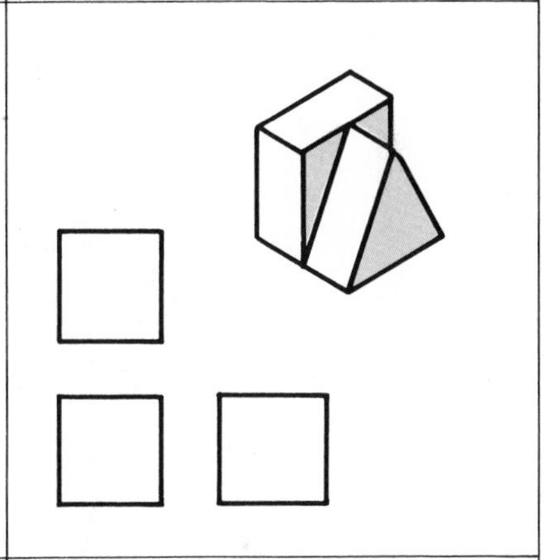

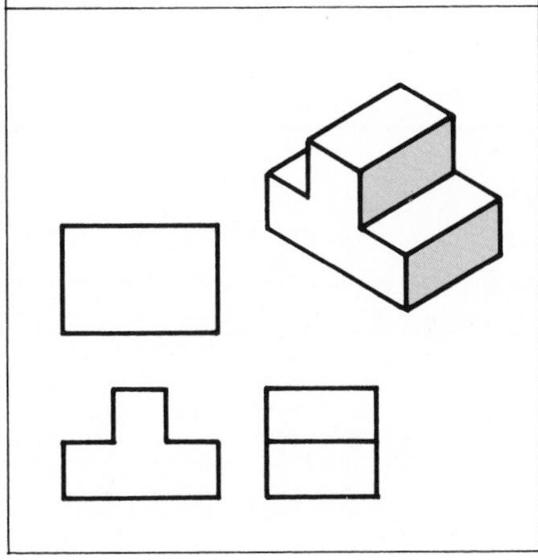

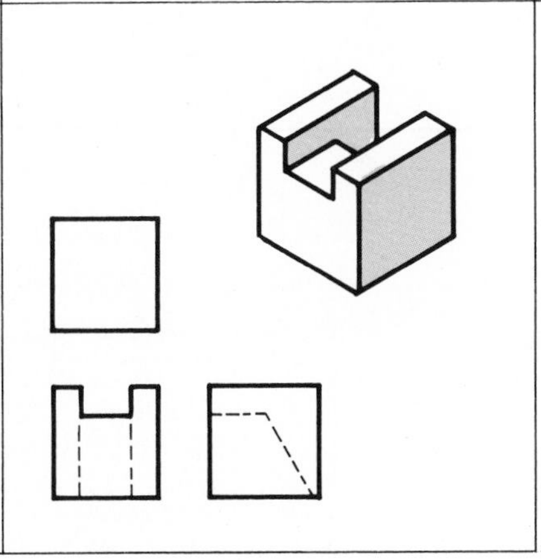

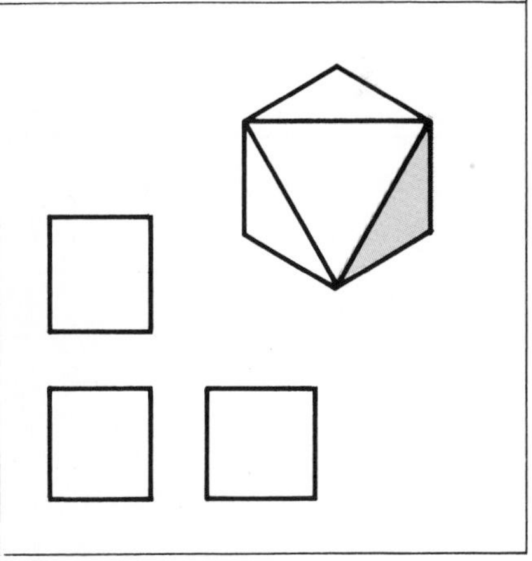

NAME ____________

DATE ____________

SCORE ____________

EXERCISE B6-1

Print #3789-4

Study Print #3789-4 found after this exercise. Then answer these questions.

QUESTIONS	ANSWERS
1. Give dimensions for *A*, *B*, *C*, and *D*.	1. *A* ________ *B* ________ *C* ________ *D* ________
2. What is this part?	2. ____________________
3. This part fits on what assembly?	3. ____________________
4. Was this part revised, and if so, on what date?	4. ____________________
5. What scale is used?	5. ____________________
6. This part is made of what material?	6. ____________________
7. What is the diameter of the counterbore?	7. ____________________
8. What is the depth of the .312 hole?	8. ____________________
9. What is the tolerance of the 1.938 dimension?	9. ____________________
10. What are the overall dimensions of this part?	10. ____________________
11. What is the tolerance of the 5/8-inch dimension?	11. ____________________
12. Prior to 6-20-80, what was dimension 1.938?	12. ____________________

NAME ____________________

DATE ____________________

SCORE ____________________

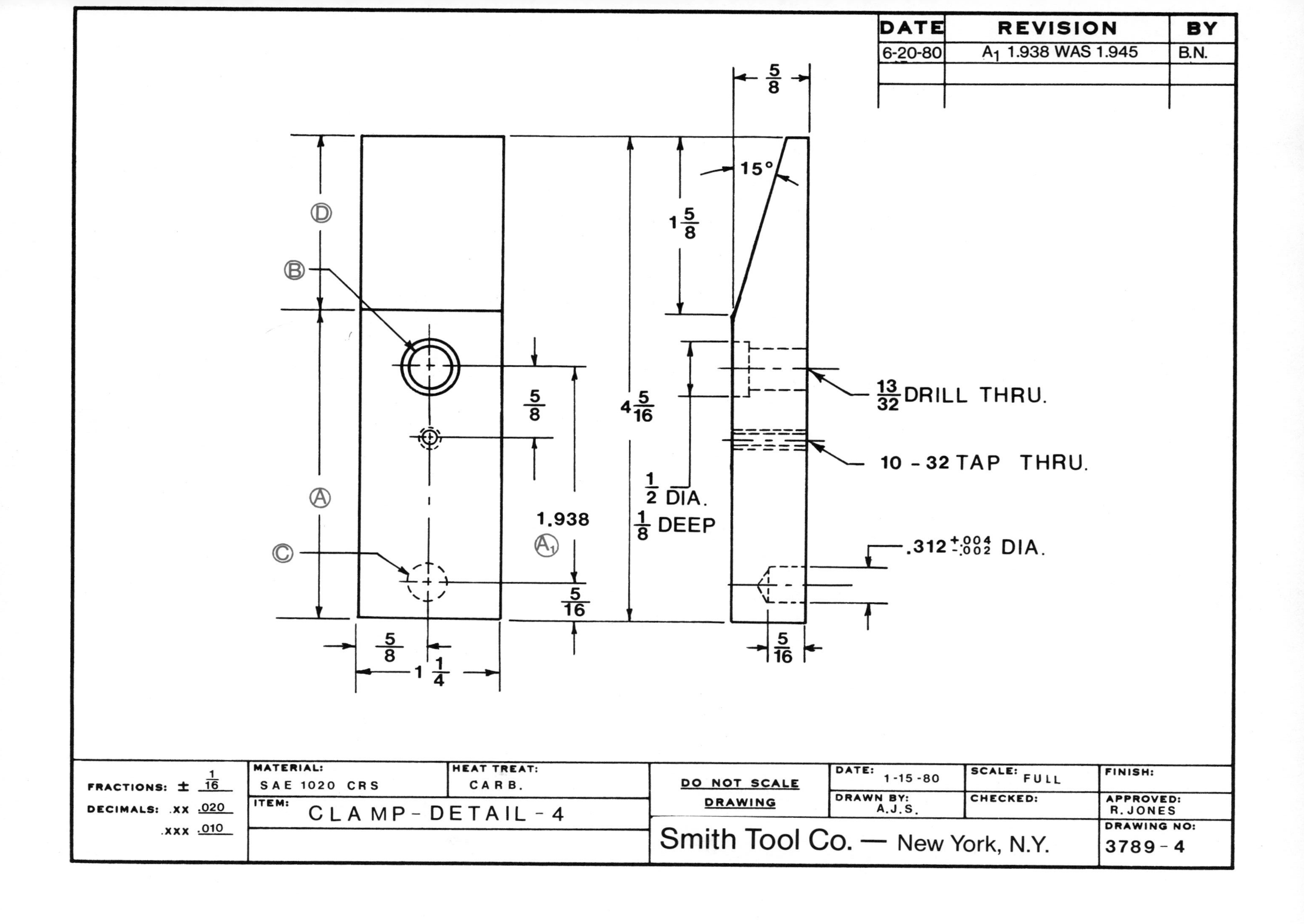
DATE
REVISION
BY
6-20-80
A1 1.938 WAS 1.945
B.N.
5/8
15°
1 5/8
4 5/16
D
B
A
C
5/8
1.938
A1
5/16
1/2 DIA.
1/8 DEEP
13/32 DRILL THRU.
10 - 32 TAP THRU.
.312 +.004 -.002 DIA.
5/8
1 1/4
5/16
FRACTIONS: ± 1/16
DECIMALS: .XX .020
.XXX .010
MATERIAL: SAE 1020 CRS
HEAT TREAT: CARB.
ITEM: CLAMP-DETAIL-4
DO NOT SCALE DRAWING
DATE: 1-15-80
SCALE: FULL
FINISH:
DRAWN BY: A.J.S.
CHECKED:
APPROVED: R. JONES
Smith Tool Co. — New York, N.Y.
DRAWING NO: 3789-4

CHAPTER 7

TOLERANCES AND SURFACE FINISHES

OBJECTIVES

After studying this chapter, you will be able to:

- Explain the purpose of tolerance dimensions.
- Describe the techniques of displaying tolerance dimensions.
- Recognize standard tolerance blocks.
- Describe the techniques of displaying surface finish.

It is not economical to produce parts with the exact specified dimensions. Therefore, the print must indicate the amount the finished part may vary from the given dimension. This allowed variation is called *tolerance.*

The tolerance may be shown with the dimension, indicated by a tolerance note, or specified in the tolerance section of the title block.

Most dimensions are subject to tolerance. Dimensions that are not subject to tolerance are those labeled *BSC* (Basic), *REF* (Reference), *MAX* (Maximum), *MIN* (Minimum), and those enclosed in a frame (5.135). (The frame is a new method of indicating a basic dimension.)

TOLERANCES

A tolerance is normally expressed in the same form as the dimension. For example, a fractional dimension will have a fractional tolerance: $5\frac{3}{16} \pm \frac{1}{16}$. A decimal dimension will have a decimal tolerance: 5.135 ± 0.002.

Angular tolerances are usually specified in degrees ($30° \pm 1°$) or in minutes ($30° \pm 30'$).

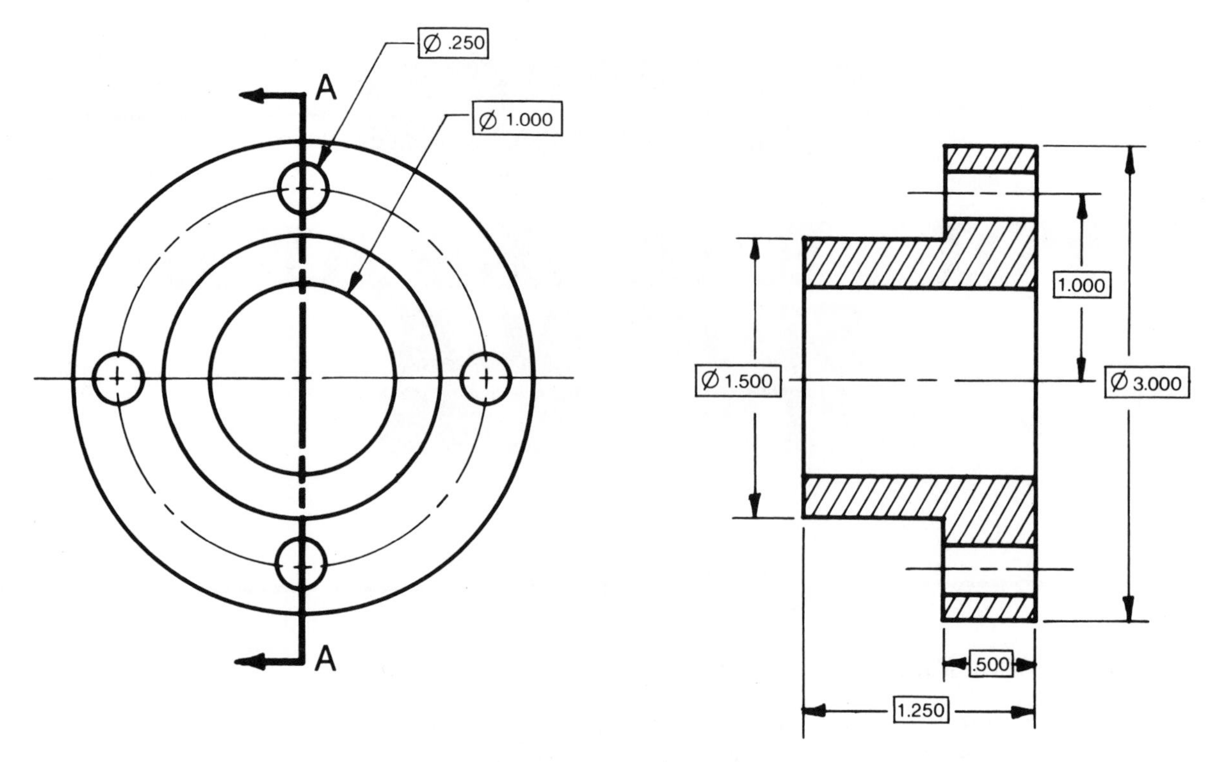

Fig. 7-1 Basic dimensions

The original part design may have only basic dimensions for size or location of features, Figure 7-1. Later, the part designer will insert the tolerances. The tolerances will indicate the amount the part features may differ from the basic dimension. See Figure 7-2.

Four main factors determine the amount of tolerance:

1. The function of the part
2. The manufacturing method or machining costs
3. Interchangeability in assembly
4. A combination of all these factors

A greater tolerance allowance will usually result in a reduction of machining costs, and will often decrease the amount of scrap. However, the greater tolerance allowance may also increase the assembly costs. On the other hand, specifying a closer tolerance than is necessary is usually a waste of money because it results in increased manufacturing costs.

TOLERANCING METHODS

Dimensional tolerances are expressed as either *limits* or as *plus and minus.* Plus and minus tolerancing may be either *bilateral* (in both directions) or *unilateral* (in one direction).

Limit Dimensions

In the *limit dimensioning method,* the maximum and minimum dimensions are specified. When dimensions are given directly, they are placed on the drawing in one of two ways:

- The high limit is placed above the low limit; or
- The high limit is placed to the right of the low limit, with a dash between the two. See Figure 7-3.

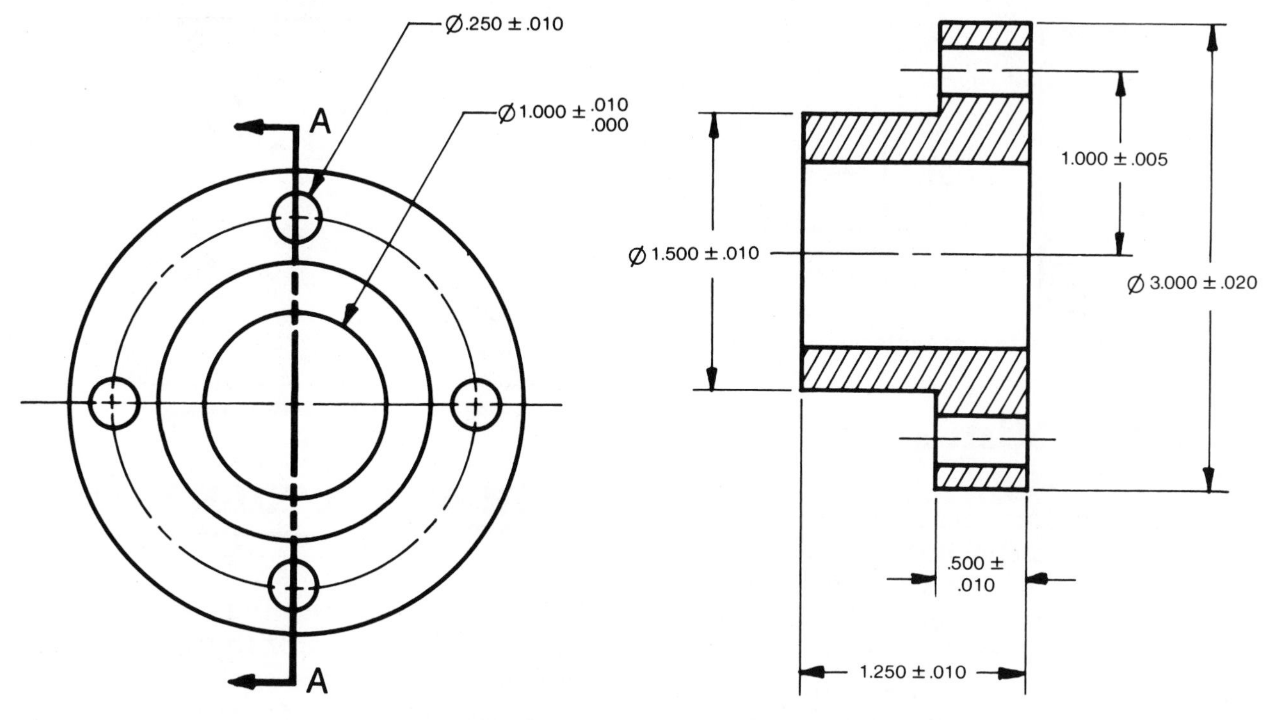

Fig. 7-2 Finish dimensions

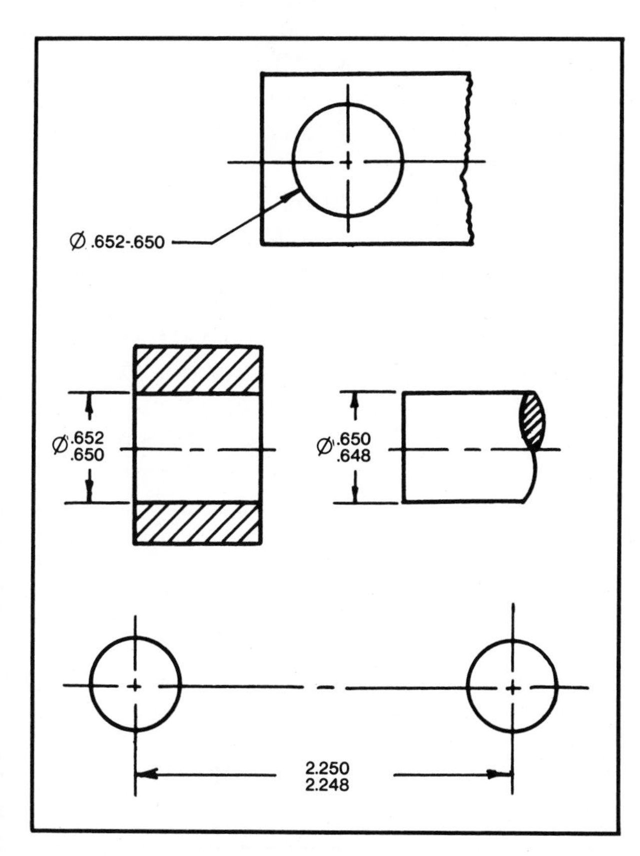

Fig. 7-3 Limit dimensions

It is not always required to state both limits. The abbreviation *MIN* (minimum) or *MAX* (maximum) is often placed after a dimension when mating part features determine the limit, Figure 7-4.

Some dimensioning systems specify that the upper limit be placed above the lower limit on an external feature. The reverse arrangement is specified for an internal feature. See Figure 7-5.

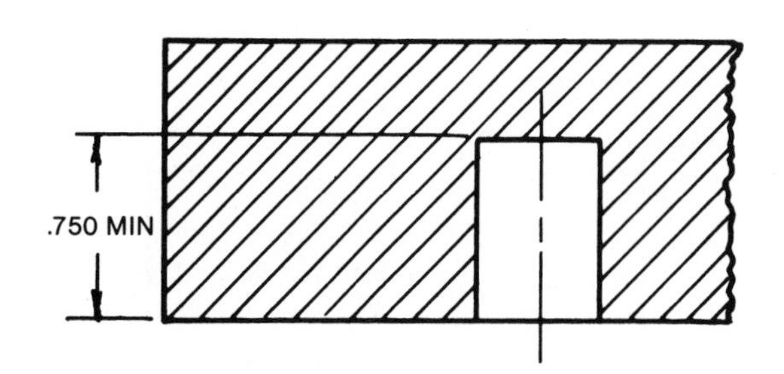

Fig. 7-4 Single limit

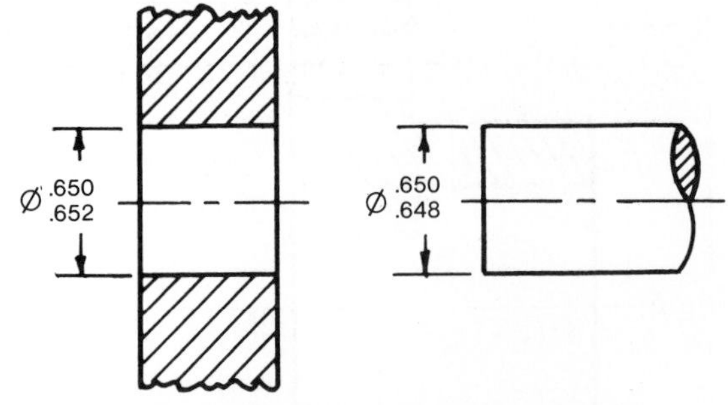

Fig. 7-5 External and internal parts

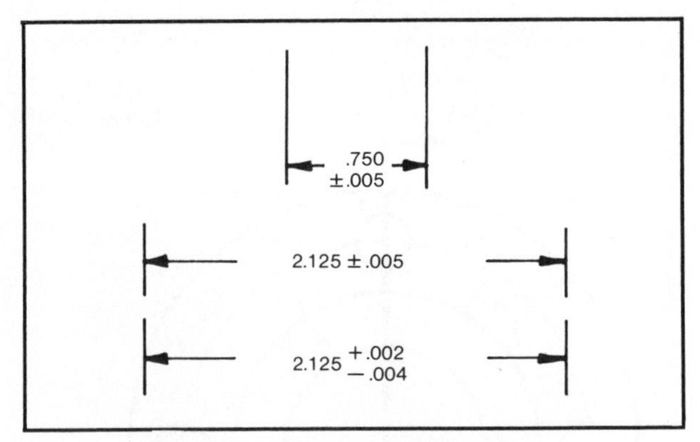

Fig. 7-7 Bilateral tolerancing

Plus and Minus Tolerancing

In the *plus and minus tolerancing method,* the dimension is given, followed by a plus and minus (±) expression of tolerance. Examples of unilateral tolerancing are shown in Figure 7-6. The tolerances are in one direction only, and can be either larger or smaller than the specified dimensions.

The bilateral tolerancing method provides for the tolerance to be plus or minus in both directions from the given dimension, Figure 7-7. The tolerance limit may vary from one direction to the other.

The specified dimension is usually the desired or ideal dimension. The tolerance indicates the amount the production department may vary from this ideal. Figure 7-8 shows three methods of stating the same dimension and tolerance. In all three cases, the ideal dimension is 2.125.

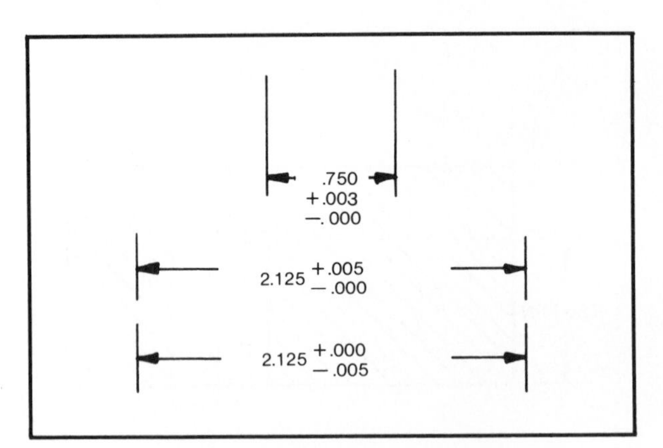

Fig. 7-6 Unilateral tolerancing

TITLE BLOCK TOLERANCE NOTATIONS

Most machining industries utilize the *decimal dimensioning system.* In accordance with this system, all measurements on their prints are in inches and decimals of an inch. Screw threads, which, by industry standards, contain fractional dimensions, are excluded from this system. Some other accessories which are manufactured to large fractional tolerances are also excluded from decimal dimensioning.

The decimal dimensioning system has been a very important change. All machine dials and measuring tools are now decimally graduated. This system will eventually eliminate the need for fraction-to-decimal conversion tables.

Most decimal dimensions end in an even figure (*i.e.,* .06, .004, .0008). This

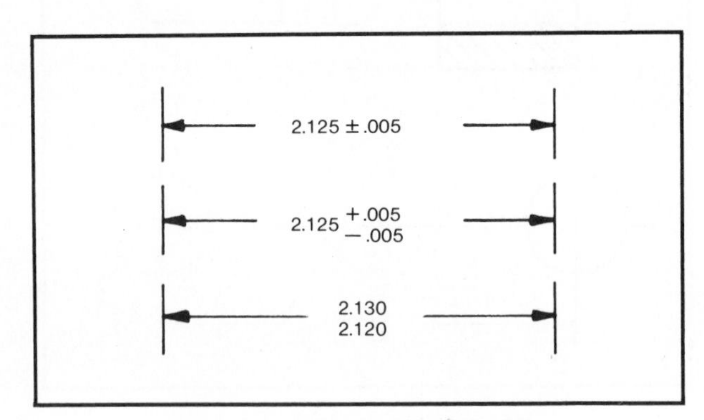

Fig. 7-8 Tolerance dimensions

practice is used so that when the dimension is halved (for example, diameter to radius) the same number of decimal places will result.

Most industries have established standard title block tolerance notations. A sample is as follows:

TOLERANCE UNLESS NOTED

.00 ± .02

.000 ± .010

.0000 ± (zero tolerance)

Angle tolerance ± 1°

Figure 7-9 shows several title block tolerance systems. The older systems are listed first. Even though the decimal system is a tremendous improvement, many older prints are still in use which contain fractions. It is important to be familiar with all systems.

TOLERANCES UNLESS
OTHERWISE SPECIFIED:
FRACTIONS .... ± 1/64
DECIMALS............... ± .003
ANGLES.................. ± 1/2°
DO NOT SCALE
DRAWING

± .008 VARIATION ALLOWED WHERE FRACTIONAL DIMENSIONS ARE USED LOCATING FINISHED SURFACES
TOLERANCE DIMENSIONS SPECIFY ACTUAL GAGE SIZES
COUNTERSINK ALL TAPPED HOLES 90° INCLUDED ANGLE TO MAJOR DIAMETER OF THREAD
- REMOVE ALL BURRS -
DESIGNED
FOR

NORMAL TOLERANCES
FRACTIONAL DIMEN. ± .015
DECIMAL DIMEN. ± .003
ANGULAR DIMEN. ± 1 DEGREE

± .020 TOL. ON TWO PLACE DECIMALS
± .010 TOL. ON THREE PLACE DECIMALS
ZERO TOL. ON FOUR PLACE DECIMALS
ANGLE TOL. ± 1°
DO NOT SCALE
DESIGNED
FOR

Fig. 7-9 Title block tolerance notations

SURFACE FINISH

When machined properly, most part surfaces are of a surface texture which will satisfy their functional requirements. Surfaces on rough parts, such as sand castings, are machined to produce a flat surface or to meet required dimensions.

Symbols are placed on drawings to indicate which surfaces are to be machined. The symbol ∫ or *V* was used in the past to indicate that some kind of finish was required. Variations of these symbols have been used when the surface texture of the finished surfaces was to be controlled.

The two most common practices are shown in Figures 7-10 and 7-11. On the print part views, the finished surface is

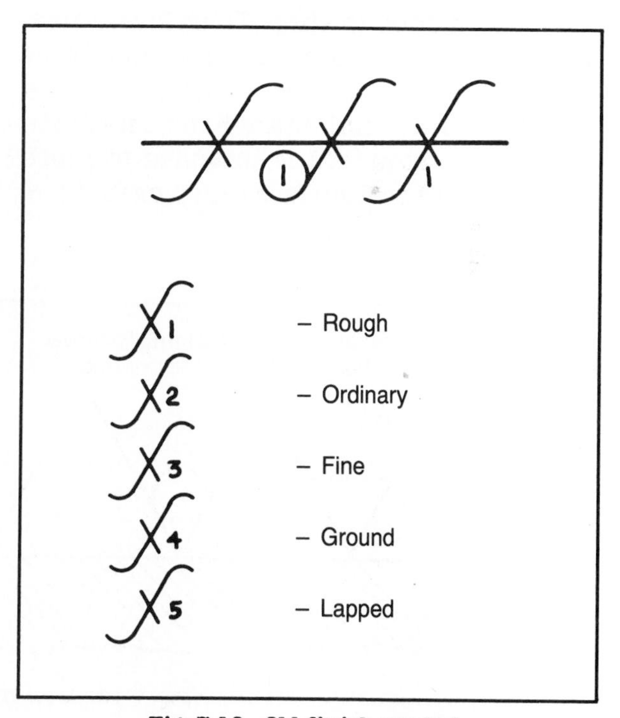

Fig. 7-10 Old finish symbol

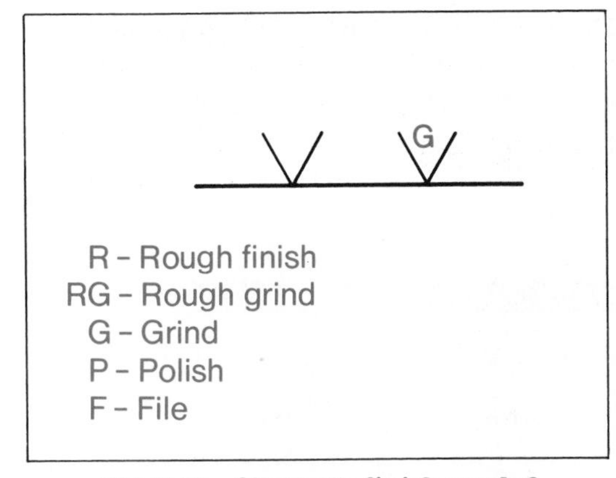

Fig. 7-11 Alternate finish symbol

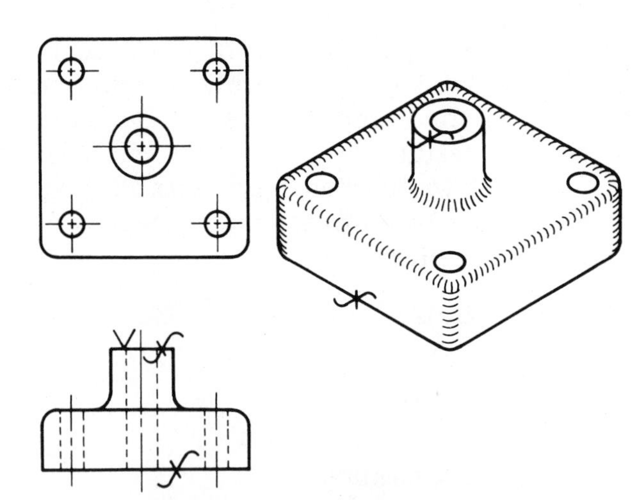

Fig. 7-12 Locating finish symbols

represented by a line with the symbol *∫* crossing the line, or by the symbol *V* touching the line. See Figure 7-12.

If the part is to be finished all over, the abbreviation *FAO* may be placed as a note on one view of the part, or near the title block of the drawing.

These two systems of indicating surface finish are still used in some industries. However, the two systems have led to confusion from shop to shop. Therefore, a new standard was adopted by the Society of Automotive Engineers (SAE) and the American National Standards Institute (ANSI). The standard is intended to more adequately control surface finish or texture.

AMERICAN NATIONAL STANDARD SURFACE TEXTURE

The American National Standard Surface Texture is a standard used to correctly indicate surface roughness, waviness, and lay. These terms are discussed in this section.

The new surface finish symbol, Figure 7-13, is placed on the drawing in a manner similar to that of the older symbols. The notations placed next to the symbol control a variety of surface

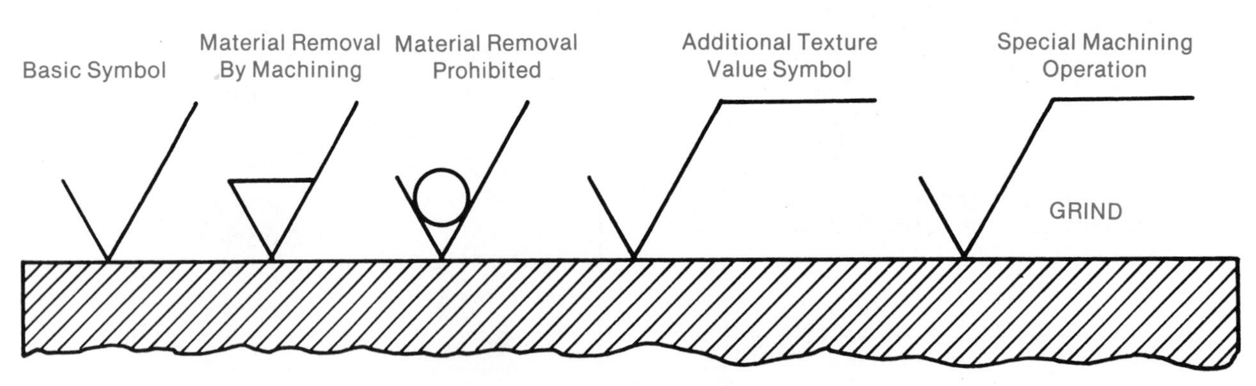

Fig. 7-13 New finish symbol standard

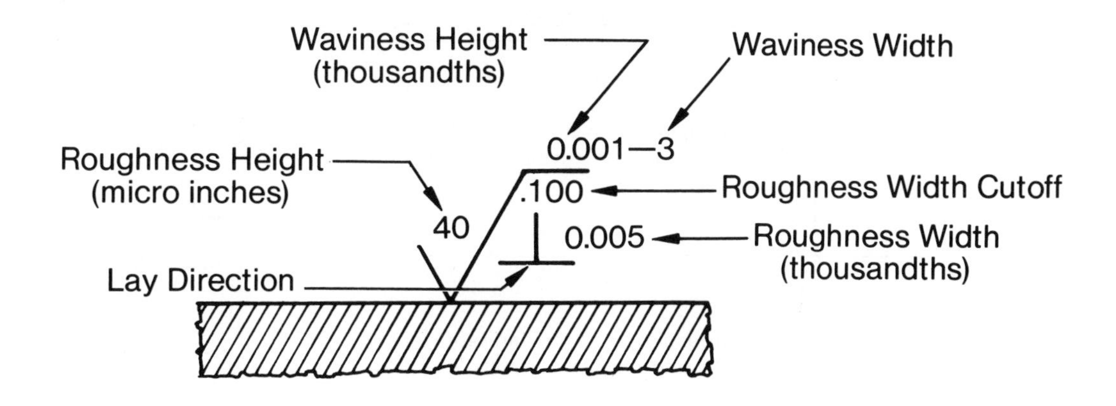

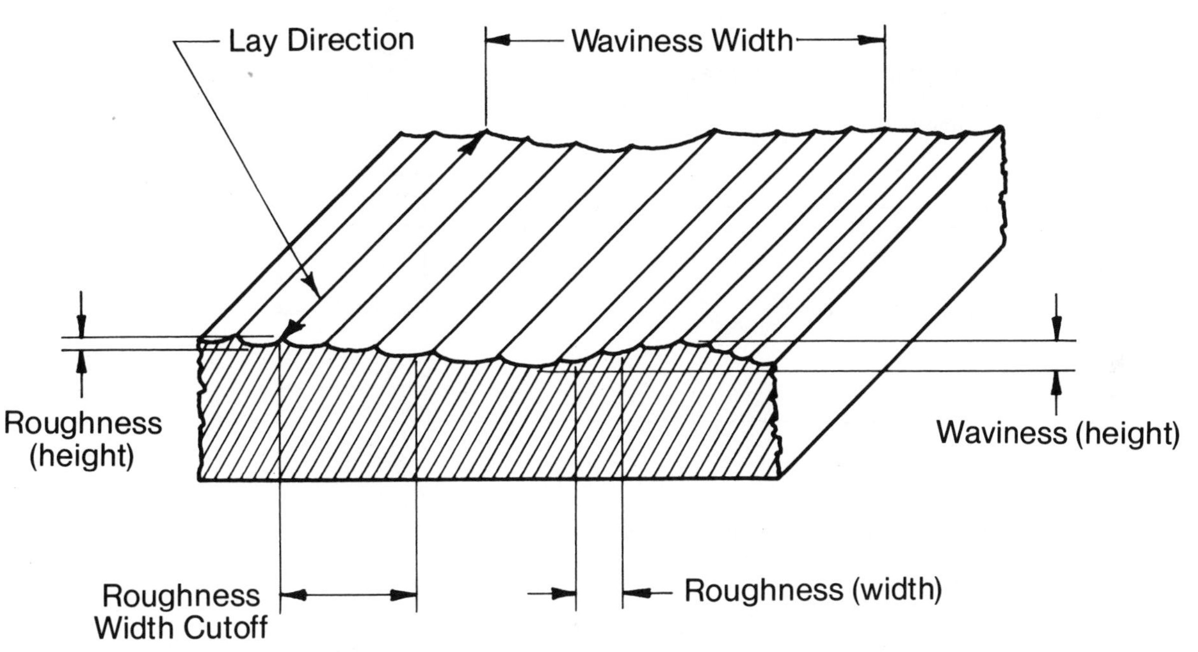

Fig. 7-14 Symbol identification and notations

irregularities. The main emphasis is on roughness, waviness, and lay. See Figure 7-14.

Roughness

Roughness consists of the finely spaced irregularities of the finished surface. The irregularities are produced by the machining operation.

The roughness height is expressed in *micro inches.* One micro inch equals one-millionth of an inch (.000001 in.). The micro inch may be abbreviated *MU in.* or *μ in.*

The indicated value is rated as the arithmetic average. It is placed to the left of the long leg of the surface symbol.

The *roughness width* is the distance between the repetitive pattern of roughness. The roughness width is measured parallel to the finished surface, in inches. The roughness width cutoff is the amount the measuring instrument will be adjusted when checking the roughness height value. The *roughness height* should always be

LAY SYMBOL	DESIGNATION	EXAMPLE
‖	Lay parallel to the line representing the surface to which the symbol is applied.	DIRECTION OF TOOL MARKS
⊥	Lay perpendicular to the line representing the surface to which the symbol is applied.	DIRECTION OF TOOL MARKS
X	Lay angular in both directions to line representing the surface to which the symbol is applied.	DIRECTION OF TOOL MARKS
M	Lay multidirectional.	
C	Lay approximately circular relative to the center of the surface to which the symbol is applied.	
R	Lay approximately radial relative to the center of the surface to which the symbol is applied.	
P	Lay particulate, non-directional, or protuberant.	

Fig. 7-15 Lay symbols (*Courtesy of ASME, extracted from ANSI Y14.36-1978*)

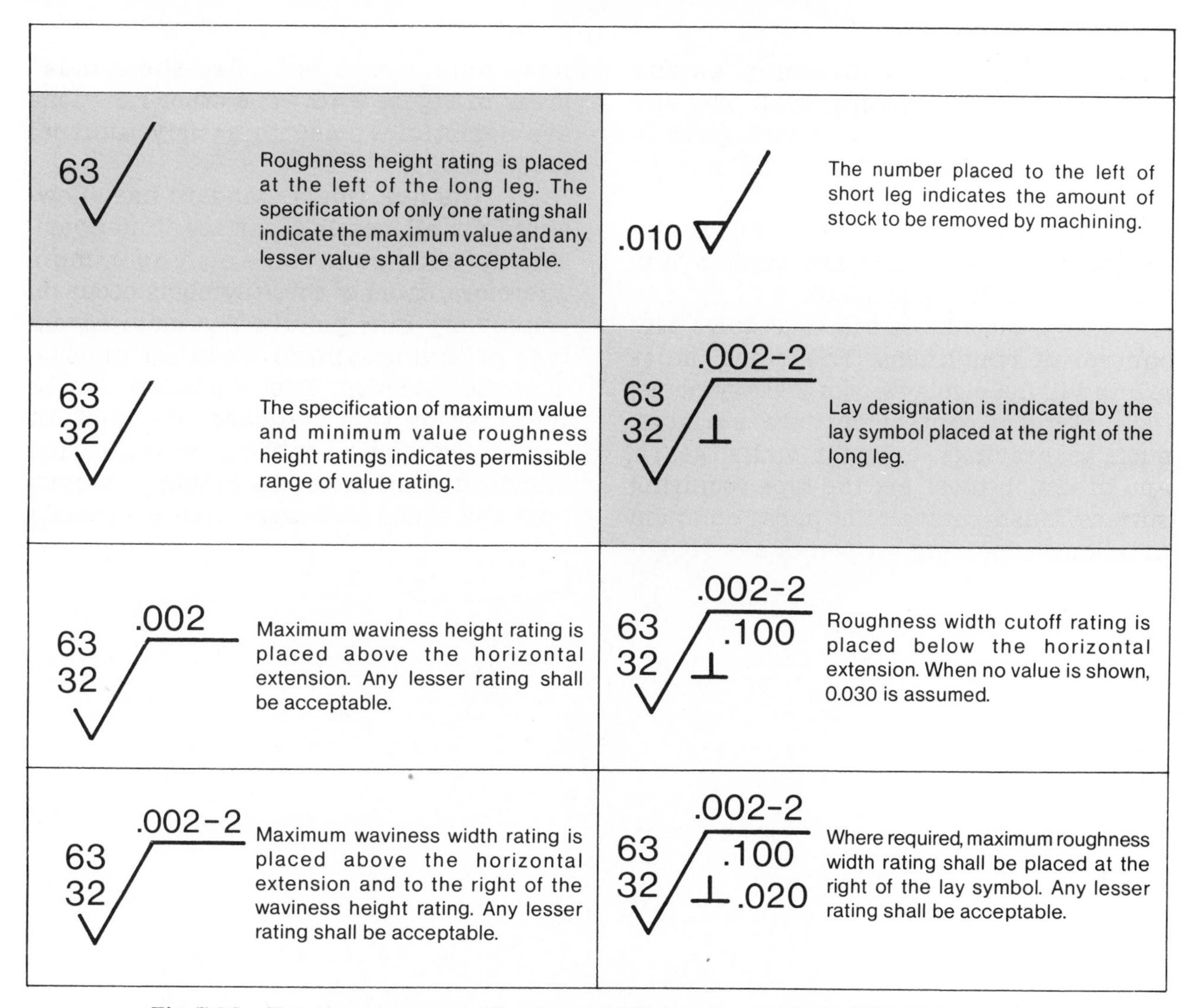

Fig. 7-16. Notation placement (*Courtesy of ASME, extracted from ANSI Y14.36-1978*)

more than the roughness width, and is measured in inches.

The standard roughness width cutoff values (inches) are .003, .010, .030, .100, .300, and 1.000. When no value is specified, the value .030 is assumed.

Waviness

Waviness refers to the surface irregularities that are spaced too far apart to be classified as roughness. Waviness is a result of machine or work deflection, vibration, chatter, heat treatment, or warping strains. Roughness, by comparison, may be defined as irregularities on a wavy surface.

The *waviness height* is defined as the peak-to-valley distance. The waviness height is rated in inches. The *waviness width* is defined as the successive wave peaks or successive wave valleys. Waviness width is also rated in inches.

Lay

The direction of the predominate surface irregularity, produced by the machining method, is called the *lay.* This is also referred to as *tool marks.* The lay is usually in the same direction as the movement of the cutting tool. The lay direction is indicated by standard symbols placed to the lower right of the surface symbol, Figure 7-15.

A step-by-step evaluation of the American National Standard Surface Texture is shown in Figure 7-16.

Few machine surfaces require any control of roughness. Therefore, most prints will not display any of these symbols.

Typical contact or wear surfaces, such as bearings, cylinder walls, gears, and precision tools, are the type requiring surface finish control. The most common notations appearing on prints are roughness height and lay. (See the shaded areas in Figure 7-16.) The other notations are sometimes used to satisfy unusual requirements.

The new finish standard has allowances for all possible surface finish variations, both common and uncommon. Therefore, some of these symbols occur on prints very infrequently. The main advantage of this system is when an unusual condition occurs. The condition can be indicated by this standard, allowing for easy interpretation by the print reader. The standard eliminates the ambiguous symbols and notes that were used previously.

REVIEW

This review is provided to serve as reinforcement study material. Fill in the appropriate word(s) to complete the sentences below.

1. Most dimensions are subject to tolerance. The exceptions are those labeled ______________________, ______________________, ______________________, ______________________, and those enclosed in (a) (an) ______________________.

2. The amount of tolerance used is subject to:

   a. ______________________ c. ______________________

   b. ______________________ d. ______________________

3. On a print, the dimension tolerance is specified in the ____________ ____________.

4. A greater tolerance will usually result in a reduction in ______________________.

5. Most machined surfaces satisfy their ______________________.

6. One micro inch is equal to ______________________ of an inch.

7. In the American National Standard Surface Texture system, the roughness height is expressed in ______________________.

NAME ______________________

DATE ______________________

SCORE ______________________

EXERCISE B7-1

Print #316088

Answer the following questions after studying Print #316088 found after this exercise.

Note that Print #316088 uses aligned dimensions.

QUESTIONS	ANSWERS
1. Give dimensions *A*, *B*, *C*, *D*, *E*, and *F*.	1. *A* ______ *B* ______ *C* ______ *D* ______ *E* ______ *F* ______
2. On what assembly does this part fit?	2. ______
3. What is the minimum diameter of the longer portion of this part?	3. ______
4. What scales are used?	4. ______
5. Of what material is this part made?	5. ______
6. What machining operation brings this shaft to its finished size?	6. ______
7. What is the tolerance for the regular dimensions on this part?	7. ______
8. In inches, what is the tap drill used for the cut threads?	8. ______
9. What kind and size of thread is used?	9. ______
10. Prior to 4-16-70, how deep was the tapped hole drilled?	10. ______
11. When the threads are rolled, what size drill is used?	11. ______
12. What are the maximum and minimum sizes for the 1.274 dimension?	12. Maximum ______ Minimum ______

NAME ______

DATE ______

SCORE ______

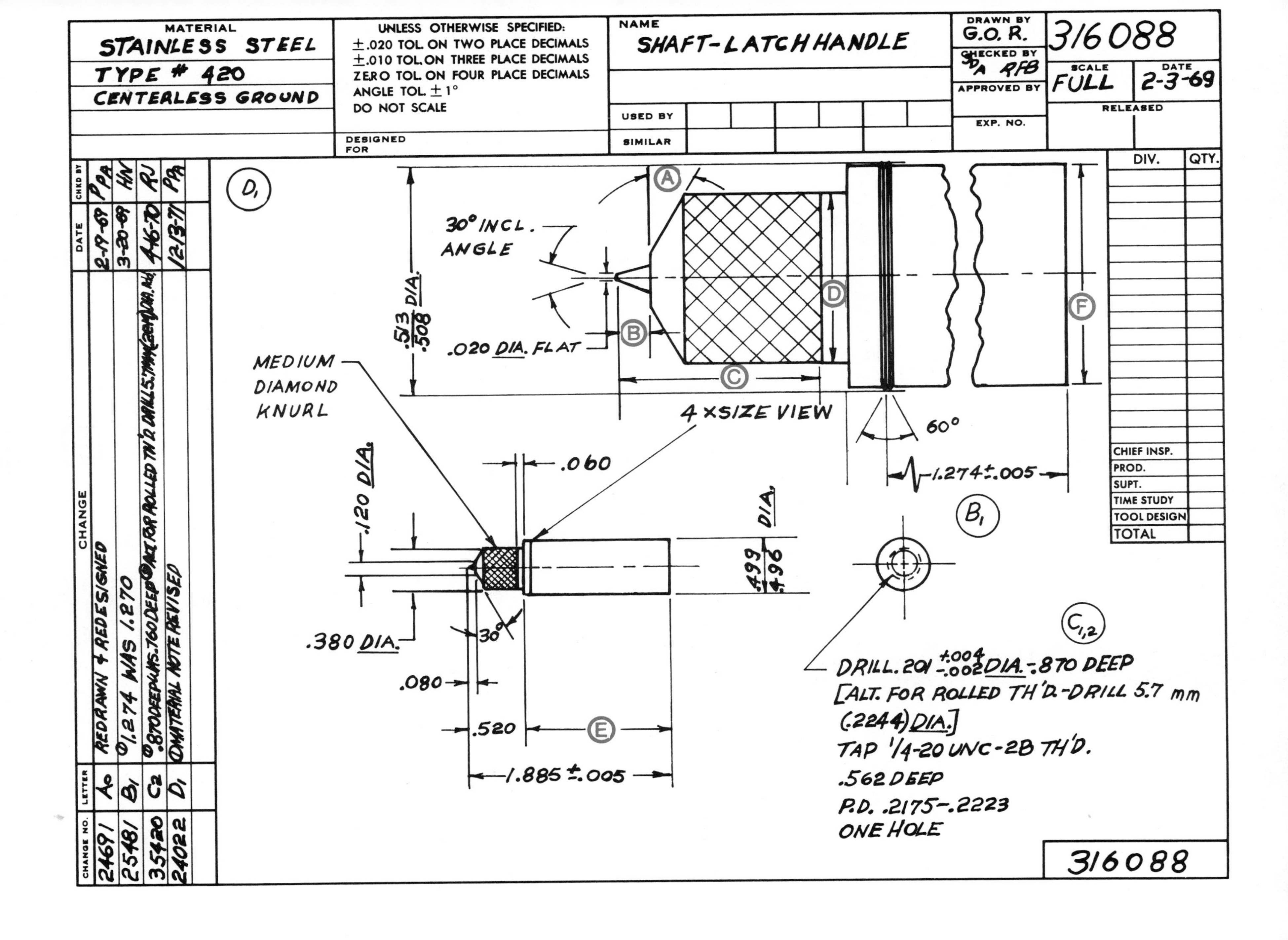
MATERIAL
STAINLESS STEEL
TYPE # 420
CENTERLESS GROUND
UNLESS OTHERWISE SPECIFIED:
±.020 TOL. ON TWO PLACE DECIMALS
±.010 TOL. ON THREE PLACE DECIMALS
ZERO TOL. ON FOUR PLACE DECIMALS
ANGLE TOL. ±1°
DO NOT SCALE
DESIGNED FOR
NAME
SHAFT-LATCH HANDLE
USED BY
SIMILAR
DRAWN BY G.O.R.
CHECKED BY RFB
APPROVED BY
EXP. NO.
316088
SCALE FULL
DATE 2-3-69
RELEASED
DIV.
QTY.
CHIEF INSP.
PROD.
SUPT.
TIME STUDY
TOOL DESIGN
TOTAL
CHANGE NO.
LETTER
CHANGE
DATE
CHKD BY
24691
A0
REDRAWN + REDESIGNED
2-19-69
PPA
25481
B1
1.274 WAS 1.270
3-20-69
HN
35420
C2
.870 DEEP-WAS .760 DEEP
4-16-70
RJ
24022
D1
MATERIAL NOTE REVISED
12/13-71
PPA
D1
30° INCL. ANGLE
.020 DIA. FLAT
.513 .508 DIA.
MEDIUM DIAMOND KNURL
4 X SIZE VIEW
60°
1.274±.005
B1
.060
.120 DIA.
.499 .496 DIA.
.380 DIA.
30°
.080
.520
E
1.885±.005
C1,2
DRILL .201 +.004 -.002 DIA. - .870 DEEP
[ALT. FOR ROLLED TH'D. - DRILL 5.7 mm (.2244) DIA.]
TAP 1/4-20 UNC-2B TH'D.
.562 DEEP
P.D. .2175-.2223
ONE HOLE
316088

CHAPTER 8

SUPPLEMENTARY INFORMATION

OBJECTIVES

After studying this chapter, you will be able to:

- Discuss the scaling of drawings.
- Describe the supplementary information shown on prints.
- Discuss the common machining notations.
- Explain the classification (class) of fits and how fit dimensions are determined.
- Describe the purpose and types of engineering specifications.

Supplementary information is provided on prints to describe the machining or assembly of the components illustrated.

Holes are to be drilled, bored, threaded, and so on, to accommodate fastening devices. The print reader must be familiar with the standard notations used to describe these and other machining operations.

SCALING

If possible, the drawing is made so that the object is shown in its full or actual size. If the object is too large to fit onto a standard size drafting paper, the object is drawn in reduced size.

Some parts are so small that a full-size drawing will not allow for a complete understanding of the part details, or the dimensions would be too crowded. In such cases, the drawing is enlarged. The most common practice for enlarging a drawing is to draw the object 2 or 4 times its actual size. When an enlarged scale is used, often the part is also shown on the print in its actual size.

The scale used for the drawing is always indicated on the print, usually in the title block. If more than one detail is shown on the print and different scales are used, then the principal scale is noted in the

Preferred		
1/1 = FULL SIZE	= 1.00 = 1.00 =	1 = 1
1/2 = HALF SIZE	= .50 = 1.00 =	½ = 1
1/4 = QUARTER SIZE	= .25 = 1.00 =	¼ = 1
1/8 = EIGHTH SIZE	= .125 = 1.00 =	⅛ = 1
2/1 = TWICE SIZE	= 2.00 = 1.00 =	2 = 1
10/1 = TEN TIMES SIZE	= 10.00 = 1.00 =	10 = 1

Fig. 8-1 Scale standards

title block. The views of the object that are drawn to a different scale must be noted with the proper scale.

Prints which have the object shown enlarged or reduced are called *scale drawings.* The ratio of object size to drawing size is referred to as the *scale.*

If the object is drawn at twice its actual size, it is called a *2 × scale drawing.* If the object is drawn at one-half its actual size, it is called a *½ scale drawing.*

To eliminate confusion, standard notations are indicated in Figure 8-1, with the preferred examples indicated.

For example, if the scale notation on the drawing is ½, this means that 1 inch on the drawing equals 2 inches on the part; the drawing is at *half size.*

A scale notation of ¹⁄₁ means that 1 inch on the drawing equals 1 inch on the part; the drawing is at *full size.* A scale notation of ²⁄₁ means that 2 inches on the print equal 1 inch on the part; the drawing is at *twice size.* See Figure 8-2.

Regardless of what scale is used on the drawing, the actual size of the object will be shown by the dimensions given. **Caution:** Do not find a dimension by actually measuring from the drawing sheet, since the drawing may not be dimensionally accurate. Drawing paper may stretch or shrink, depending upon the weather conditions. Use only the dimensions as given on the print.

When a dimension has been changed, it may not be practical to redraw the detail concerned. It is better to draw a wavy line under the dimension (1.376) to indicate that this dimension may not be drawn to scale.

FILLETS AND ROUNDS

Parts which are made by the casting or forging process are designed to eliminate both sharp inside and outside corners. These rounded corners provide greater strength to the part. They also eliminate problems in the casting process.

The term *fillet* is used for the rounded inside corners. The term *rounds* is applied to the rounded outside corners. Fillets and rounds are specified on the prints as a *radius* and on the older prints, the letter *R* is placed after the dimension. See Figure 8-3. According to the new stan-

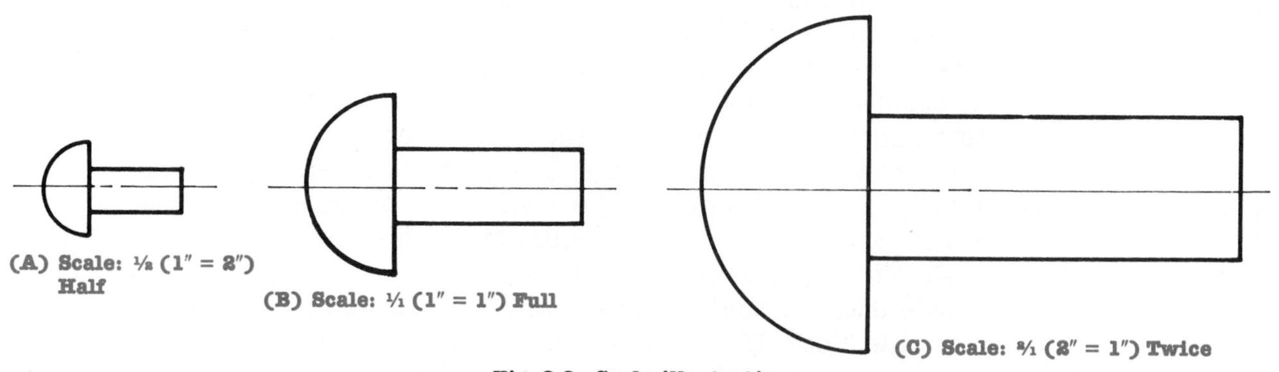

Fig. 8-2 Scale illustrations

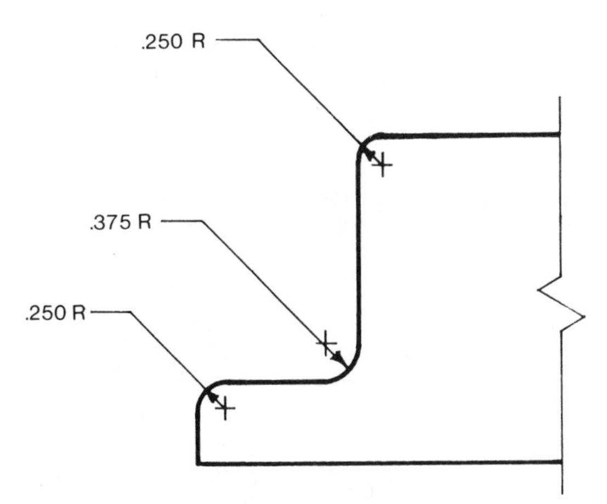

Fig. 8-3 Fillets and rounds

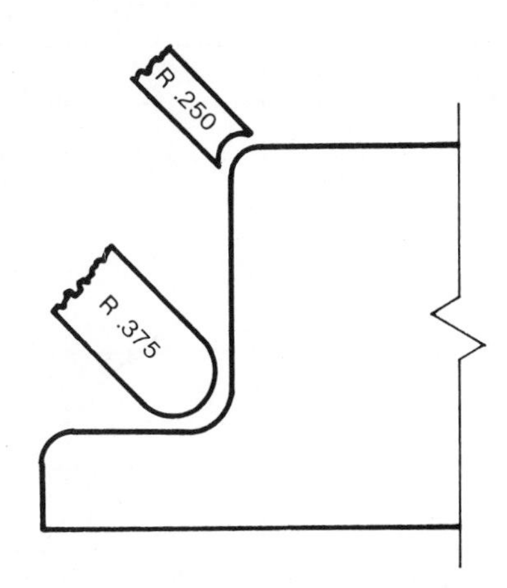

Fig. 8-4 Fillet and round gages

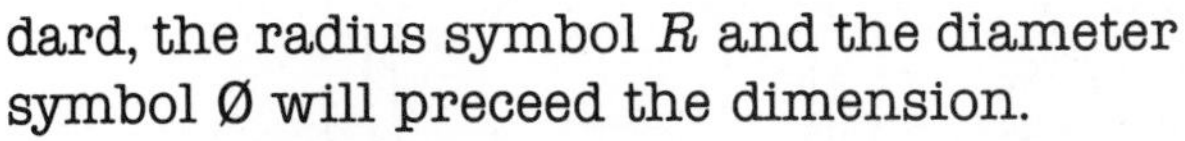
dard, the radius symbol *R* and the diameter symbol Ø will preceed the dimension.

Standard templates with machined inside and outside radii (plural of radius) are used to measure fillets and rounds on the part. See Figure 8-4.

Runouts

When a fillet meets a round surface, the extension of the fillet is called a *runout.* This is shown on the print as a short curve with its radius equal to the tangent points projecting from the other view. The shapes of the intersecting surfaces determine whether the curve will turn inward or outward. See Figure 8-5.

Objects with fillets or rounds do not have precise corners. Therefore, the object is drawn as if the surfaces meet at a definite place. This practice eliminates any misleading information about the shape of the object. Examples are shown in Figures 8-6A and 8-6B.

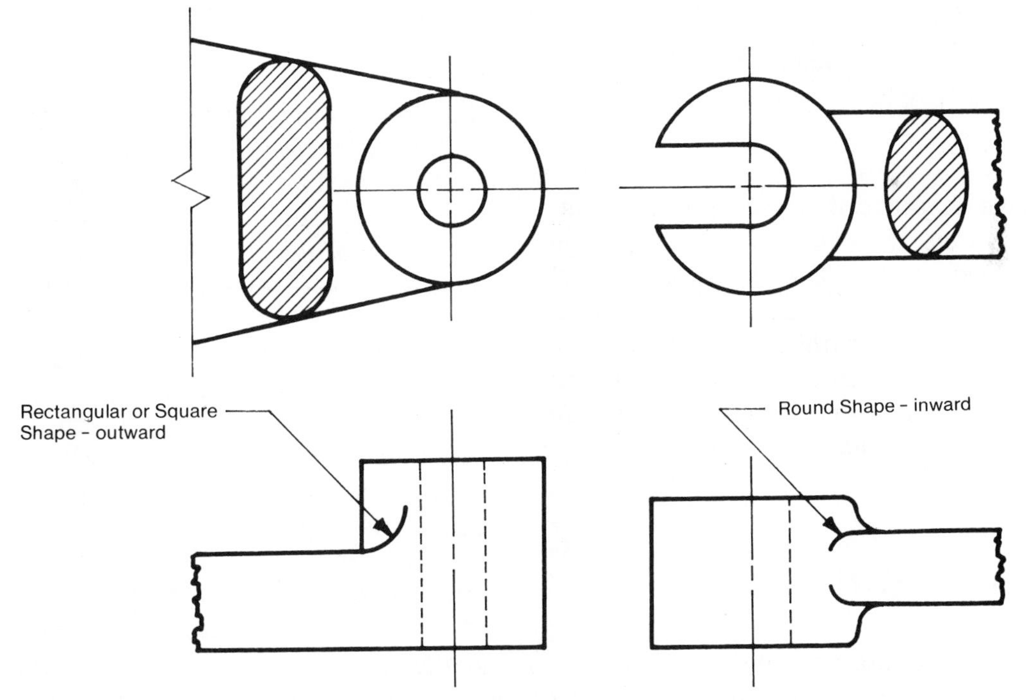

Fig. 8-5 Runout shown on prints

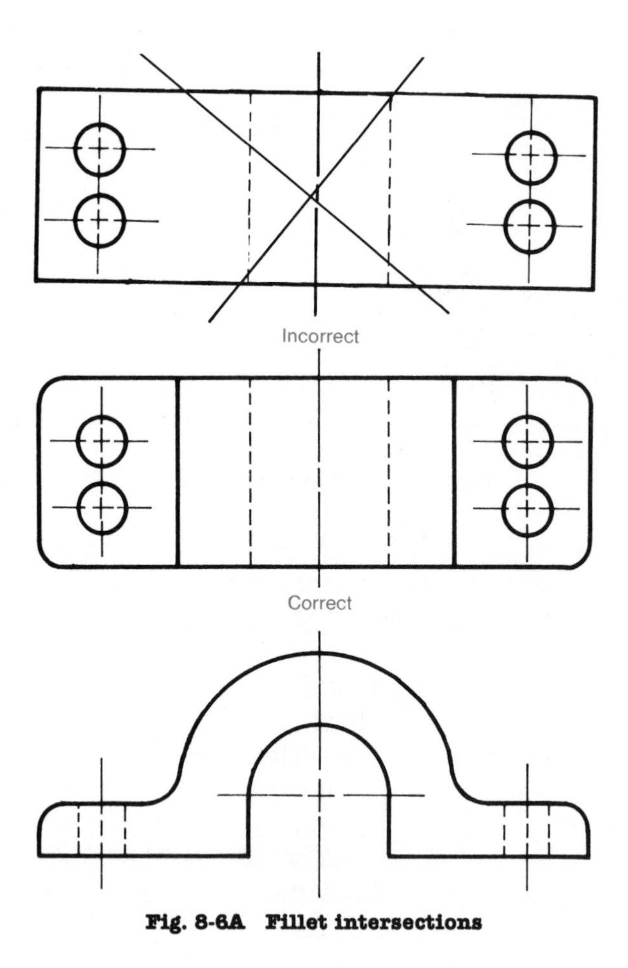

Fig. 8-6A Fillet intersections

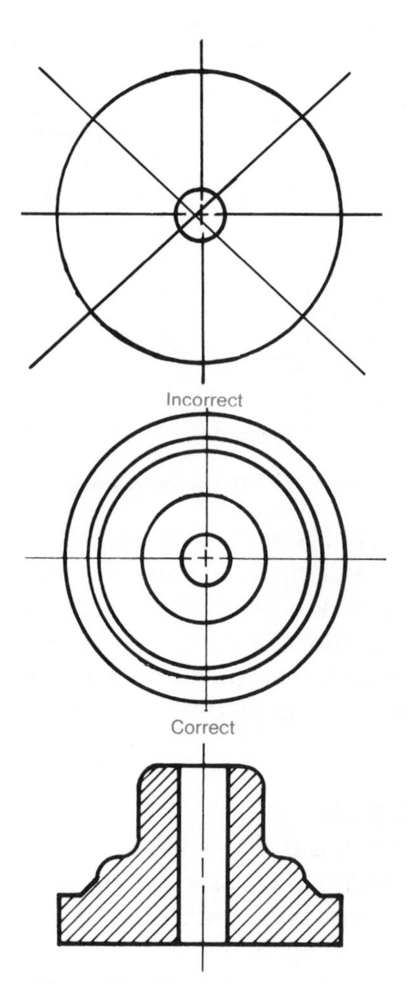

Fig. 8-6B Fillet intersections

COMMON MACHINING NOTATIONS

It is important to understand machining operations and how they are presented on prints. This knowledge will help in accurately reading industrial prints. The most common of these operations are discussed in this section. Others are described in later chapters or in the Glossary of this book.

Drilled and Reamed Holes

The most common machined hole is one that is *drilled.* This type of hole requires less accuracy or finish. The drilled hole is used mainly for the passage of bolts and other fastener devices.

When a more exact hole is required, the hole is finished to a more precise size with a *reamer.* Larger holes are usually bored to exact size by machining with a boring tool.

These types of holes are drawn in a similar manner on the print. However, the type of machining is indicated on the print. The standard method of notation is shown on a print in the following sequence:

1. The diameter of the hole
2. The machining operation
3. The number of holes if more than one are of the same size

Figure 8-7 illustrates drilled and reamed holes.

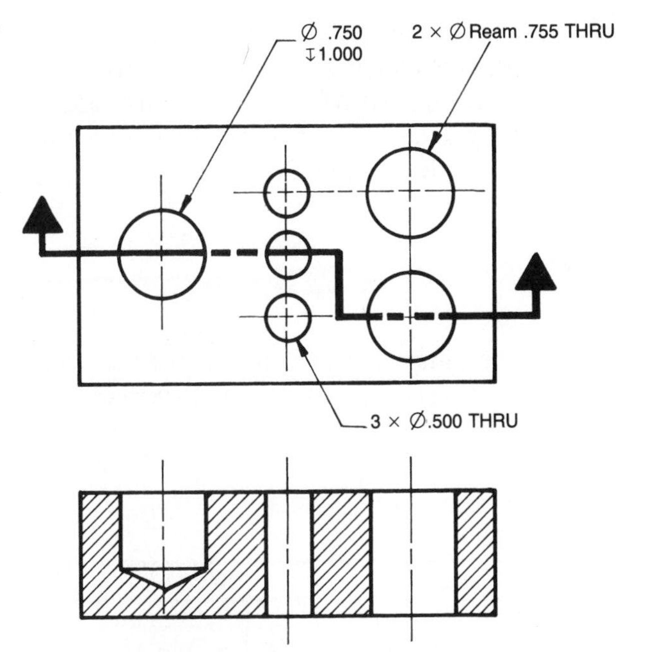

Fig. 8-7 Drilled and reamed holes

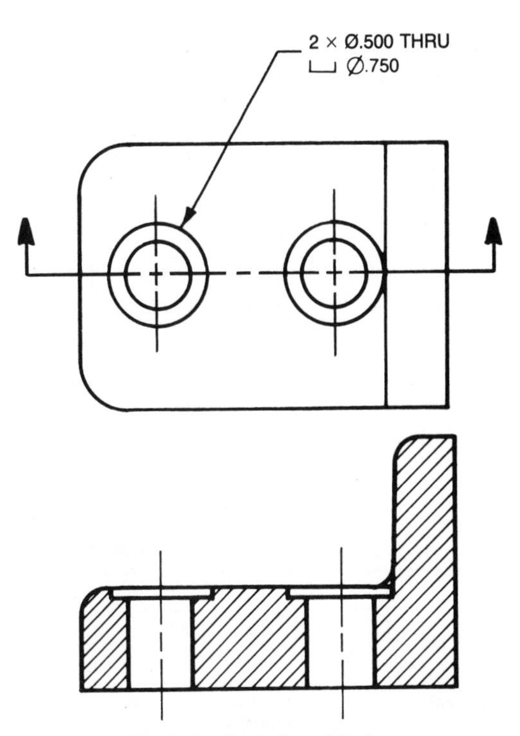

Fig. 8-9 Spot-faced holes

Counterbored, Countersunk, and Spot-faced Holes

Sometimes a portion of a drilled hole may be enlarged to accommodate the head of the fastener that will pass through the hole. This will allow the head of the fastener, when tightened, to be flush with or below the machined surface. The type of head on the fastener determines this secondary machining of the drilled hole.

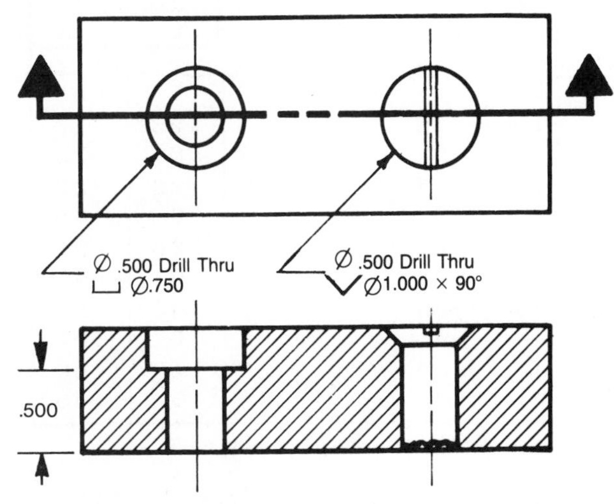

Fig. 8-8 Counterbored and countersunk holes

Fillister or Hexagon head machine screws require *counterboring* operation. A flat head machine screw requires *countersinking* the drilled hole. The notations for these operations are indicated in Figure 8-8.

A bolt head in contact with a rough unfinished surface requires a surface machining operation called *spot facing*. Spot facing is performed around the top of the drilled hole to allow the entire base of the bolt head to contact the part surface. The diameter of the spot face is slightly larger than the head of the bolt. The diameter will be specified on the print. An example is provided in Figure 8-9.

Chamfer

The outside corners of machined surfaces may have sharp edges which can easily injure workers when the parts are handled. Sharp corners may also cause problems in the assembly operations. To avoid such problems, a standard machine

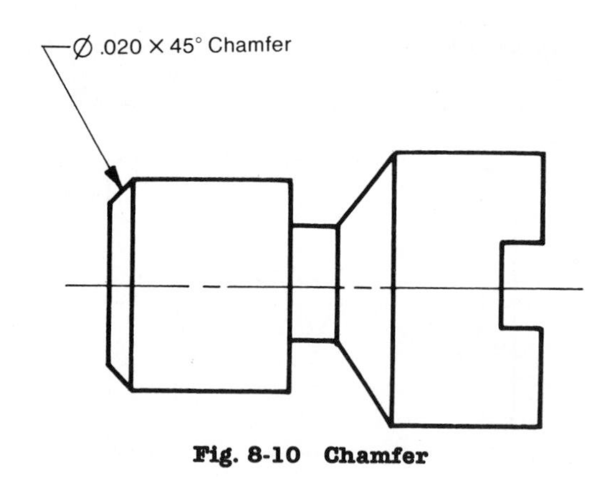

Fig. 8-10 Chamfer

shop practice is to machine off sharp corners, usually at an angle of 45°. This process is called *chamfering*. The process is specified on the print with the size and chamfer angle, Figure 8-10.

CLASS OF FITS

Fit is the general term used to indicate the degree of tightness between two

American National Standard Running and Sliding Fits (ANSI B4.1—1967, R1974)

Nominal Size Range, Inches	Class RC1			Class RC2			Class RC3			Class RC4		
	Clearance*	Standard Tolerance Limits		Clearance*	Standard Tolerance Limits		Clearance*	Standard Tolerance Limits		Clearance*	Standard Tolerance Limits	
		Hole H5	Shaft g4		Hole H6	Shaft g5		Hole H7	Shaft f6		Hole H8	Shaft f7
Over To	Values shown below are in thousandths of an inch											
0–0.12	0.1	+0.2	−0.1	0.1	+0.25	−0.1	0.3	+0.4	−0.3	0.3	+0.6	−0.3
	0.45	0	−0.25	0.55	0	−0.3	0.95	0	−0.55	1.3	0	−0.7
0.12–0.24	0.15	+0.2	−0.15	0.15	+0.3	−0.15	0.4	+0.5	−0.4	0.4	+0.7	−0.4
	0.5	0	−0.3	0.65	0	−0.35	1.12	0	−0.7	1.6	0	−0.9
0.24–0.40	0.2	+0.25	−0.2	0.2	+0.4	−0.2	0.5	+0.6	−0.5	0.5	+0.9	−0.5
	0.6	0	−0.35	0.85	0	−0.45	1.5	0	−0.9	2.0	0	−1.1
0.40–0.71	0.25	+0.3	−0.25	0.25	+0.4	−0.25	0.6	+0.7	−0.6	0.6	+1.0	−0.6
	0.75	0	−0.45	0.95	0	−0.55	1.7	0	−1.0	2.3	0	−1.3
0.71–1.19	0.3	+0.4	−0.3	0.3	+0.5	−0.3	0.8	+0.8	−0.8	0.8	+1.2	−0.8
	0.95	0	−0.55	1.2	0	−0.7	2.1	0	−1.3	2.8	0	−1.6
1.19–1.97	0.4	+0.4	−0.4	0.4	+0.6	−0.4	1.0	+1.0	−1.0	1.0	+1.6	−1.0
	1.1	0	−0.7	1.4	0	−0.8	2.6	0	−1.6	3.6	0	−2.0
1.97–3.15	0.4	+0.5	−0.4	0.4	+0.7	−0.4	1.2	+1.2	−1.2	1.2	+1.8	−1.2
	1.2	0	−0.7	1.6	0	−0.9	3.1	0	−1.9	4.2	0	−2.4
3.15–4.73	0.5	+0.6	−0.5	0.5	+0.9	−0.5	1.4	+1.4	−1.4	1.4	+2.2	−1.4
	1.5	0	−0.9	2.0	0	−1.1	3.7	0	−2.3	5.0	0	−2.8
4.73–7.09	0.6	+0.7	−0.6	0.6	+1.0	−0.6	1.6	+1.6	−1.6	1.6	+2.5	−1.6
	1.8	0	−1.1	2.3	0	−1.3	4.2	0	−2.6	5.7	0	−3.2
7.09–9.85	0.6	+0.8	−0.6	0.6	+1.2	−0.6	2.0	+1.8	−2.0	2.0	+2.8	−2.0
	2.0	0	−1.2	2.6	0	−1.4	5.0	0	−3.2	6.6	0	−3.8
9.85–12.41	0.8	+0.9	−0.8	0.8	+1.2	−0.8	2.5	+2.0	−2.5	2.5	+3.0	−2.5
	2.3	0	−1.4	2.9	0	−1.7	5.7	0	−3.7	7.5	0	−4.5
12.41–15.75	1.0	+1.0	−1.0	1.0	+1.4	−1.0	3.0	+2.2	−3.0	3.0	+3.5	−3.0
	2.7	0	−1.7	3.4	0	−2.0	6.6	0	−4.4	8.7	0	−5.2
15.75–19.69	1.2	+1.0	−1.2	1.2	+1.6	−1.2	4.0	+2.5	−4.0	4.0	+4.0	−4.0
	3.0	0	−2.0	3.8	0	−2.2	8.1	0	−5.6	10.5	0	−6.5

Fig. 8-11 Table of standard fits (*Courtesy ASME, extracted from ANSI B4.1—1964 [1974]*)

American National Standard Running and Sliding Fits (ANSI B4.1 — 1967, R1974)

Nominal Size Range, Inches	Class RC5			Class RC6			Class RC7			Class RC8			Class RC9		
	Clearance*	Standard Tolerance Limits		Clearance*	Standard Tolerance Limits		Clearance*	Standard Tolerance Limits		Clearance*	Standard Tolerance Limits		Clearance*	Standard Tolerance Limits	
		Hole H8	Shaft e7		Hole H9	Shaft e8		Hole H9	Shaft d8		Hole H10	Shaft c9		Hole H11	Shaft
Over To	Values shown below are in thousandths of an inch														
0–0.12	0.6	+0.6	− 0.6	0.6	+1.0	− 0.6	1.0	+1.0	− 1.0	2.5	+ 1.6	− 2.5	4.0	+ 2.5	− 4.0
	1.6	0	− 1.0	2.2	0	− 1.2	2.6	0	− 1.6	5.1	0	− 3.5	8.1	0	− 5.6
0.12–0.24	0.8	+0.7	− 0.8	0.8	+1.2	− 0.8	1.2	+1.2	− 1.2	2.8	+ 1.8	− 2.8	4.5	+ 3.0	− 4.5
	2.0	0	− 1.3	2.7	0	− 1.5	3.1	0	− 1.9	5.8	0	− 4.0	9.0	0	− 6.0
0.24–0.40	1.0	+0.9	− 1.0	1.0	+1.4	− 1.0	1.6	+1.4	− 1.6	3.0	+ 2.2	− 3.0	5.0	+ 3.5	− 5.0
	2.5	0	− 1.6	3.3	0	− 1.9	3.9	0	− 2.5	6.6	0	− 4.4	10.7	0	− 7.2
0.40–0.71	1.2	+1.0	− 1.2	1.2	+1.6	− 1.2	2.0	+1.6	− 2.0	3.5	+ 2.8	− 3.5	6.0	+ 4.0	− 6.0
	2.9	0	− 1.9	3.8	0	− 2.2	4.6	0	− 3.0	7.9	0	− 5.1	12.8	0	− 8.8
0.71–1.19	1.6	+1.2	− 1.6	1.6	+2.0	− 1.6	2.5	+2.0	− 2.5	4.5	+ 3.5	− 4.5	7.0	+ 5.0	− 7.0
	3.6	0	− 2.4	4.8	0	− 2.8	5.7	0	− 3.7	10.0	0	− 6.5	15.5	0	−10.5
1.19–1.97	2.0	+1.6	− 2.0	2.0	+2.5	− 2.0	3.0	+2.5	− 3.0	5.0	+ 4.0	− 5.0	8.0	+ 6.0	− 8.0
	4.6	0	− 3.0	6.1	0	− 3.6	7.1	0	− 4.6	11.5	0	− 7.5	18.0	0	−12.0
1.97–3.15	2.5	+1.8	− 2.5	2.5	+3.0	− 2.5	4.0	+3.0	− 4.0	6.0	+ 4.5	− 6.0	9.0	+ 7.0	− 9.0
	5.5	0	− 3.7	7.3	0	− 4.3	8.8	0	− 5.8	13.5	0	− 9.0	20.5	0	−13.5
3.15–4.73	3.0	+2.2	− 3.0	3.0	+3.5	− 3.0	5.0	+3.5	− 5.0	7.0	+ 5.0	− 7.0	10.0	+ 9.0	−10.0
	6.6	0	− 4.4	8.7	0	− 5.2	10.7	0	− 7.2	15.5	0	−10.5	24.0	0	−15.0
4.73–7.09	3.5	+2.5	− 3.5	3.5	+4.0	− 3.5	6.0	+4.0	− 6.0	8.0	+ 6.0	− 8.0	12.0	+10.0	−12.0
	7.6	0	− 5.1	10.0	0	− 6.0	12.5	0	− 8.5	18.0	0	−12.0	28.0	0	−18.0
7.09–9.85	4.0	+2.8	− 4.0	4.0	+4.5	− 4.0	7.0	+4.5	− 7.0	10.0	+ 7.0	−10.0	15.0	+12.0	−15.0
	8.6	0	− 5.8	11.3	0	− 6.8	14.3	0	− 9.8	21.5	0	−14.5	34.0	0	−22.0
9.85–12.41	5.0	+3.0	− 5.0	5.0	+5.0	− 5.0	8.0	+5.0	− 8.0	12.0	+ 8.0	−12.0	18.0	+12.0	−18.0
	10.0	0	− 7.0	13.0	0	− 8.0	16.0	0	−11.0	25.0	0	−17.0	38.0	0	−26.0
12.41–15.75	6.0	+3.5	− 6.0	6.0	+6.0	− 6.0	10.0	+6.0	−10.0	14.0	+ 9.0	−14.0	22.0	+14.0	−22.0
	11.7	0	− 8.2	15.5	0	− 9.5	19.5	0	−13.5	29.0	0	−20.0	45.0	0	−31.0
15.75–19.69	8.0	+4.0	− 8.0	8.0	+6.0	− 8.0	12.0	+6.0	−12.0	16.0	+10.0	−16.0	25.0	+16.0	−25.0
	14.5	0	−10.5	18.0	0	−12.0	22.0	0	−16.0	32.0	0	−22.0	51.0	0	−35.0

All data above heavy lines are in accord with ABC agreements. Symbols H5, g4, etc. are hole and shaft designations in ABC system. Limits for sizes above 19.69 inches are also given in the ANSI Standard.

*Pairs of values shown represent minimum and maximum amounts of clearance resulting from application of standard tolerance limits.

Fig. 8-11 ***(Continued)***

mating parts. Many factors must be considered in the selection of fits for particular applications. Such factors include the bearing load, operating speed, lubrication, temperature, and materials.

A preferred standard has been established to satisfy various conditions. The standard applies to fits between plain (nonthreaded) cylindrical parts. This standard is called the *ANSI Standard Limits and Fits (ANSI B4.1–1967, Revised 1974).*

The information provided on the numerous tables of this standard is used by designers and engineers to determine the proper clearances between mating parts. The calculated clearances are shown on the prints as the dimensional tolerances. A sample table from the ANSI standard is shown in Figure 8-11.

Preferred Basic Sizes

Decimal			Fractional					
0.010	2.00	8.50	1/64	0.015625	2 1/4	2.2500	9 1/2	9.5000
0.012	2.20	9.00	1/32	0.03125	2 1/2	2.5000	10	10.0000
0.016	2.40	9.50	1/16	0.0625	2 3/4	2.7500	10 1/2	10.5000
0.020	2.60	10.00	3/32	0.09375	3	3.0000	11	11.0000
0.025	2.80	10.50	1/8	0.1250	3 1/4	3.2500	11 1/2	11.5000
0.032	3.00	11.00	5/32	0.15625	3 1/2	3.5000	12	12.0000
0.040	3.20	11.50	3/16	0.1875	3 3/4	3.7500	12 1/2	12.5000
0.05	3.40	12.00	1/4	0.2500	4	4.0000	13	13.0000
0.06	3.60	12.50	5/16	0.3125	4 1/4	4.2500	13 1/2	13.5000
0.08	3.80	13.00	3/8	0.3750	4 1/2	4.5000	14	14.0000
0.10	4.00	13.50	7/16	0.4375	4 3/4	4.7500	14 1/2	14.5000
0.12	4.20	14.00	1/2	0.5000	5	5.0000	15	15.0000
0.16	4.40	14.50	9/16	0.5625	5 1/4	5.2500	15 1/2	15.5000
0.20	4.60	15.00	5/8	0.6250	5 1/2	5.5000	16	16.0000
0.24	4.80	15.50	11/16	0.6875	5 3/4	5.7500	16 1/2	16.5000
0.30	5.00	16.00	3/4	0.7500	6	6.0000	17	17.0000
0.40	5.20	16.50	7/8	0.8750	6 1/2	6.5000	17 1/2	17.5000
0.50	5.40	17.00	1	1.0000	7	7.0000	18	18.0000
0.60	5.60	17.50	1 1/4	1.2500	7 1/2	7.5000	18 1/2	18.5000
0.80	5.80	18.00	1 1/2	1.5000	8	8.0000	19	19.0000
1.00	6.00	18.50	1 3/4	1.7500	8 1/2	8.5000	19 1/2	19.5000
1.20	6.50	19.00	2	2.0000	9	9.0000	20	20.0000
1.40	7.00	19.50	....	....	...	......	....	....
1.60	7.50	20.00	All dimensions are given in inches.					
1.80	8.00	.....						

Fig. 8-12 Table of preferred basic sizes (*Courtesy ASME, extracted from ANSI B4.1—1967 [1974]*)

The following information and examples are intended to provide a basic understanding of the classification of fits. This section is not all-inclusive of the subject matter.

Preferred Basic Sizes

The basic sizes of mating parts are usually selected from the fractional or decimal series in the table shown in Figure 8-12. This process simplifies machining, tooling, and material inventory.

Standard Fits

Standard types of fits have been established. The standards ensure uniform operation of mating parts throughout the range of sizes. The type of fit is determined by the service requirements of the designed equipment. The type of fit is indicated in the tables by letter symbols, as follows.

- RC—Running or Sliding Fit
- LC—Locational Clearance Fit
- LT—Transition Locational Fit
- LN—Locating Interference Fit
- FN—Force and Shrink Fit

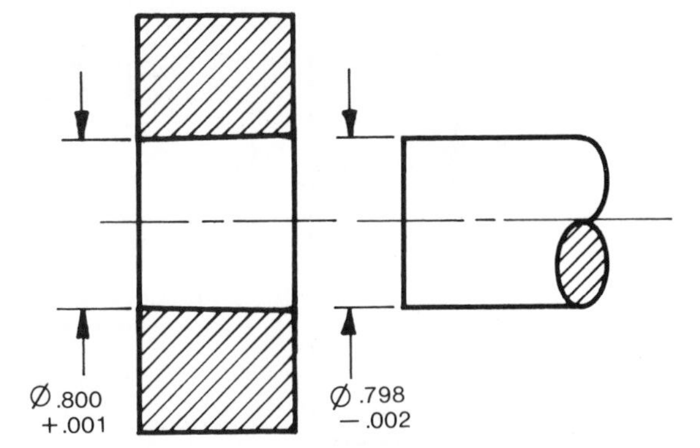

Fig. 8-13 Basic hole system (shaft is changed for fit)

Numbers are also used with the letter symbol to indicate a class of fit. For example, the notation *RC4* represents a class 4 running fit. These letter or number symbols will rarely be indicated on prints. Dimensional sizes with their tolerances, as listed in the tables, will be shown.

Description of Fits

Some basic types of fits and their functions are described here. A complete listing of all fits can be found in engineering drafting manuals.

- **RC1 Close Sliding Fits** are intended for the accurate location of parts which must assemble without perceptible play.

- **RC2 Sliding Fits** are intended for accurate location, but with greater maximum clearance than class RC1. Parts made to this fit move and turn easily, but are not intended to run freely. In the larger sizes they may bind with small temperature changes.

- **RC3 Precision Running Fits** are about the closest fits which can be expected to run freely. They are intended for precision work at low speeds and light bearing pressures. They are not suitable where higher temperatures are likely to be encountered.

- **RC4 Close Running Fits** are intended mainly for running fits on accurate machinery with moderate surface speeds and bearing pressures, and where accurate locations and minimum play are desired.

There are additional standard RC fits from class 5 through class 9. Similar classifications of fits exist for the LC, LT, LN, and FN standards.

Cylindrical Fit Dimensions

It is necessary to know whether a basic hole system or a basic shaft system will be used. This information should be established before calculating the tolerance of cylindrical mating parts for the application involved.

A *basic hole system* is one in which the hole size is basic and the allowance for fit is applied to the shaft. The dimensions on a print using the basic hole system will indicate the minimum hole size. The indicated shaft size will be the hole size minus the amount of clearance needed for the class of fit used. See Figure 8-13.

The basic hole system is the system used most often on industrial prints. It is used most often because it allows for economies in the machining operations. The holes can be machined by standard-size drills, reamers, and so on. They can also be inspected by standard gaging devices.

The *basic shaft system* is used when a standard shaft size will be used and the allowance will be applied to the hole. The minimum shaft size will be indicated. The hole size will be equal to the minimum shaft size minus the fit clearance, Figure 8-14. The basic shaft system is used only when a standard shaft size is used.

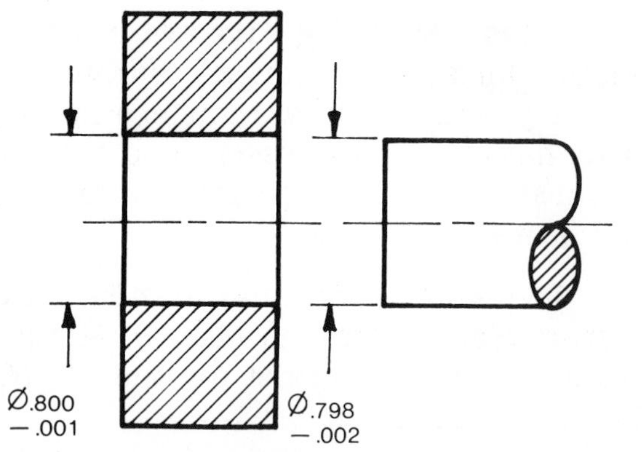

Fig. 8-14 Basic shaft system (hole is changed for fit)

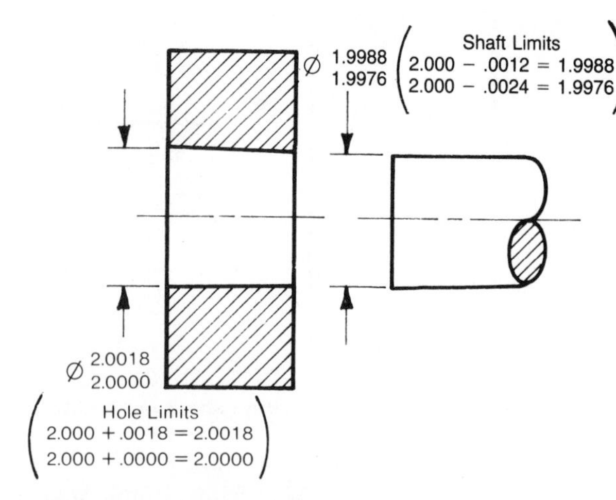

Fig. 8-15 Basic hole size calculations

Selecting Proper Fit Limits from Tables

The following example illustrates how designers and engineers use fit tables.

The shaft size is 2.000 inches, and will be a close running fit, RC4. From the Running and Sliding Fits table (Figure 8-11):

1. Locate the size range from the column on the left.
2. From the vertical columns find the class of fit (RC4). In this column locate the hole and shaft tolerance limits.
3. To the hole size (2.000), add .0018 for the maximum size, and add .0000 for the minimum size. For the shaft, subtract .0012 for the maximum size, and subtract .0024 for the minimum size. See Figure 8-15.

ENG. SPEC. 415

11% SILICON DIE CAST ALUMINUM

CASTING SPECIFICATION	
SILICON	10–12%
COPPER	.6 MAX.
MAGNESIUM	.1 MAX.
MANGANESE	.8 MAX.
ZINC	.5 MAX.
IRON	.4–.8
NICKEL	.5 MAX.
CHROMIUM	.3 MAX.
TITANIUM	.3 MAX.
OTHERS	.2 MAX.

PURCHASE SPECIFICATION	
SILICON	10–12%
COPPER	.6 MAX.
MAGNESIUM	.1 MAX.
MANGANESE	.3 MAX.
ZINC	.5 MAX.
IRON	.4–.8
NICKEL	.5 MAX.
CHROMIUM	.3 MAX.
TITANIUM	.3 MAX.
OTHERS	.2 MAX.

20 POUND BARS PUT UP IN BUNDLES OF 50–100 BARS.

TAG EACH BUNDLE WITH A *GREEN* TAG, SHOWING ALLOY #415 AND HEAT NUMBER.

FURNISH THREE COPIES OF CERTIFIED ANALYSIS WITH EACH SHIPMENT.

#415	CASTING AND PURCHASING SPEC.	12/15/75	R. JONES

Fig. 8-16 Engineering specification

ENGINEERING SPECIFICATIONS

In recent years manufacturing processes have become more complicated. To identify these processes thoroughly on a print (as a Note:) would require considerable print space. This would leave less room for the drawings and dimensioning. It would also clutter up the print, reducing the ability of the print reader to visualize the details of the part.

To eliminate this problem, many companies have established standards to satisfy their particular requirements. These standards are catalogued separately from the print. They are identified as *Engineering Specifications* (Eng. Spec.).

Each specification is numbered. When a special manufacturing process is to be performed on a part, the print will carry the correct engineering specification number.

Engineering specifications may be used for heat treating, plating, painting, thread cutting, or other special operations that may vary from the industry standard. If an engineering specification is indicated as a note on a print, the producer of the part shown on the print must have a copy of the engineering specification to complete all the requirements of the print.

The engineering specification shown in Figure 8-16 concerns an alumi-

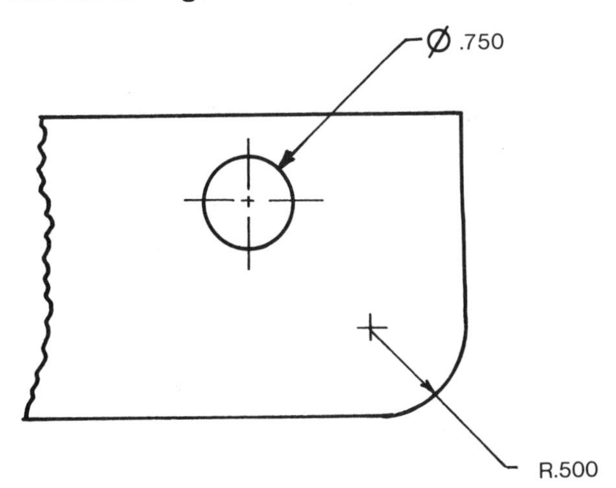

Fig. 8-17 Radius and diameter notations

Ø = Diameter

S Ø = Spherical diameter

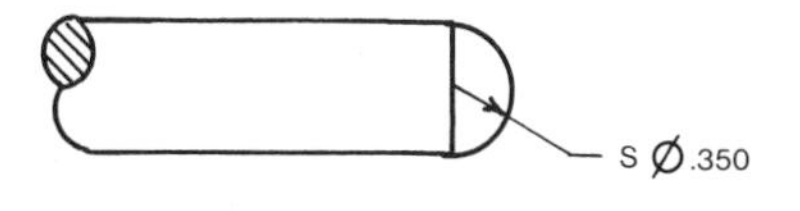

R = Radius

SR = Spherical radius

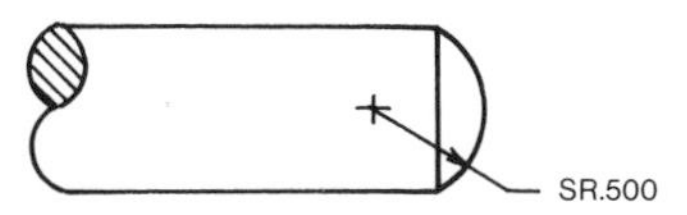

Fig. 8-18 Special diameter and radius notations

num alloy. The sheet specifies the limits of each alloy element that would best suit a particular application.

Some industries use the engineering specification notations on the prints to safeguard their trade secrets. This is a common practice when parts are made of exotic metal alloys.

NEW MACHINING SYMBOLS

New machining symbols and dimensions notations have been introduced in the ANSI Y14.5—1982 Dimensioning and Tolerancing Standard. These new practices have been incorporated into this textbook.

The new standard uses special notations for different types of radii and diameters, Figure 8-17 and Figure 8-18. Repetitive features are identified by *X*, Figure 8-19.

New symbols are also used to replace frequently used terms, Figure 8-20. Examples of their usage are illustrated in Figure 8-21 through Figure 8-27.

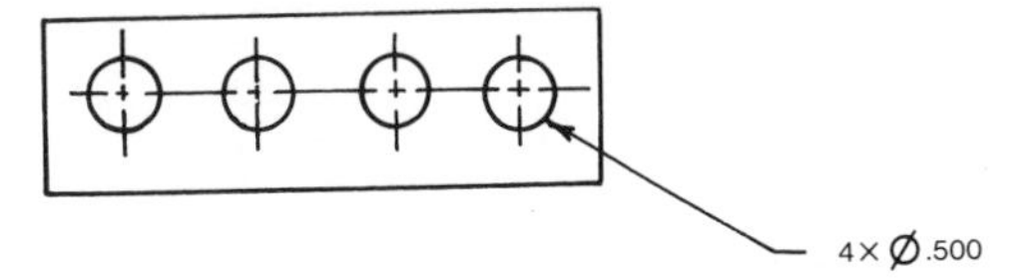

Fig. 8-19 Repetitive features

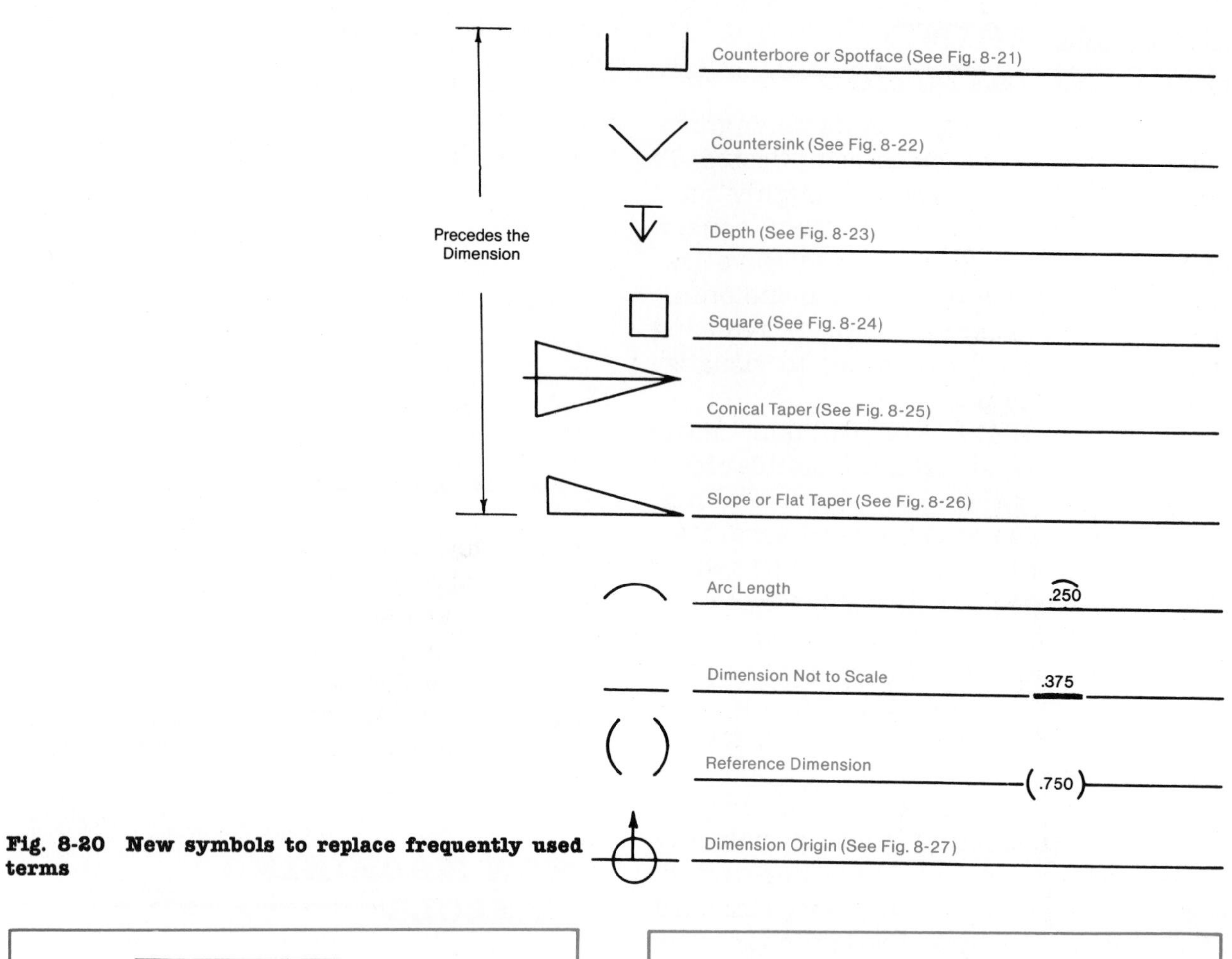

Fig. 8-20 New symbols to replace frequently used terms

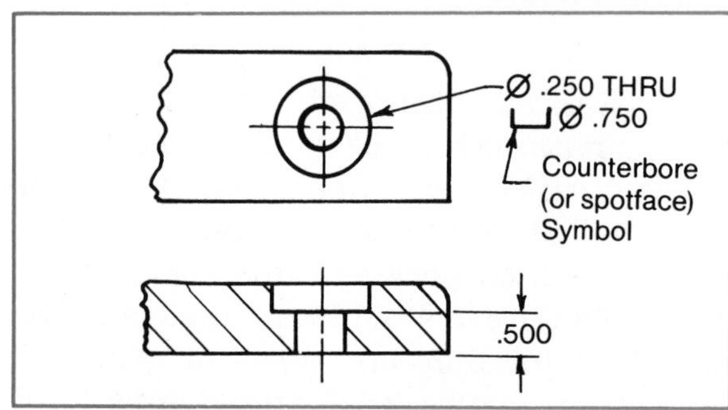

Fig. 8-21 Counterbore or spot-face symbol

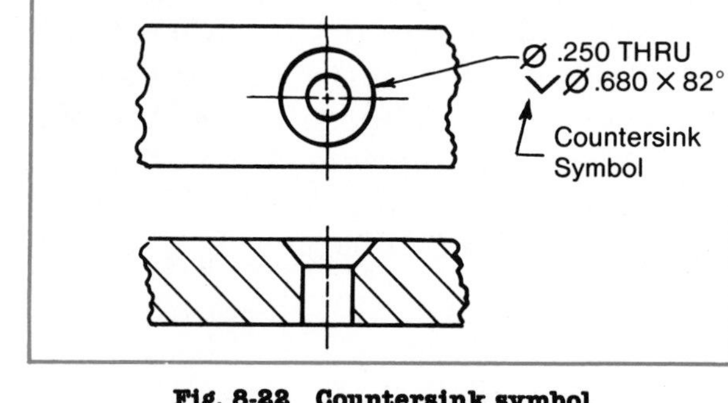

Fig. 8-22 Countersink symbol

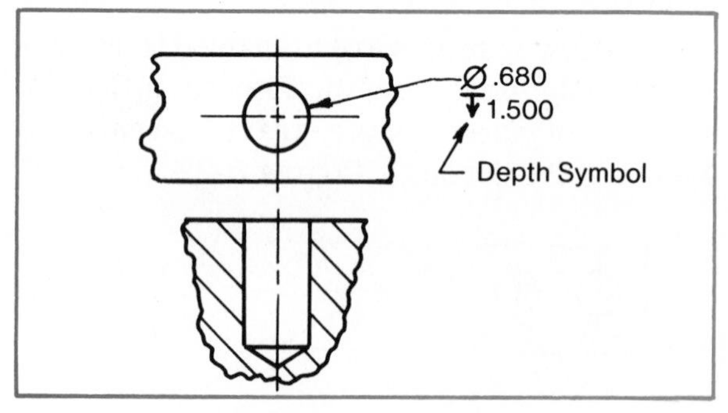

Fig. 8-23 Depth symbol

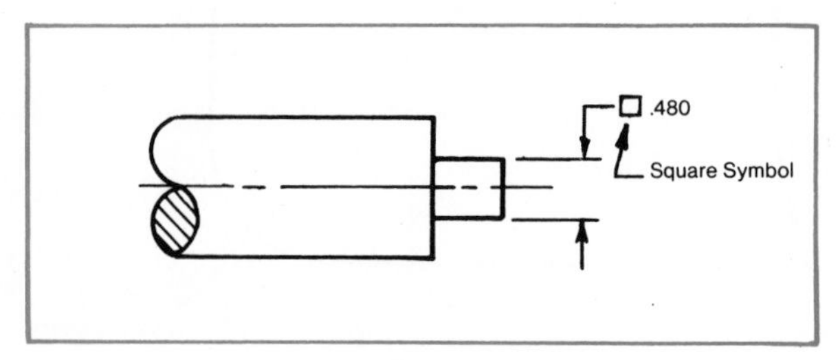

Fig. 8-24 Square symbol

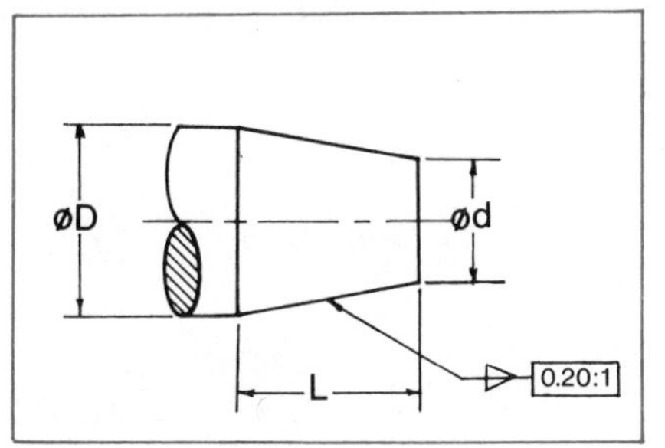

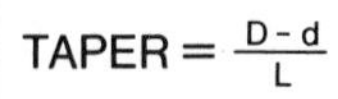

Fig. 8-25 Specifying a conical taper

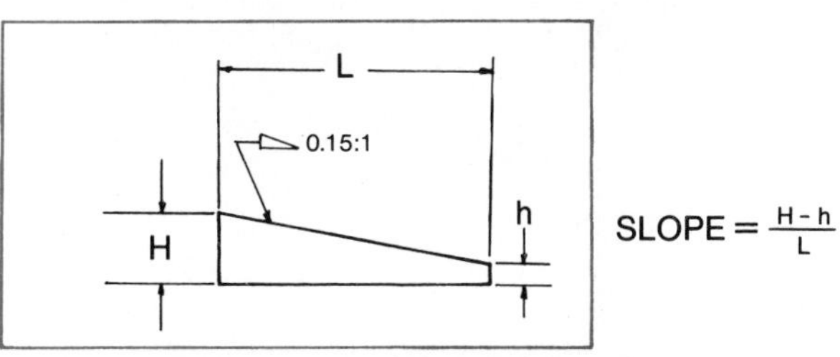

Fig. 8-26 Specifying a flat taper

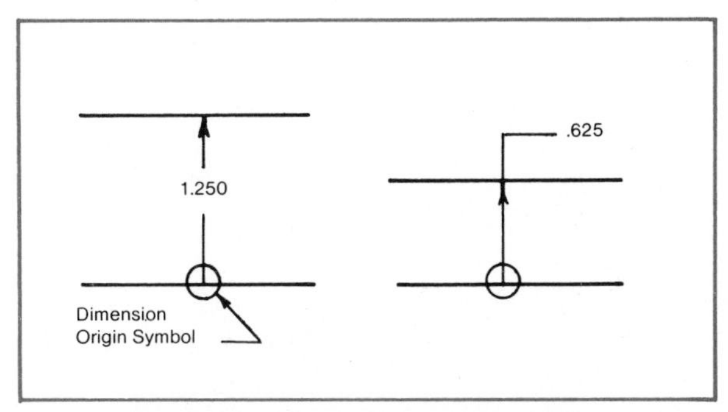

Fig. 8-27 Dimension origin symbol

REVIEW

This review is provided to serve as reinforcement study material. Fill in the appropriate word(s) to complete the sentences below.

1. The scale used on a print is usually specified in the ______________.

2. If the object drawn is large, the print will usually have (a) (an) ______________________________ scale.

3. If the object drawn is small, the print will usually have (a) (an) ______________________________ scale.

4. When a dimension has been changed, the changed dimension will be shown with (a) (an) ______________________________ under the dimension.

5. According to the new standard, fillet and round dimensions are preceded by (a) (an) ______________________________, which specifies that the dimensions have a radius value.

6. The class of fits (rarely shown on prints) is satisfied by the actual ______________________________ shown on the print.

7. Common machining notations are usually spelled out immediately preceding the ______________________________.

NAME ____________________

DATE ____________________

SCORE ____________________

EXERCISE B8-1

Print #701

Answer these questions after studying Print #701 found after this exercise.

QUESTIONS	ANSWERS
1. Give dimensions 1 through 9.	1. (1) ____ (4) ____ (7) ____ (2) ____ (5) ____ (8) ____ (3) ____ (6) ____ (9) ____
2. The fillets on this part would be what decimal inch size? a. .250 c. .512 b. .375 d. .125	2. ____________
3. The rounds on this part would be what decimal inch size? a. .250 c. .375 b. .512 d. .125	3. ____________
4. If the object were drawn to 2″ = 1″ scale, what would be the actual drawing length of the dimension indicated by (2)? a. .750 c. 3.000 b. 1.500 d. 6.000	4. ____________
5. If the object were drawn to 1″ = 2″ scale, what would be the actual drawing length of the dimension indicated by (2)? a. .750 c. 3.000 b. 1.500 d. 6.000	5. ____________
6. What detail number refers to a counterbored hole? a. 5 c. 7 b. 6 d. 8	6. ____________
7. What detail number refers to a countersunk hole? a. 5 c. 7 b. 6 d. 8	7. ____________
8. What detail number refers to the spot-faced holes? a. 5 c. 7 b. 6 d. 8	8. ____________

(continued)

(Exercise B8-1 continued)

9. How are the top and bottom of the pad containing the counterbored holes machined? 9. ____________

   a. Milled c. Broached
   b. Ground d. Planed

10. Why is the .232 hole specified as reamed? 10. ____________

   a. To reduce machining costs
   b. To accurately locate the hole
   c. It is the only hole this size
   d. To maintain size tolerance

11. What would be the maximum height of the center pad? 11. ____________

   a. 1.248 c. 1.496
   b. 1.250 d. 1.252

12. What would be the minimum width of the center pad? 12. ____________

   a. 1.500 c. 1.496
   b. 1.502 d. 1.498

13. This part is made of what material? (Refer to Figure 8-16.) 13. ____________

   a. Brass c. Cast iron
   b. Aluminum d. Aluminum alloy

14. What would be the proper way of specifying a counterbored hole? 14. ____________

   a. .750 c'bore, .500 drill, 2 holes, .250 deep
   b. 2 holes, .500 drill, .750 c'bore, .250 deep
   c. .500 drill, .750 c'bore, .250 deep, 2 holes
   d. .500 drill, .250 deep, .750 c'bore, 2 holes

15. What is the difference between the maximum and minimum diameter of the counterbored holes? 15. ____________

   a. .006 b. .003
   b. .002 d. .004

16. What would be the overall length of the bracket? 16. ____________

   a. 3.000 c. 3.750
   b. 4.000 d. 3.500

NAME ____________

DATE ____________

SCORE ____________

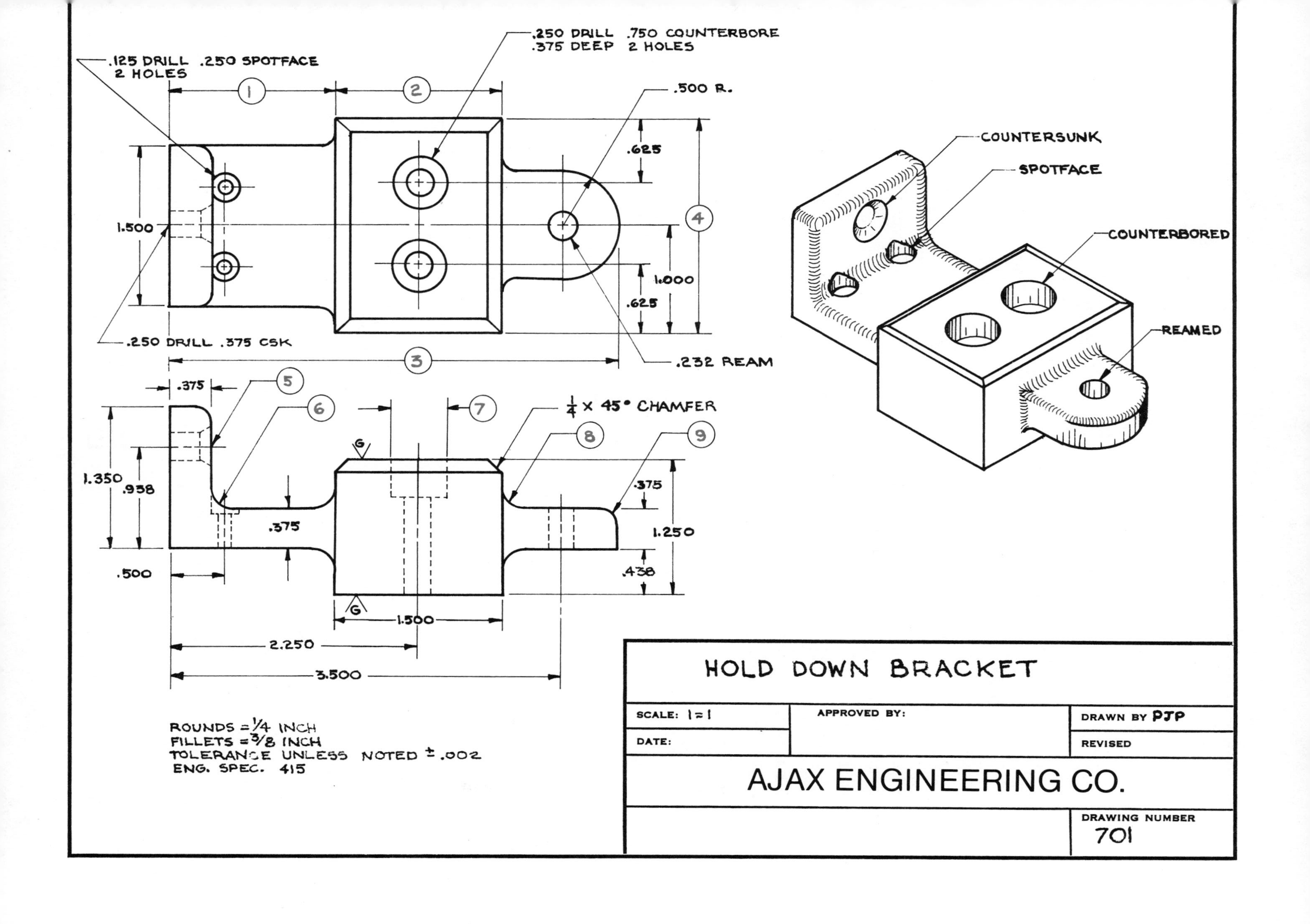

.125 DRILL .250 SPOTFACE
2 HOLES
.250 DRILL .750 COUNTERBORE
.375 DEEP 2 HOLES
.500 R.
.625
1.500
1.000
.625
.250 DRILL .375 CSK
.232 REAM
.375
¼ X 45° CHAMFER
1.350
.938
.375
.375
1.250
.500
.438
1.500
2.250
3.500
G
G
1
2
3
4
5
6
7
8
9
COUNTERSUNK
SPOTFACE
COUNTERBORED
REAMED
ROUNDS = 1/4 INCH
FILLETS = 3/8 INCH
TOLERANCE UNLESS NOTED ± .002
ENG. SPEC. 415
HOLD DOWN BRACKET
SCALE: 1 = 1
DATE:
APPROVED BY:
DRAWN BY PJP
REVISED
AJAX ENGINEERING CO.
DRAWING NUMBER
701

CHAPTER 9

THREADS AND FASTENING DEVICES

OBJECTIVES

After studying this chapter, you will be able to:

- Identify the common methods of representing threads on an industrial print.
- Discuss the difference between bolt threads and pipe threads.
- Explain the meaning of the thread notations on a print.
- Describe the common thread forms.
- Explain the metric thread notations.
- Describe the different types of threaded fasteners.
- Describe the different types of nonthreaded fasteners.

Before the development of the welding process, machine and structural parts were held together by threaded or nonthreaded fasteners. Nonthreaded fasteners, usually rivets, were used with parts that were to be permanently assembled.

Currently, most permanently assembled parts are welded rather than riveted. However, threaded fasteners are used when parts are to be disassembled for maintenance or repair. Threads may also be used for purposes of adjustment and transmission of power.

Standards which specify bolt size, number of threads, and other technical data have been established to fill the needs of many different thread uses. These standards are listed in engineering tables and charts. They are also indicated on the part or assembly prints.

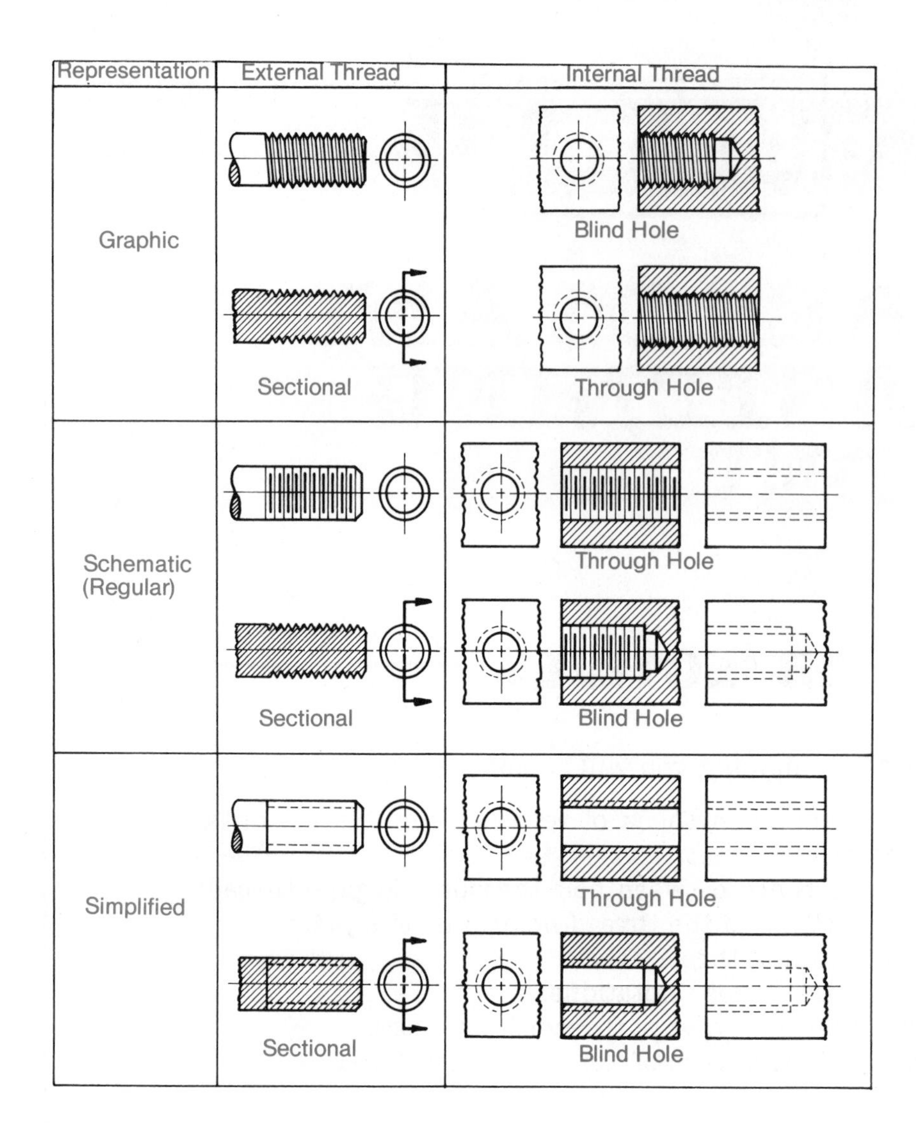

Fig. 9-1 Thread representation methods

THREAD REPRESENTATION

Three methods are in general use to indicate forms of screw threads on industrial prints. These methods are the *graphic*, the *schematic*, and the *simplified representations*, Figure 9-1.

The graphic representation closely resembles what the eye would actually see when viewing the threads. This method is very effective in showing the threads. However, it is time consuming for the drafter. Therefore, graphic representation is used only when it is necessary to identify the thread from other details of the part.

The schematic representation requires much less drawing time. This method is very effective for showing threaded details of a part. Schematic representation is not used for hidden internal threads or in sectional views.

The simplified representation is the easiest to draw. This is the method most often used on engineering drawings. In drawings which contain many invisible

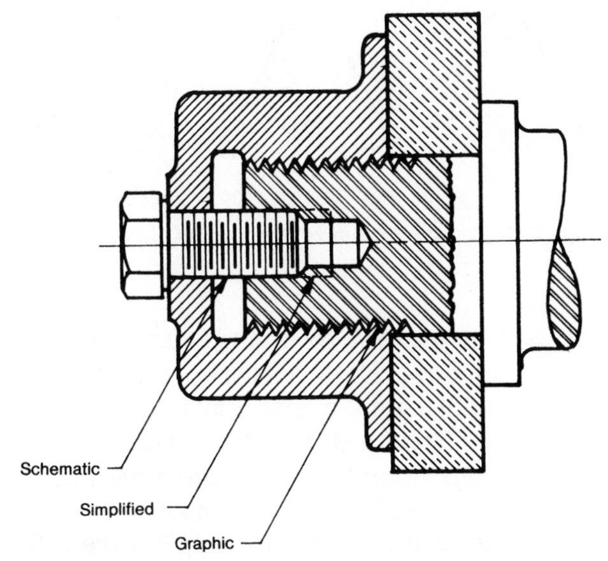

Fig. 9-2 Three thread representation methods on the same drawing

surfaces, however, the schematic method is used to avoid confusion with other details of the part.

Occasionally all three thread representation methods may be used on the same drawing. See Figure 9-2.

Pipe threads which are cut at a taper are usually shown on the drawing using the simplified method. Pipe threads are drawn at a slight angle because they are cut on the part at a taper, Figure 9-3.

THREAD DIMENSIONS—

A standard specification notation procedure has been established to indicate thread dimensions on industrial prints.

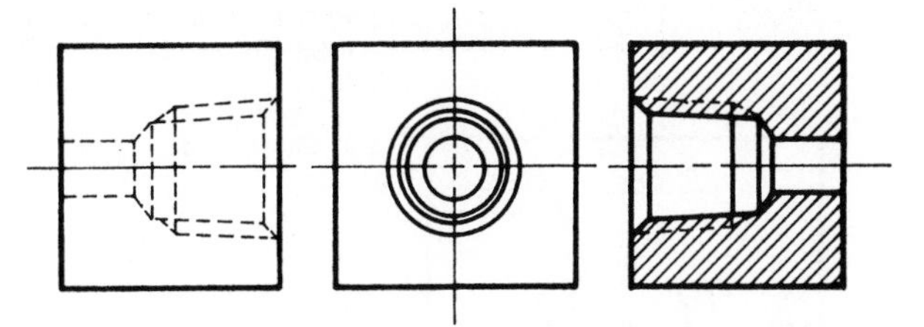

Fig. 9-3 Pipe thread—simplified representation

This standard is used regardless of the thread representation method.

The specification notation is placed close to the thread, with a leader and arrow pointing to the threads. As shown in Figure 9-4, the proper order of listing the most important thread features is as follows:

1. The diameter of the thread (nominal size)
2. The number of threads per inch of length
3. The thread type (Unified National)
4. The thread series (coarse—C, fine—F, extra fine—EF)
5. The class of fit (looseness or tightness of mating threads). The three main classes of fits are *class 1* (loose), *class 2* (standard), and *class 3* (tightest).
6. When used, the notation *A* (external thread) or *B* (internal thread) is placed next to the class of fit notation.

Unless otherwise specified, all threads are assumed to be right-hand thread (turn clockwise to tighten). Left-hand threads are always specified *LH*, following the class of fit notation.

Other notations for special thread fits may be indicated immediately below

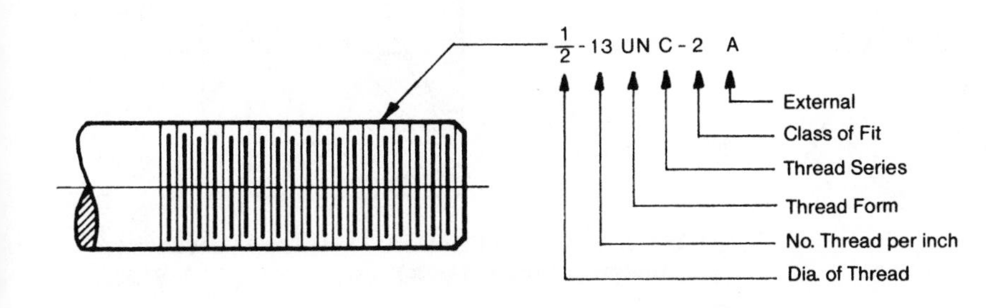

Fig. 9-4 Print thread notations

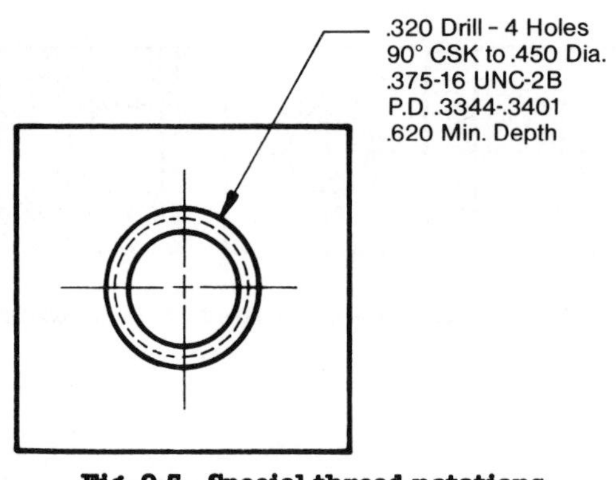

Fig. 9-5 Special thread notations

the standard notations, as shown in Figure 9-5. Such notations include pitch diameter (*PD*), size limits, full thread length, and other specifications.

Other terms occasionally used on industrial prints are explained in Table 9-1.

THREAD FORMS

Before the 1960s, the *American National* thread (N) was the most widely used thread form in the world, Figure 9-6. The next most common thread was the *British Standard Whitworth,* Figure 9-7. Because of their different thread angles (American 60°, British 55°), these two thread forms would not fit together properly.

To solve this problem, the *Unified National Standard* thread (UN) was adopted. To establish this thread form,

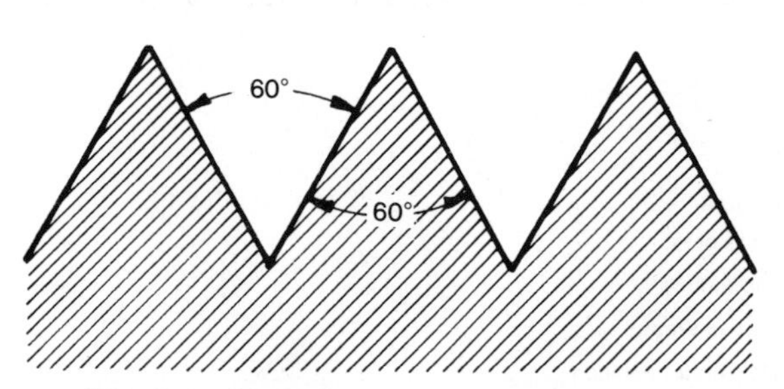

Fig. 9-6 American National thread form

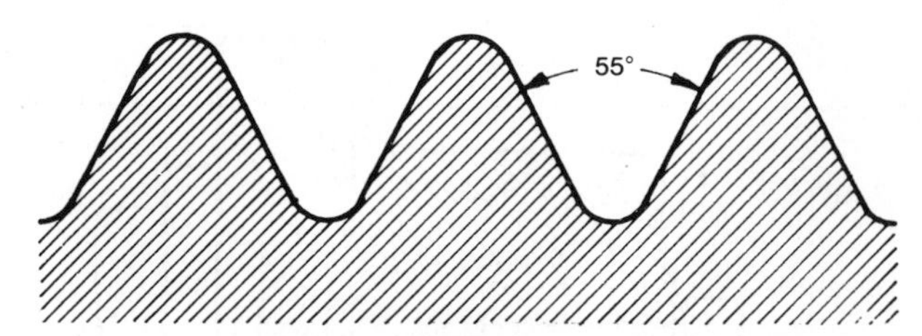

Fig. 9-7 British Standard Whitworth thread form

the British agreed to change their thread angle to 60°. This allowed the interchange of American and British threaded fasteners, Figure 9-8. All threads in the inch measurement system are now specified as Unified National Standard.

Other forms of thread, Figure 9-9, are used to meet special engineering requirements. When these forms are used, they are identified by name on the industrial print.

PIPE THREADS

Pipe threads are available in two standard forms: *regular* and *dryseal.*

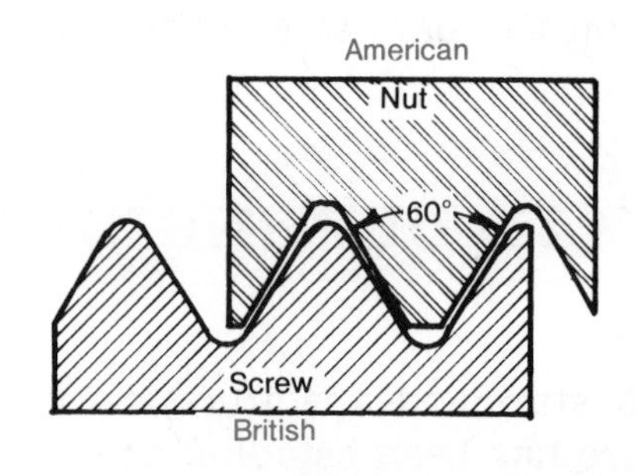

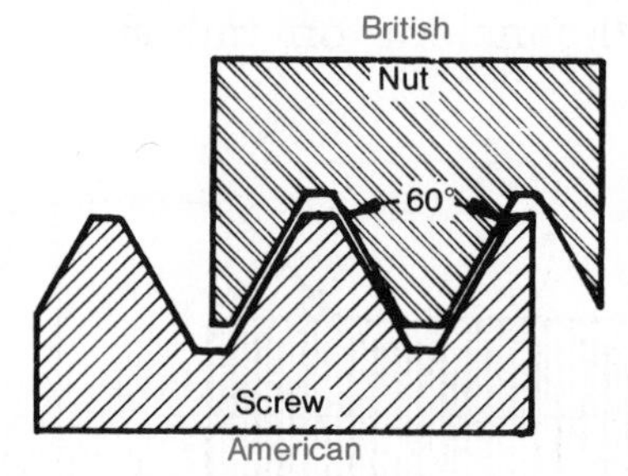

Fig. 9-8 Unified National Standard thread form (American and British)

Table 9-1 Screw Thread Tolerances

It is generally recognized that in mass production, it is impossible to reproduce in exact detail, the theoretically perfect product as laid out on the drawing board. The slight variation between the theoretically perfect product and each unit of the actual product is called the TOLERANCE.

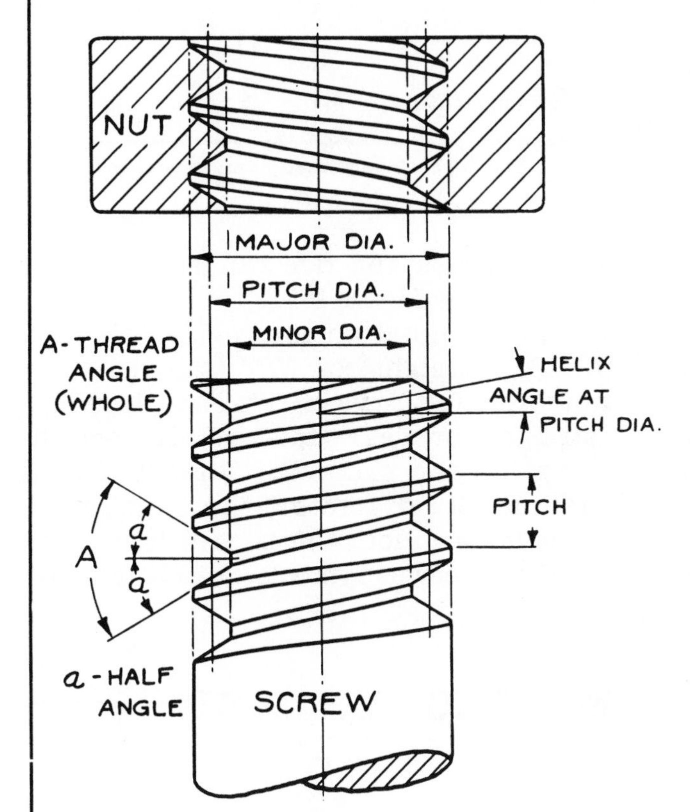

ALLOWANCE—An intentional difference in correlated dimensions of mating parts. It is the minimum clearance or the maximum interference between such parts.

ANGLE OF THREAD—The angle included between the flanks of the thread measured in an axial plane.

HALF ANGLE OF THREAD—The angle included between a flank of the thread and the normal (90°) to the axis, measured in an axial plane.

LEAD OF THREAD—The distance a screw thread advances axially in one turn. On a single-thread screw the lead and pitch are identical. On a double thread the lead is 2 x pitch, on a triple thread the lead is 3 x pitch, etc.

MAJOR DIAMETER—The largest diameter of a straight thread.

MINOR DIAMETER—The smallest diameter of a straight thread.

PITCH—The distance from a point on a screw thread to a corresponding point on the next thread measured parallel to the axis.

The pitch in inches =

$$\frac{1}{\text{Number of threads per inch}}$$

PITCH DIAMETER— On a straight screw thread, the diameter of an imaginary cylinder, the surface of which would pass through the threads at such points as to make equal the width of the threads and the width of the spaces cut by the surface of the cylinder.

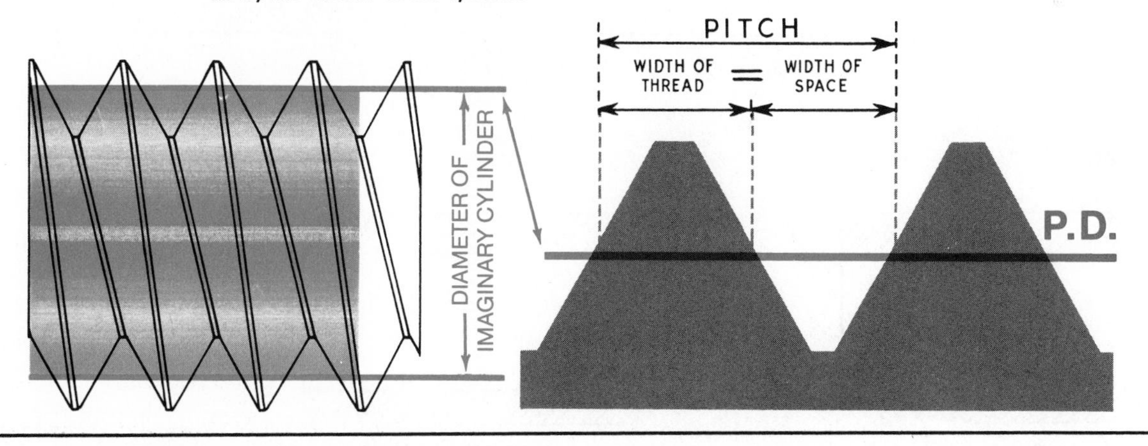

(Courtesy of Greenfield Tap & Die, A Division of TRW Corp.)

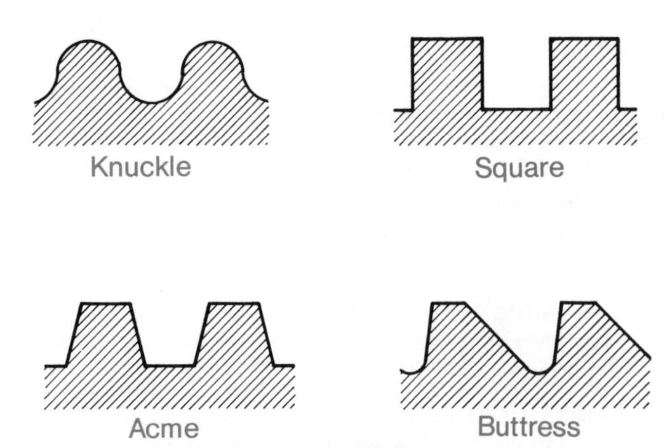

Fig. 9-9 Special thread forms

Nominal Pipe Sizes	Threads Per Inch
$\frac{1}{16}$, $\frac{1}{8}$	27
$\frac{1}{4}$, $\frac{3}{8}$	28
$\frac{1}{2}$, $\frac{3}{4}$	14
1, 1$\frac{1}{4}$, 1$\frac{1}{2}$, 2	11$\frac{1}{2}$
2$\frac{1}{2}$ and larger	8

Fig. 9-11 Pipe thread number ranges

Regular pipe threads are designed with a root (bottom) and crest (top) clearance. Thread compound or sealer is required to obtain a watertight joint. The regular thread is used on most plumbing systems and plumbing hardware.

No clearance is allowed on dryseal pipe threads. Therefore, a sealer is not required. Pipes with dryseal threads are used in automotive, refrigeration, and hydraulic applications.

Both the regular and dryseal forms are available in tapered or straight threads, Figure 9-10. The tapered threads (¾-inch taper per foot standard) are usually used in the plumbing trade. Straight threads are most often used in automotive and hydraulic systems. Pipe has only one thread series and five ranges of thread per inch. See Figure 9-11.

The notation for regular threads is *NPS* (straight) or *NPT* (tapered). The notation for dryseal threads is *NPSF* (straight) or *NPTF* (tapered). The specification on the prints for threads should include the diameter, the number of threads per inch, the thread form, and the thread series symbols. See Figure 9-12.

METRIC THREADS

The profile of metric threads is similar to the Unified inch thread. However, metric thread diameters are specified in millimeter (mm) sizes. Unified inch and metric threads are not interchangeable.

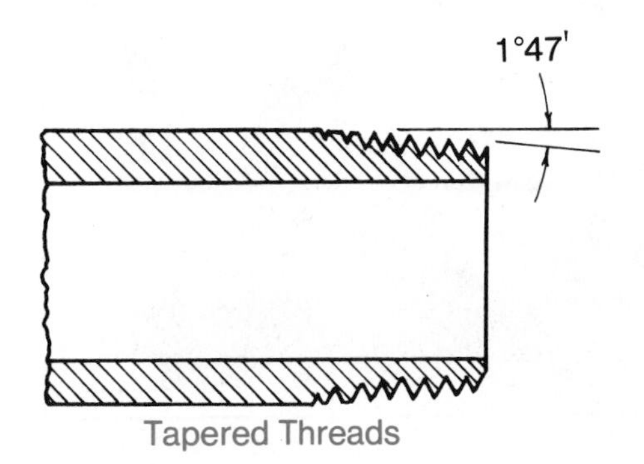

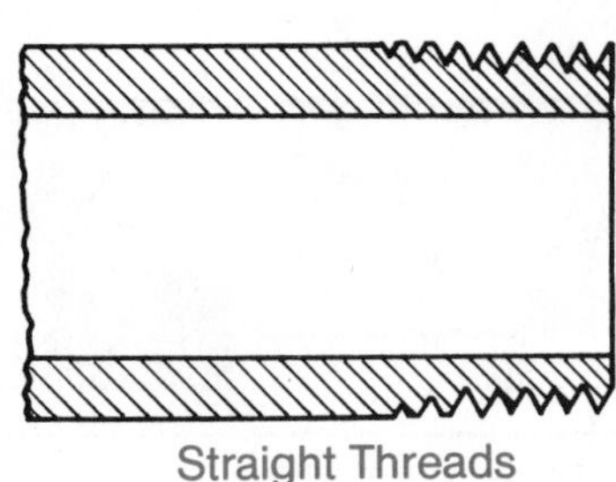

Fig. 9-10 Pipe threads

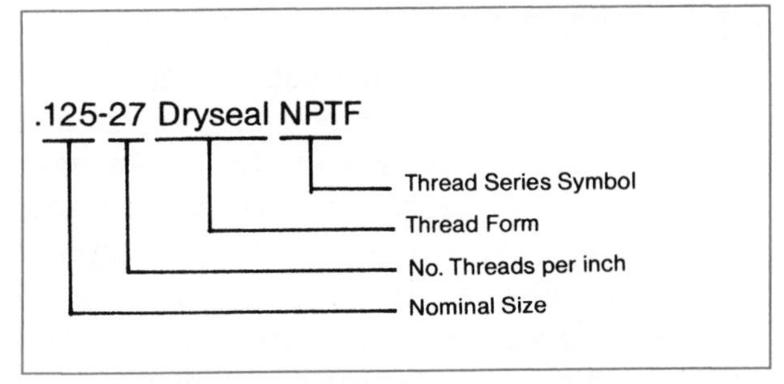

Fig. 9-12 Pipe thread notations

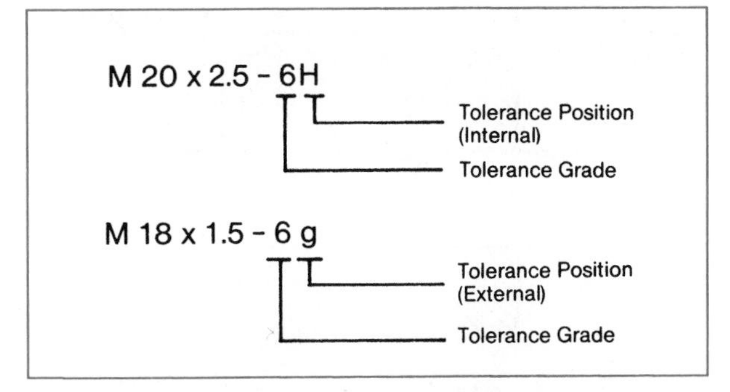

Fig. 9-14 Additional metric thread notations

The letter *M* precedes all metric thread notations. The thread series includes both coarse and fine threads. Coarse threads are used for general applications. Fine threads are used for precision work.

Metric fine-thread notations also include the pitch (threads per millimeter of length) after the nominal size. The sign × is used to separate the two notations, Figure 9-13.

Tolerance grade (the class of fit on Unified inch threads) may also be indicated on metric threads by a numbering system from 3 through 9. Number 3 is the tightest fit (fine). Number 6 is medium, and number 9 is coarse.

The term *H* (internal thread) or *g* (external thread) is shown next to the tolerance grade. See Figure 9-14.

The method of indicating specifications for metric threads is similar to that used for Unified threads. However, it is important to remember that the measuring system for Unified threads is the inch, while metric threads are measured in millimeters.

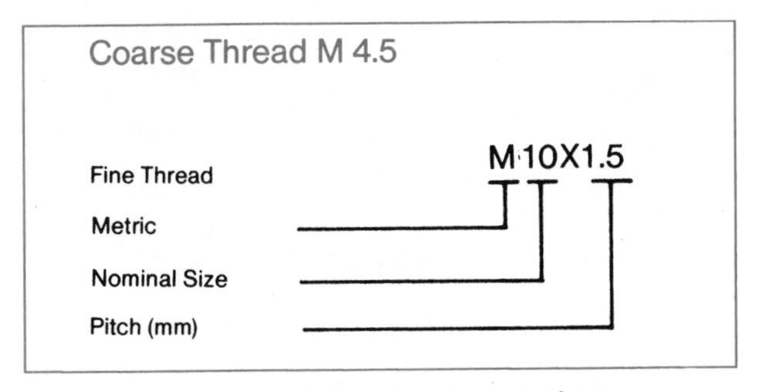

Fig. 9-13 Metric thread notations

One inch equals 25.4 millimeters. Therefore, a Unified thread of .500–20 UN F–2B would be very similar to the metric thread *M 12.7–6H*. These two threads would be interchangeable. However, standard metric threads are made only in whole or fractional multiples of the millimeter. Thus, the *M 12.7 × 1.27–6H* would not be a standard metric thread. The closest metric standard would be *M 12.5 × 1.5–6H*. (This thread, however, would not be interchangeable with the Unified inch thread.)

THREADED FASTENERS

The most common types of threaded fasteners are *set screws, cap screws, machine screws, bolts, studs,* and *nuts.* Knowledge of these fasteners is very helpful in reading industrial prints.

Set Screws

Set screws are common devices used to prevent rotary motion between two mating parts which are not subject to heavy stresses. Their most common

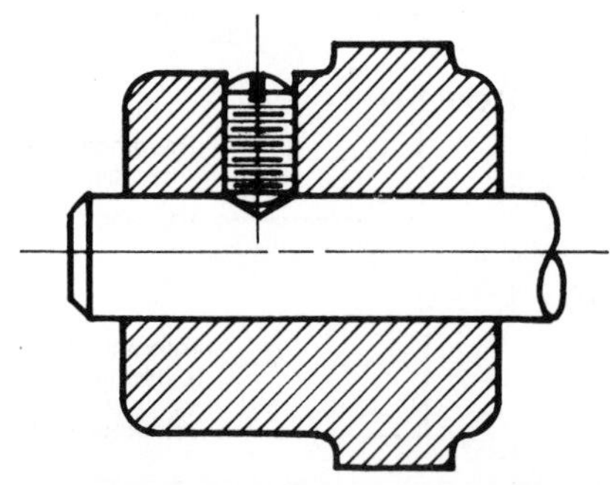

Fig. 9-15 Set screw assembly

application is to secure a pulley to a shaft, Figure 9-15.

The different types of set screws are illustrated in Figure 9-16. The square head set screw has been replaced by the headless type. This change was made because of the potential danger of the projecting head on the revolving shaft.

Different types of points, Figure 9-17, are used to meet a variety of needs. The specification on the print should include the diameter, threads per inch, class of fit, and length. The type of head and type of point should be shown below that information, as in the following example:

.375–24 UN F–2A–.50
Hex Socket Oval Point Set Screw

Cap Screws

Cap screws are used to fasten together two parts, when one part has a drilled clearance hole and the other part is threaded to receive the screw threads.

The various types of cap screw heads are shown in Figure 9-18. Their diameter sizes range from .25 inch to 1.25 inches. Their lengths run from .50 inch to approximately 3.00 inches. The specification should include the diameter, thread form, class of fit, length, and type of head.

Machine Screws

Machine screws are usually smaller than cap screws; they are used for finer work. The threads on machine screws run the entire length of the screw body.

Machine screw sizes under .250 inch are specified by the appropriate number from 1 through 12 (see the thread table in the Appendix). Those sizes above .250 inch are usually in fractional inch sizes.

The types of machine screw heads are shown in Figure 9-19. Machine screw specifications on industrial prints are similar to those for cap screws.

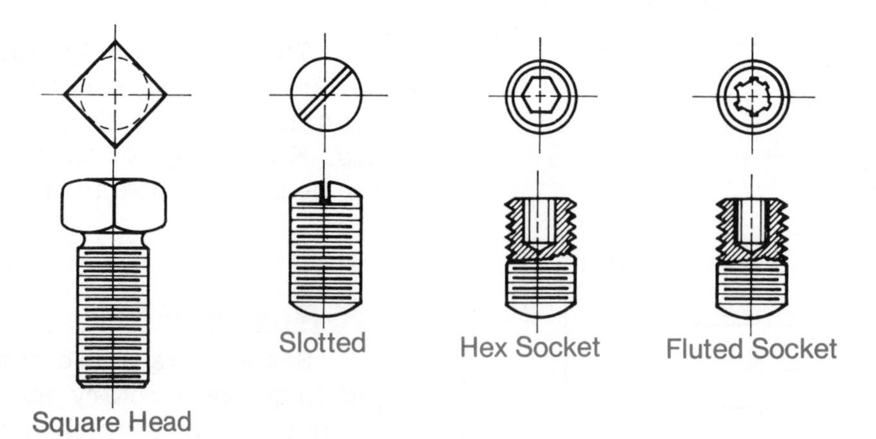

Fig. 9-16 Types of set screws

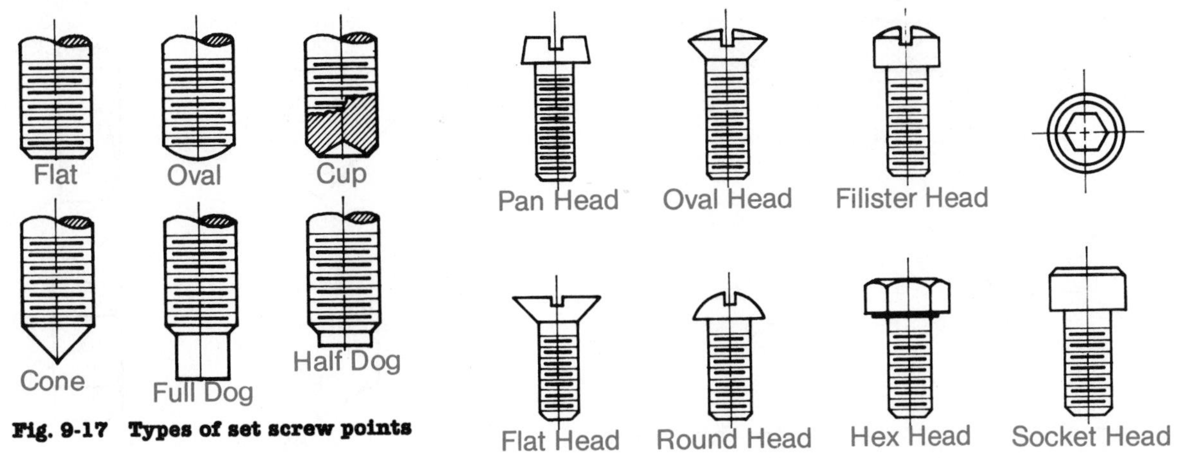

Fig. 9-17 Types of set screw points

Fig. 9-18 Cap screws

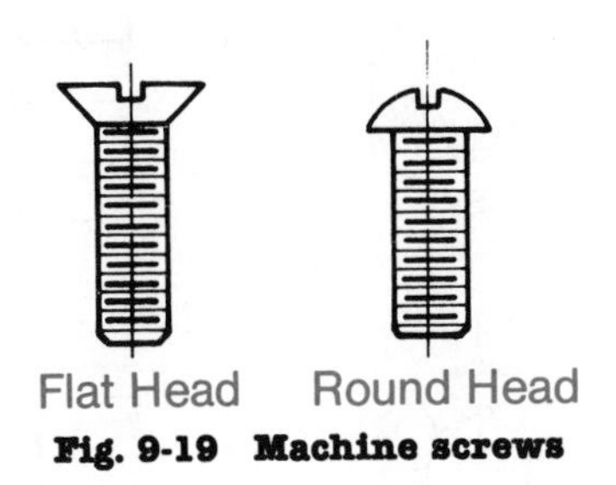

Fig. 9-19 Machine screws

Bolts

There are two series of bolts, *regular* and *heavy duty.* Regular bolts are used for general-purpose work. Heavy-duty bolts are used when more working stress is required.

Bolts are classified as *finished, semifinished,* and *unfinished.* The finished classification means that the bolt is manufactured to a higher quality and tolerance. The strength requirements of the assembly determine what type of bolt will be used. Fine-thread bolts are usually stronger than coarse-thread bolts. Fine-thread bolts are used when a limited thread engagement is available or when extra strength is required.

Numerous head types are available to satisfy different design requirements, Figure 9-20. On a print, bolt specifica-

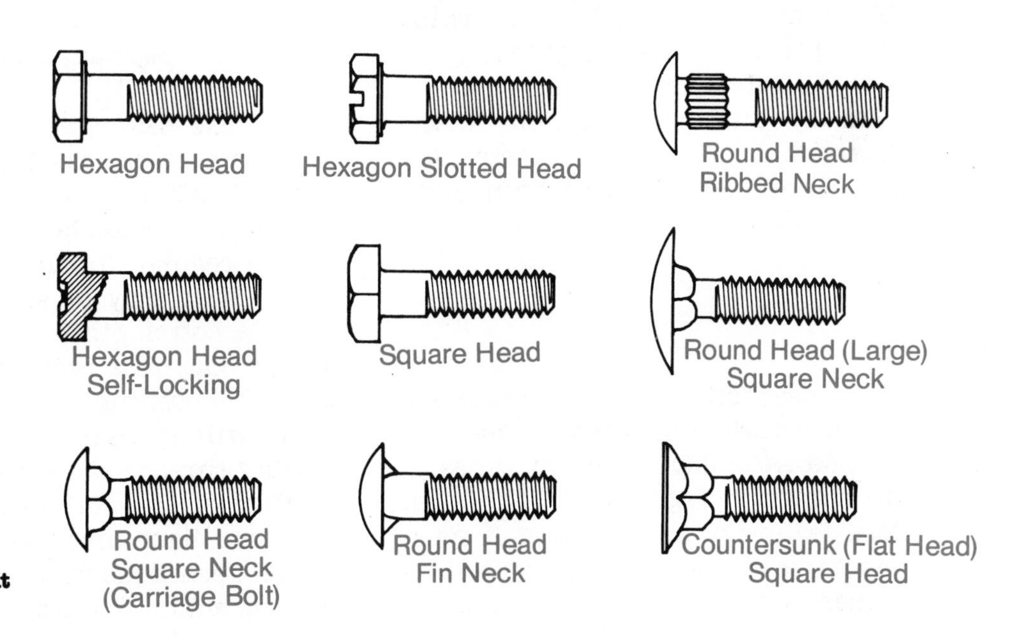

Fig. 9-20 Types of bolt heads

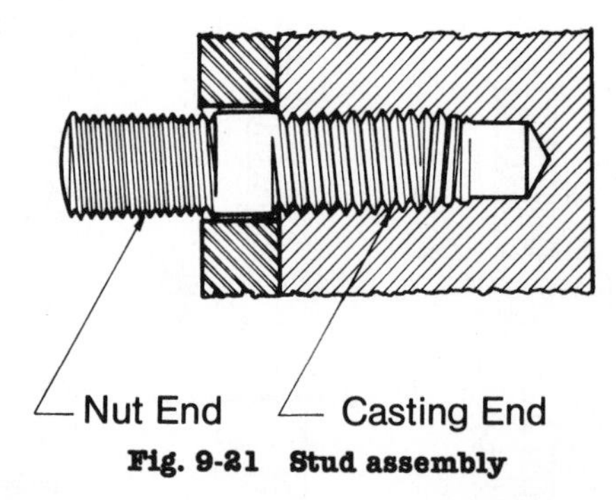

Fig. 9-21 Stud assembly

tions should be stated as in the following arrangement:

.500–13 UN C–2A × 275
Finished Hex H D Bolt

Studs

Stud bolts, or studs, are special bolts threaded on both ends. Studs are used on parts which must be removed frequently, such as engine cylinder heads. See Figure 9-21.

One end of the bolt, usually having a coarse thread, is screwed into a tapped hole with a class 3 fit. The other end, usually with a fine thread and a class 2 fit, fits into the removable portion of the assembly. A nut is used on the projecting portion of the stud. When tightened, the nut holds the two parts together.

The specifications on the print must indicate the stud bolt diameter, the type and length of thread at both ends, and the overall length.

Nuts

Nuts used for threaded fasteners are available in a variety of types to meet the needs of the assembly. Different types of nuts are shown in Figure 9-22. The most important feature of a nut is that it must have the same thread type and pitch as the fastener to which it will be attached.

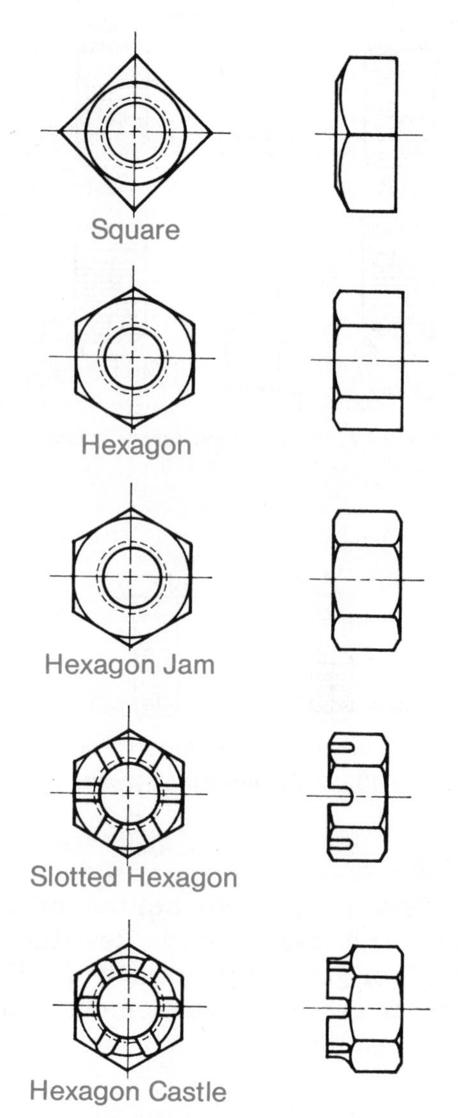

Fig. 9-22 Common nuts

The *hexagon nut* is the most common type used today. The dimensional size must closely match the head of the bolt it will fit. Like bolts, hexagon nuts are classified as regular, finished, and semifinished.

Nuts may be slotted on one end to allow a cotter pin to pass through a hole in the bolt. This reduces the possibility of the nut loosening. Other machining operations will increase the friction between mating threads. This will reduce the possibility of the nut backing off the bolt threads. Each nut manufacturer has a unique technique for producing this thread locking action (lock nuts).

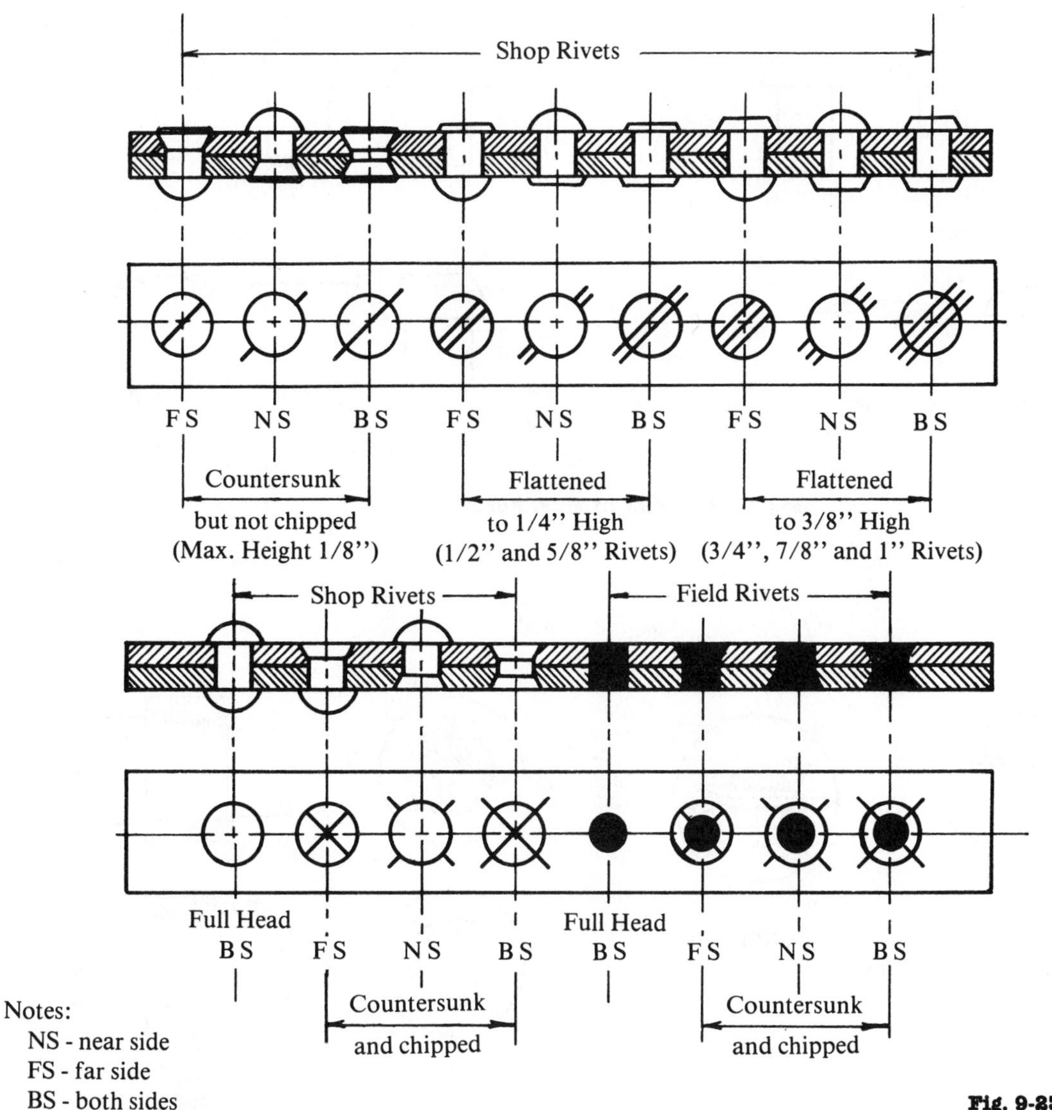

Fig. 9-23 Rivet symbols

NONTHREADED FASTENERS

Before the development of welding, *rivets* (malleable metal pins) were used as fasteners in most permanently fastened assemblies, sheet metal, and steel plates. Rivets are still used in many applications where welding would be impractical.

Rivets are shown on industrial prints by circle symbols, Figure 9-23. Each type of symbol represents a certain type of rivet. The specification note states the rivet size, type of head, and length. Open circle symbols are used for shop-installed rivets. Solid circles represent field-installed rivets.

Different types of rivet heads are illustrated in Figure 9-24.

Pins, keys, and *washers* are other types of nonthreaded fasteners. Fasteners, pins, keys, and washers are made of many different materials. They are also made with various degrees of hardness to satisfy design requirements.

Pins

Cotter pins are used to lock slotted nuts on bolts. They may also be used on

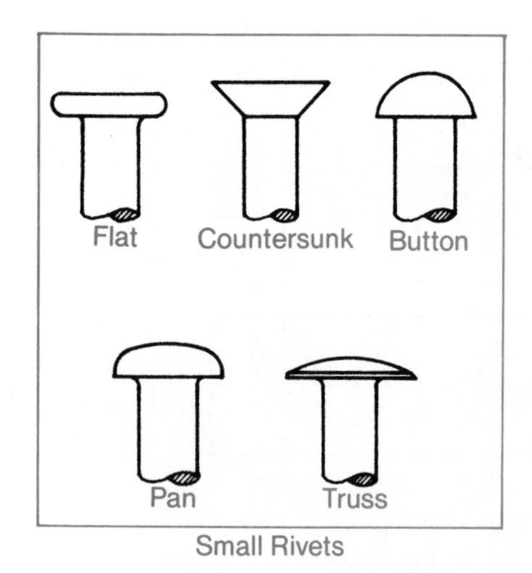

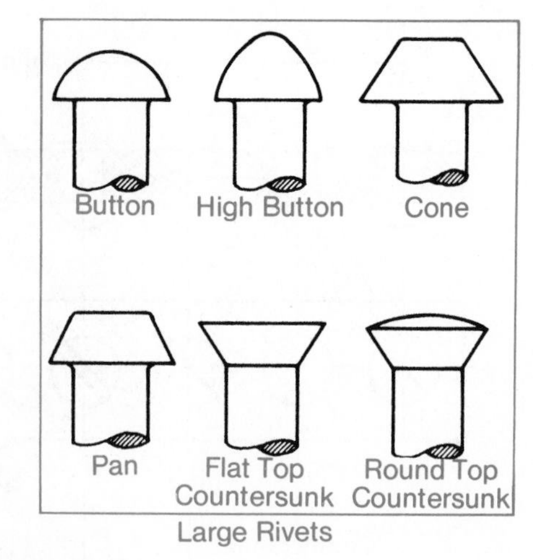

Fig. 9-24 Types of rivet heads

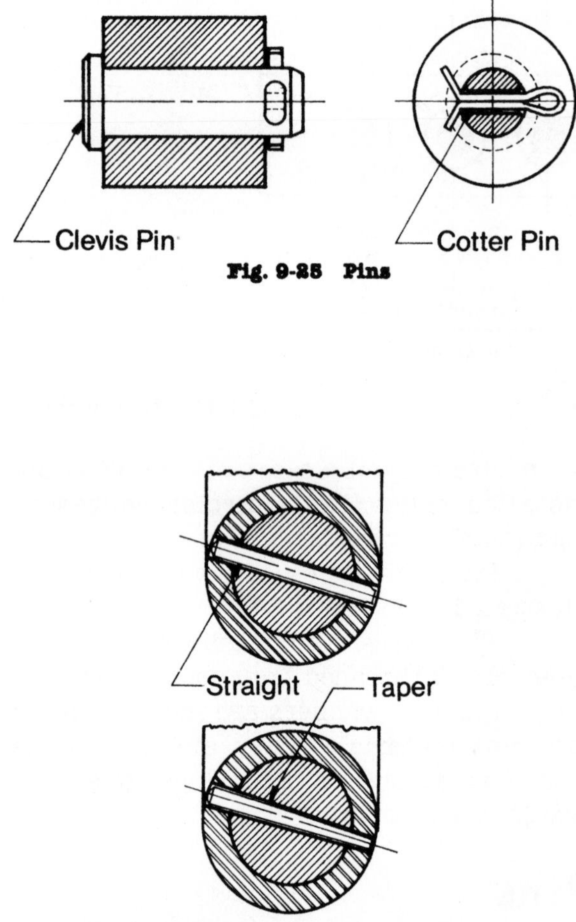

Fig. 9-25 Pins

Fig. 9-26 Straight pins and tapered keys

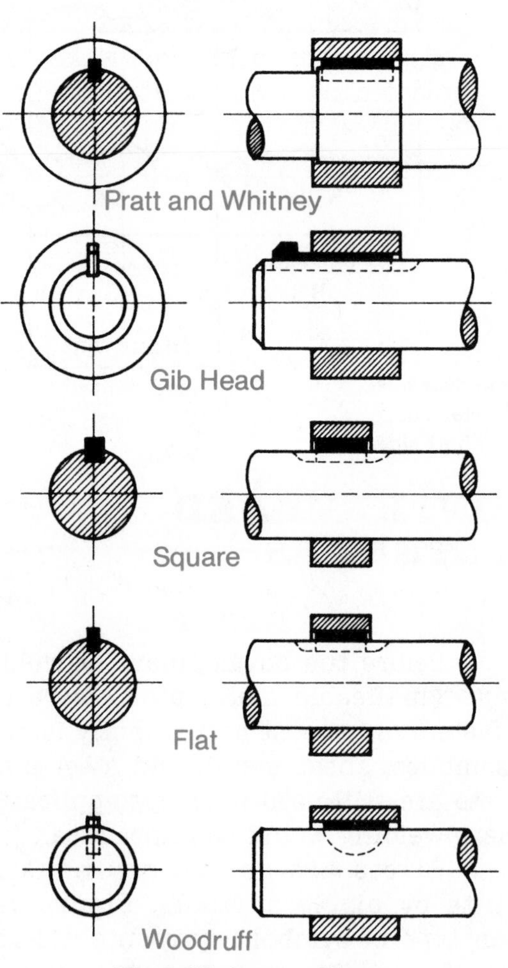

Fig. 9-27 Types of keys

movable links or rods, Figure 9-25. Grooved pins, straight pins, or tapered pins, Figure 9-26, may be used to more securely retain parts together.

Keys

Keys are used to provide positive drive to rotating parts. A key will keep a gear or pulley tight to a shaft while rotating. The type of key used is determined by the amount of force or work the machine must do. See Figure 9-27.

Specifications on the print should include the width, height, and length dimensions. Pratt and Whitney and Woodruff keys use code numbers referring to a definite key size, as listed in tables in standard machining handbooks.

The proper positions of key slots are indicated on prints. Slot dimensions are determined from machining handbook key tables.

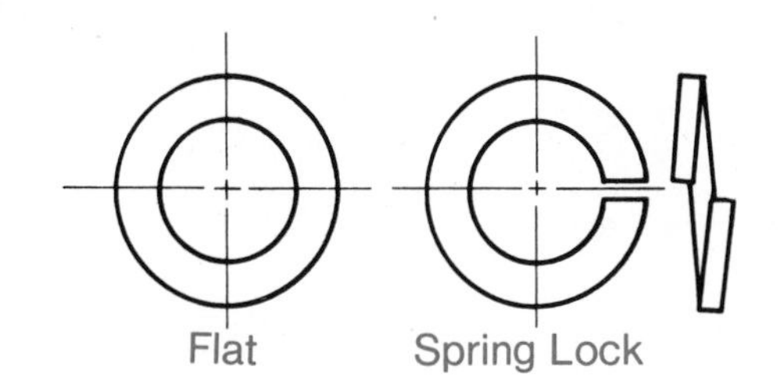

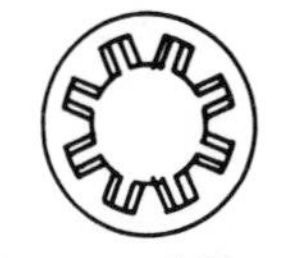

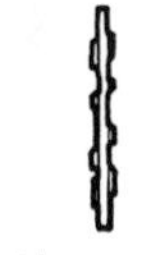

Fig. 9-28 Washers

Washers

There are many types of washers. The most common types are the flat, spring lock, and tooth or star washers. Each type of washer is available in standard sizes to fit standard bolt diameters. See Figure 9-28.

REVIEW

This review is provided to serve as reinforcement study material. Fill in the appropriate word(s) to complete the sentences below.

1. The three methods of representing threads on a print are the ____________________, ____________________, and ____________________ representations.

2. The meanings of the first five thread notations are:

a. ____________________

b. ____________________

c. ____________________

d. ____________________

e. ____________________

3. The notation *A* at the end of the thread notation means __________.

4. The notation *B* at the end of the thread notation means __________.

5. All threads in the inch measuring system are now ______________.

6. Metric thread diameters are specified in ____________________.

7. The letter ____________________ precedes all metric thread notations.

8. Three common types of threaded fasteners are ________________, ____________________, and ____________________.

NAME ________________

DATE ________________

SCORE ________________

EXERCISE B9-1

Print #EWD-48

Answer the following questions after studying Print #EWD-48 found after this exercise.

QUESTIONS	ANSWERS
1. What is this part?	1. ________
2. Does this part have a size?	2. ________
3. What does the ℄ mean?	3. ________
4. This part is made of what material?	4. ________
5. What is the indicated scale?	5. ________
6. What is the thickness of either end?	6. ________
7. What is the tolerance of the wrench size?	7. ________
8. What is the hardness range?	8. ________
9. What is the part number?	9. ________
10. How many changes were made since 1-80?	10. ________

NAME ________

DATE ________

SCORE ________

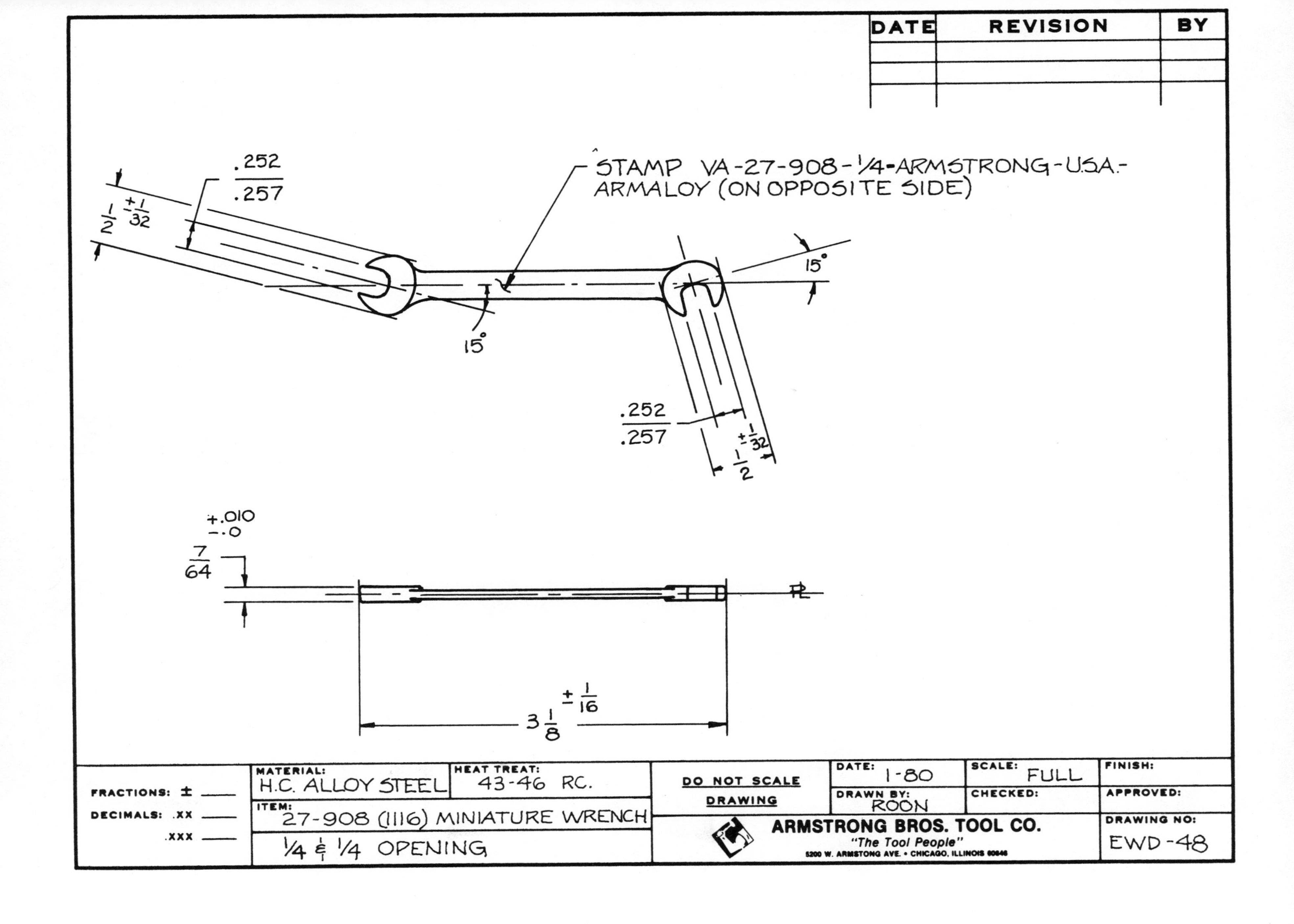
DATE
REVISION
BY
STAMP VA-27-908-1/4-ARMSTRONG-USA.-
ARMALOY (ON OPPOSITE SIDE)
.252
.257
1/2 ±1/32
15°
15°
.252
.257
±1/32
1/2
+.010
-.0
7/64
℄
3 1/8 ±1/16
FRACTIONS: ±
DECIMALS: .XX
.XXX
MATERIAL:
H.C. ALLOY STEEL
HEAT TREAT:
43-46 RC.
ITEM:
27-908 (1116) MINIATURE WRENCH
1/4 & 1/4 OPENING
DO NOT SCALE
DRAWING
DATE:
1-80
SCALE:
FULL
FINISH:
DRAWN BY:
ROON
CHECKED:
APPROVED:
ARMSTRONG BROS. TOOL CO.
"The Tool People"
5200 W. ARMSTONG AVE. • CHICAGO, ILLINOIS 60646
DRAWING NO:
EWD-48

CHAPTER 10

GEARS AND SPLINES

OBJECTIVES

After studying this chapter, you will be able to:

- Explain the basic gear theory.
- List the factors that determine which type of gear to use.
- Explain the meaning of the common gear terms used on industrial prints.
- Identify the required print gear data.
- Explain the required basic gear rack and spline theory.
- Describe how to calculate additional gear data from handbook tables.

The advent of the mechanical age brought about the need for transmission of large quantities of power from one rotating shaft to another. This chapter describes the most common methods of achieving this function, and how the required information is shown on industrial prints.

BASIC GEAR THEORY

When a cylinder mounted on a rotating shaft makes contact with a cylinder mounted on another shaft, the power of the first shaft will be transferred to the second shaft. In Figure 10-1, when shaft *A* rotates in the direction indicated by the arrow, shaft *B* will rotate in the opposite direction. As the power from shaft *A* is increased, a certain amount of slippage will occur between the two cylinders. This slippage can be reduced by increasing the friction between the cylinders. Friction can be increased by pushing together the two cylinders more tightly, or by adding a coarse material between the cylinders.

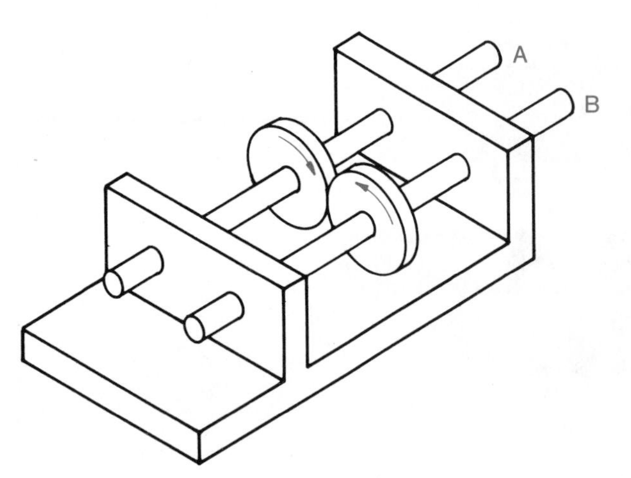

Fig. 10-1 Cylinders in contact

This procedure is quite similar to the technique of placing sand between automobile tires and an icy road pavement.

The most effective method of eliminating slippage is to machine teeth on cylinder *A*. These teeth should interlock with similarly machined teeth on cylinder *B*. See Figure 10-2. This principle of producing rotary motion by the use of interlocking teeth is used on all types of gears. This is the basic gear theory.

TYPES OF GEARS

Many different types of gears are available, Figure 10-3. The type used on a mechanism is determined by the position of the shafts transmitting the power.

1. When the gear shafts are parallel, a spur or some variation of spur gears is used.

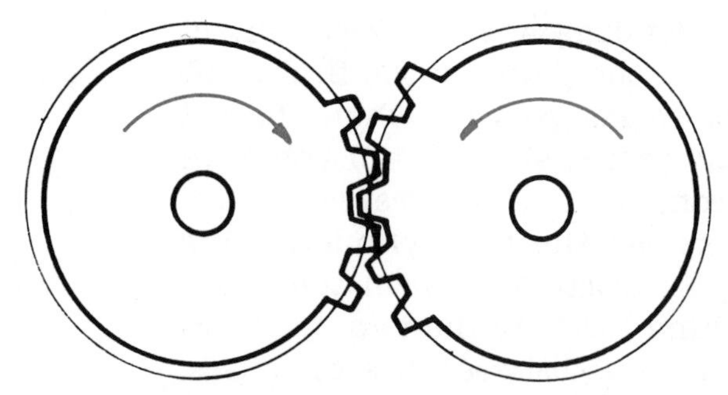

Fig. 10-2 Cylinders with gear teeth

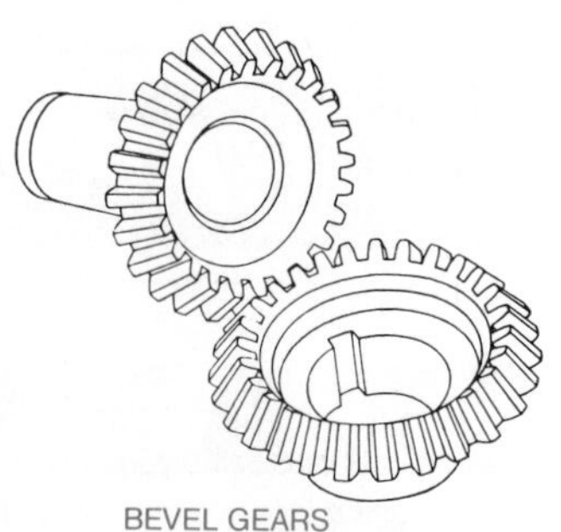

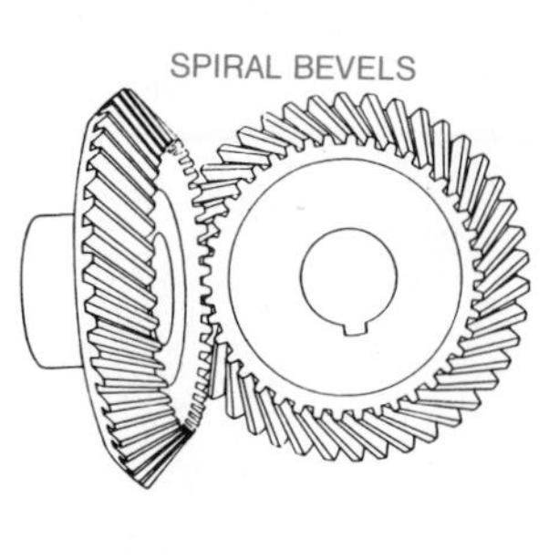

HELICAL GEARS

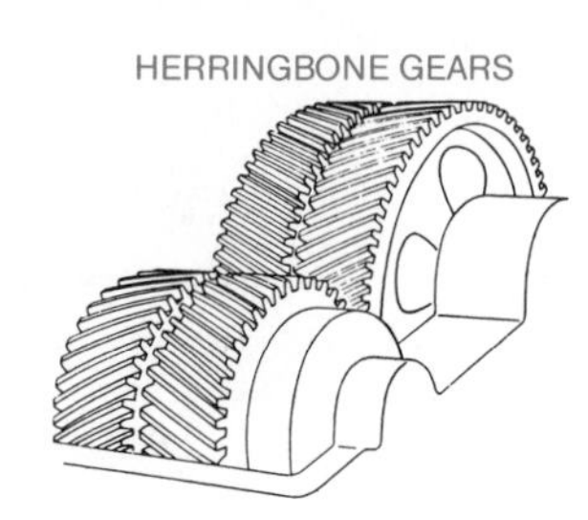

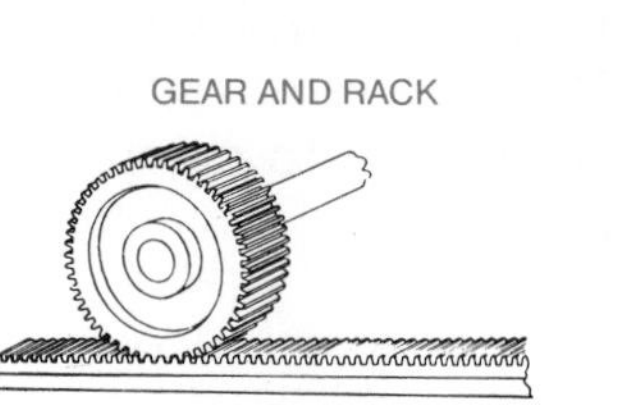

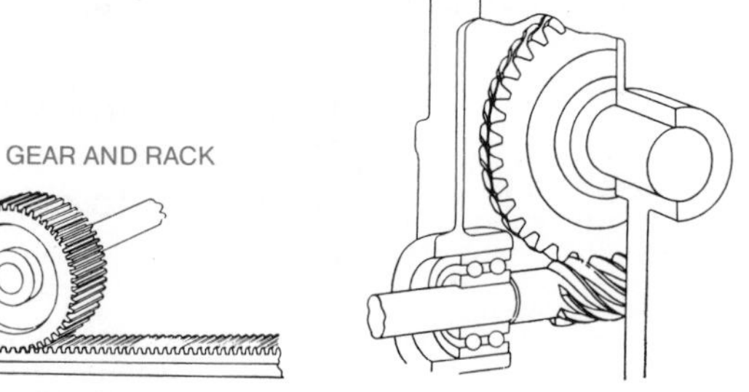

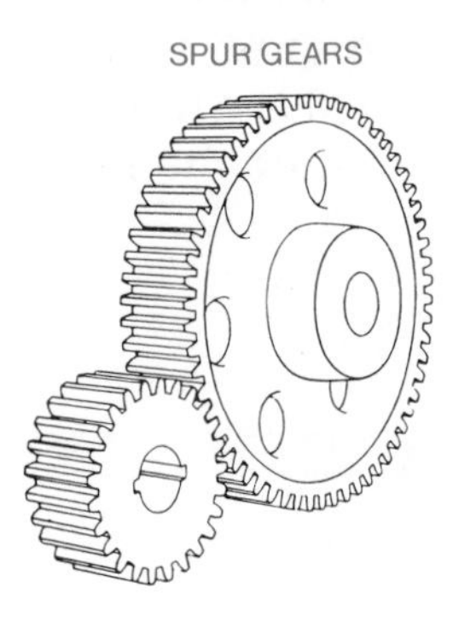

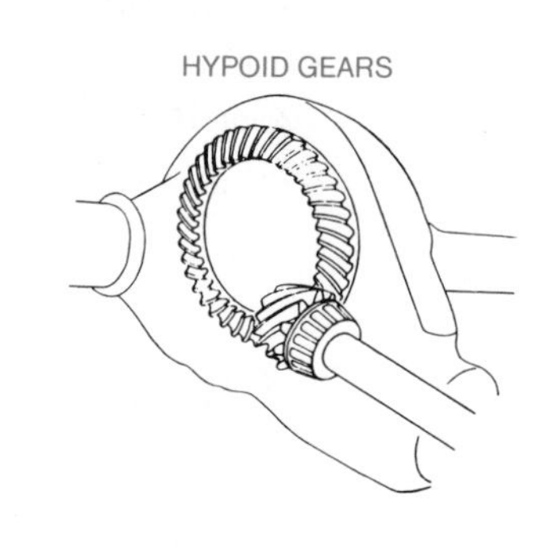

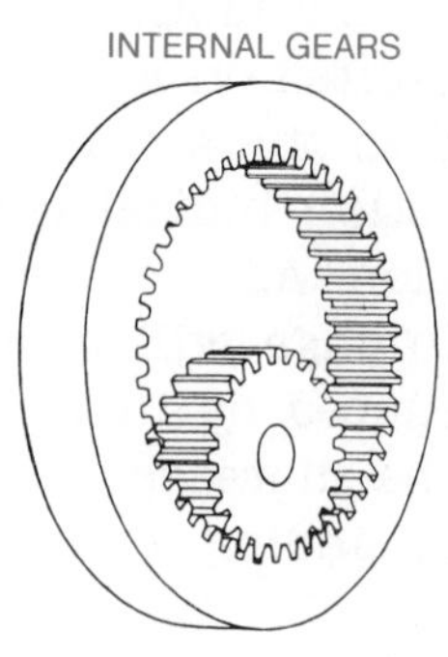

Fig. 10-3 Different types of gears (*Courtesy of Brown & Sharpe Mfg. Co.*)

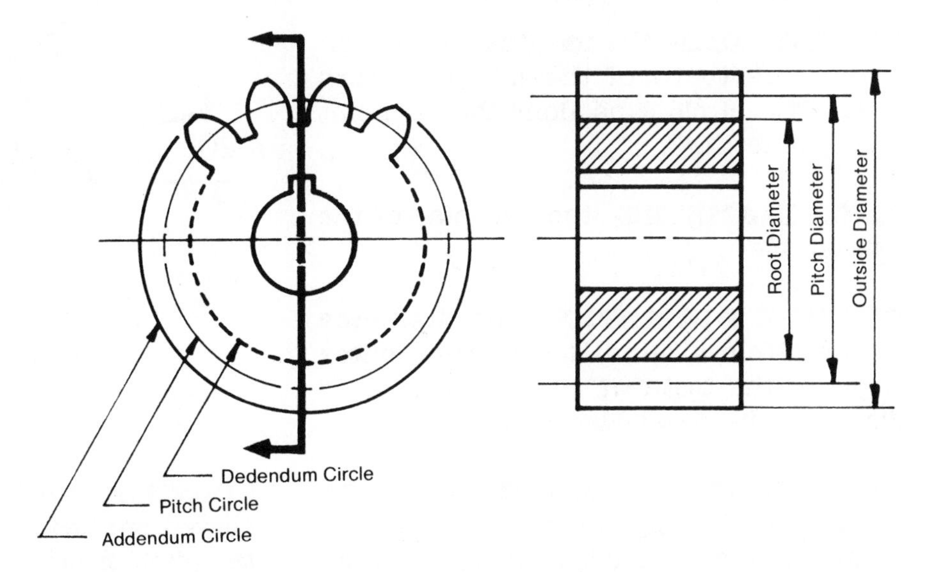

Fig. 10-4 Main gear terms

2. When the gear shafts are not parallel or meet at an angle, bevel gears are used.
3. When the gear shafts are not parallel or do not intersect, worm gears, or another type of gears, are used.

GEAR NOMENCLATURE (TERMINOLOGY)

The spur and bevel gears are the most commonly used types of gears. These and other types are discussed in this chapter to illustrate the information and terms most likely to appear on industrial prints.

SPUR GEARS

Spur gears are gears with teeth cut parallel to the axis of the gear blank. The *pitch circle* is the imaginary circle representing the contact surface of one of the cylinders. (See Figure 10-1.) The *gear size* is normally the diameter of the pitch circle. That portion of the gear tooth that is added or projects beyond the pitch circle is called the *addendum*. That portion of the gear tooth that is deducted or inside of the pitch circle is called the *dedendum*.

More complete definitions of gear parts are listed here. These parts are illustrated in Figures 10-4 and 10-5.

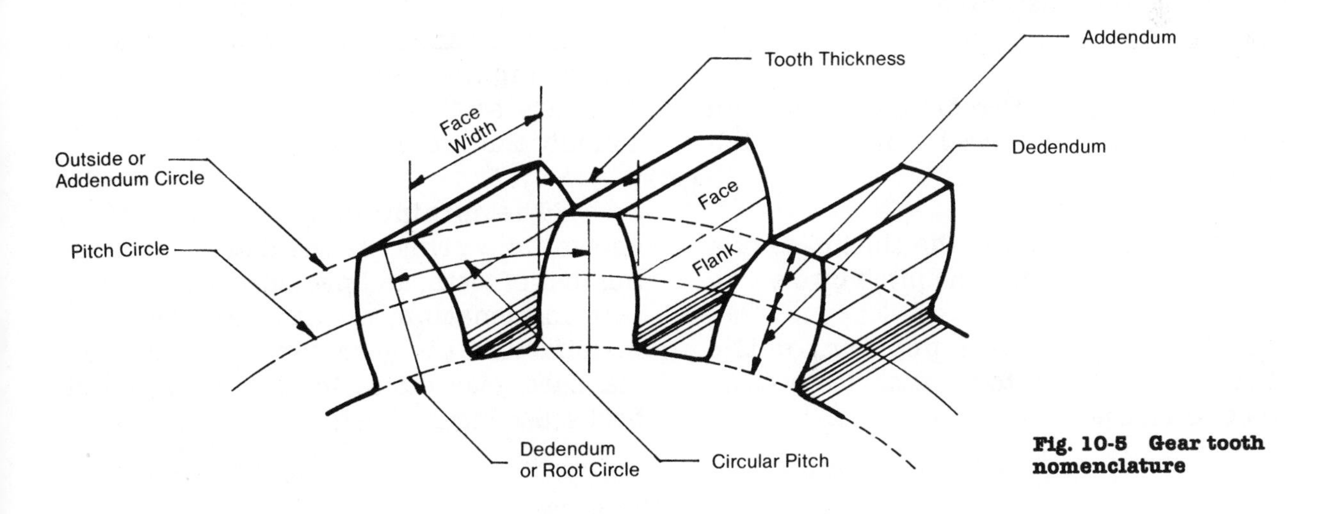

Fig. 10-5 Gear tooth nomenclature

PITCH CIRCLE—the imaginary circle on which the teeth of the mating gears mesh. This circle runs along the midpoint of all the teeth.

PITCH DIAMETER—the diameter of the pitch circle.

OUTSIDE DIAMETER—the diameter of the circle around the extreme outer edges of the teeth or equal to the pitch diameter plus two addendums.

DEDENDUM, OR ROOT, CIRCLE—the circle formed along the roots of the gear teeth whose diameter is equal to the pitch circle minus two dedendums.

ADDENDUM—that portion of the gear teeth which is added to the pitch diameter.

DEDENDUM—that portion of the gear teeth which is deducted from the pitch diameter.

CLEARANCE—the amount that the dedendum exceeds the length of the addendum. This amount allows a clearance between the top of one tooth and the bottom of the mating tooth when gears are in mesh.

CIRCULAR PITCH—the distance from the center of one tooth to the center of the next tooth, measured along the pitch circle.

TOOTH SPACE—the distance between adjacent teeth, measured along the pitch circle.

TOOTH THICKNESS—the thickness of a tooth, measured along the pitch circle.

TOOTH FACE—that portion of the curved surface of a tooth that lies outside the pitch circle.

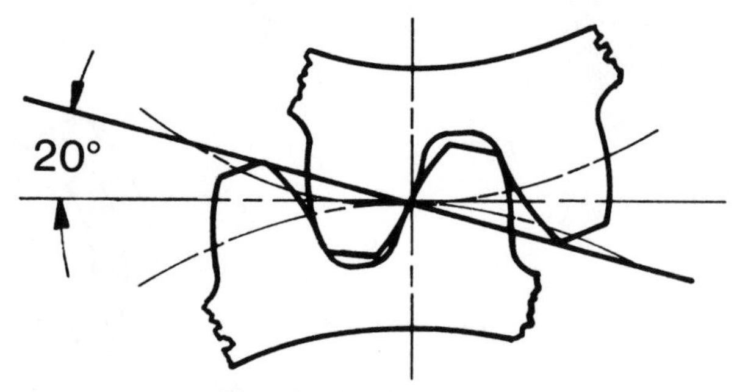

Fig. 10-6 Pressure angle

TOOTH FLANK—that portion of the curved surface of a tooth that lies inside the pitch circle.

DIAMETRAL PITCH—diametral pitch refers to tooth size. It is the number of teeth on a gear per inch of pitch diameter. For example, an *8 pitch gear* has 8 teeth for every inch of pitch diameter. The term diametral pitch (*DP*) followed by the appropriate number of teeth will always be indicated on a gear print.

As is discussed in Chapter 9, a nut must have the same pitch (threads per inch of length) as the bolt or they will not fit together properly. A similar situation is true for gears. All mating gears must have the same diametral pitch number or they will not mesh properly.

PRESSURE ANGLE—the angle between the line of action and the line perpendicular to the center line of the two mating gears, Figure 10-6. The most common pressure angles are 14½° or 20°. Occasionally the pressure angle is 25°.

The minimum gear information shown on a print are the diametral pitch, number of teeth, and pressure angle. Given this information, all data necessary to manufacture the gear can be found from standard gear tables in machining handbooks. See Figure 10-7.

RULES AND FORMULAS FOR SPUR GEARS*				
	TO GET:	HAVING:	RULE:	USE FORMULA:
1	Addendum	Diametral pitch	Divide 1 by the diametral pitch	$A = \frac{1}{P_d}$
2	Addendum	Circular pitch	Multiply circular pitch by .318	$A = P \times .318$
3	Circular pitch	Diametral pitch	Divide 3.1416 by diametral pitch	$P = \frac{3.1416}{P_d}$
4	Circular pitch	Pitch diameter & no. of teeth	Divide pitch diameter by product of .3183 and no. of teeth	$P = \frac{D}{.3183 \times N}$
5	Clearance	Circular pitch	Multiply circular pitch by .05	$C = .05 \times P$
6	Clearance	Thickness of tooth	One-tenth thickness of tooth at pitch line	$C = \frac{T}{10}$
7	Clearance	Diametral pitch	Divide .157 by diametral pitch	$C = \frac{.157}{P_d}$
8	Dedendum	Diametral pitch	Divide 1.157 by diametral pitch	$S = \frac{1.157}{P_d}$
9	Dedendum	Circular pitch	Multiply circular pitch by .368	$S = P \times .368$
10	Diametral pitch	Circular pitch	Divide 3.1416 by circular pitch	$P_d = \frac{3.1416}{P}$
11	Diametral pitch	Pitch diameter & no. of teeth	Divide number of teeth by pitch diameter	$P_d = \frac{N}{D}$
12	Diametral pitch	Outside diameter & no. of teeth	Divide number of teeth plus 2 by the outside diameter	$P_d = \frac{N + 2}{O}$
13	Number of teeth	Pitch diameter & diametral pitch	Multiply pitch diameter by the diametral pitch	$N = D \times P_d$
14	Number of teeth	Outside diameter & diametral pitch	Multiply outside diameter by diametral pitch and subtract 2	$N = O \times P_d - 2$
15	Number of teeth	Pitch diameter & circular pitch	Divide product of pitch diameter and 3.1416 by the circular pitch	$N = \frac{3.1416 \times D}{P}$
16	Outside diameter	Number of teeth & addendum	Multiply addendum by the number of teeth plus 2	$O = A \times (N + 2)$
17	Outside diameter	Number of teeth & diametral pitch	Divide number of teeth plus 2 by the diametral pitch	$O = \frac{N + 2}{P_d}$
18	Outside diameter	Pitch diameter & number of teeth	Pitch diameter plus the quotient of 2 divided by the diametral pitch	$O = D + \frac{2}{P_d}$
19	Pitch diameter	Number of teeth & diametral pitch	Divide number of teeth by the diametral pitch	$D = \frac{N}{P_d}$

(continued)

*Symbols Used in Spur Gear Calculations

A = Addendum
C = Clearance
D = Pitch Diameter
N = Number of Teeth
O = Outside Diameter
P = Circular Pitch
P_d = Diametral Pitch
S = Dedendum
T = Thickness of Tooth
W = Whole Depth

Fig. 10-7 Spur gear table

	RULES AND FORMULAS FOR SPUR GEARS *(continued)*			
	TO GET:	HAVING:	RULE:	USE FORMULA:
20	Pitch diameter	Number of teeth & outside diameter	Divide product of outside diameter and no. of teeth by no. of teeth plus 2	$D = \frac{O \times N}{N + 2}$
21	Pitch diameter	Outside diameter & diametral pitch	Outside diameter minus the quotient of 2 divided by the diametral pitch	$D = O - \frac{2}{P_d}$
22	Pitch diameter	Addendum & no. of teeth	Multiply the number of teeth by the addendum	$D = N \times A$
23	Thickness of tooth	Diametral pitch	Divide 1.5708 by the diametral pitch	$T = \frac{1.5708}{P_d}$
24	Whole depth	Diametral pitch	Divide 2.157 by the diametral pitch	$W = \frac{2.157}{P_d}$
25	Whole depth	Circular pitch	Multiply the circular pitch by .6866	$W = .6866 \times P$

Fig. 10-7 ***(Continued)***

TABLE EXAMPLES

The following examples illustrate how to use the information in standard gear tables. When the diametral pitch and number of teeth are listed on the print, other data regarding the gear may be found. As you read these examples, refer to Figure 10-7.

- **Given:** Diametral Pitch (P_d) = 4
  Number of Teeth (N) = 22

- **To find:** Outside diameter (O)

Use Rule 17: $O = \frac{N + 2}{P_d}$

$$O = \frac{22 + 2}{4} = \frac{24}{4} = 6$$

- **To find:** Addendum (A)

Use Rule 1: $A = \frac{1}{P_d}$

$$A = \frac{1}{4} = .25$$

More than one rule can be used to find each item. Select the rule which contains the information given on the print plus the desired data.

Some prints include a gear data block which includes all information necessary to machine the gear teeth. This practice eliminates possible errors that can occur when calculations are done in the shop. See Figure 10-8.

HELICAL SPUR GEARS

Not all spur gear teeth are machined parallel to the gear shafts. When

GEAR DATA	
Number of Teeth	88
Diametral Pitch	16
Pressure Angle	20
Pitch Diameter (REF)	5.500
Addendum (REF)	.0625
Dedendum (REF)	.072
Whole Depth (REF)	.134
Circular Tooth Thickness (REF)	.098
Center Distance	4.125
Measuring Pin Diameter (REF)	.105
Diameter Over Pins (REF)	5.657
Part Number Of Mating Gear	85740
Number Of Teeth On Mating Gear	44

Fig. 10-8 Spur gear data

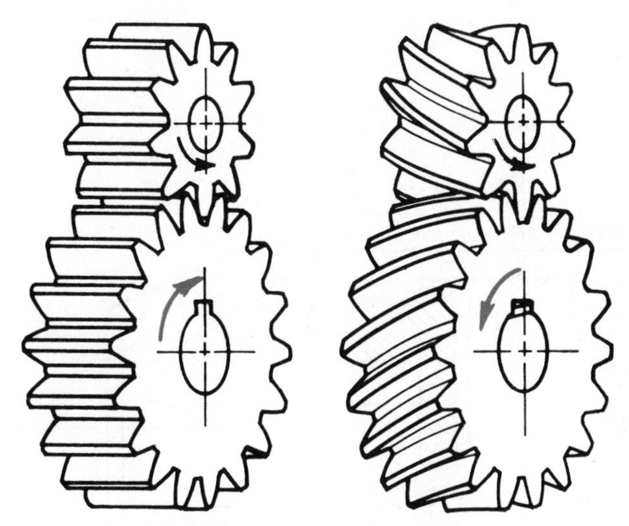

Fig. 10-9 Spur gears **Fig. 10-10 Helical spur gears**

the teeth are machined at an angle to the gear shafts, they are called *helical spur gears.*

Spur gears, Figure 10-9, are used most commonly to connect two parallel shafts which rotate in opposite directions. The teeth are parallel to the axis of the gear. Therefore, spur gears can be used as sliding gears to change the speed mechanism in gear boxes.

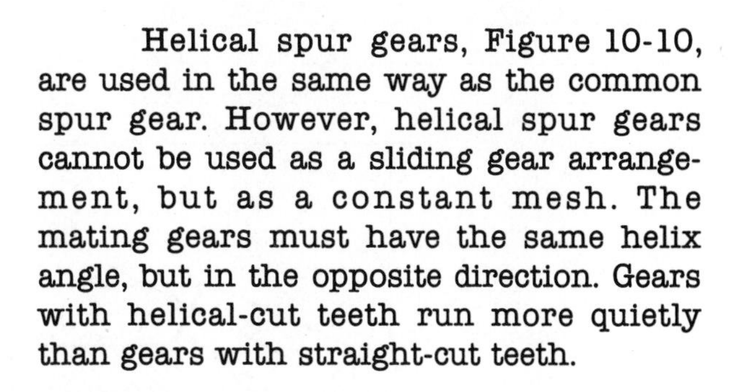

Helical spur gears, Figure 10-10, are used in the same way as the common spur gear. However, helical spur gears cannot be used as a sliding gear arrangement, but as a constant mesh. The mating gears must have the same helix angle, but in the opposite direction. Gears with helical-cut teeth run more quietly than gears with straight-cut teeth.

BEVEL GEARS

The *bevel gear* is a commonly used gear. It is found in many power-driven mechanisms. A spur gear is defined as two wheels mounted on parallel shafts, with teeth cut on the wheel contact edges. In contrast, the mounting shafts of bevel gears are not parallel.

The bevel gear is similar to two cones mounted on non-parallel shafts, with teeth cut on the conical contact edges. The shafts of bevel gears are usually positioned at a right angle. See Figure 10-11.

Bevel gear teeth can also be machined at an angle; these are called

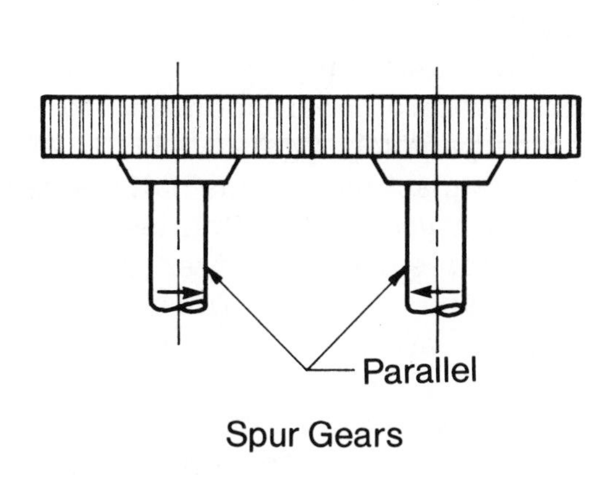

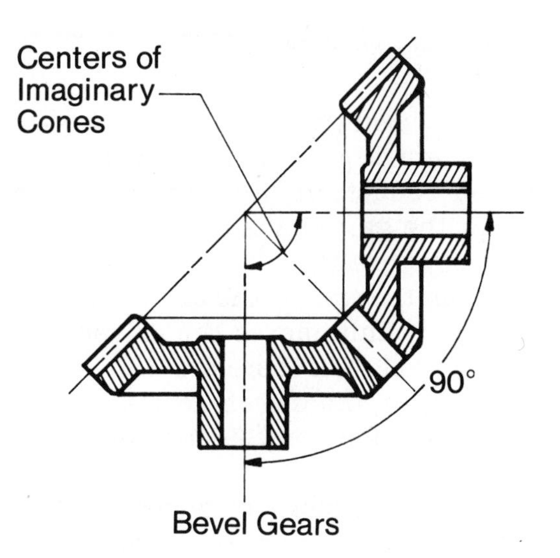

Fig. 10-11 Spur and bevel gears

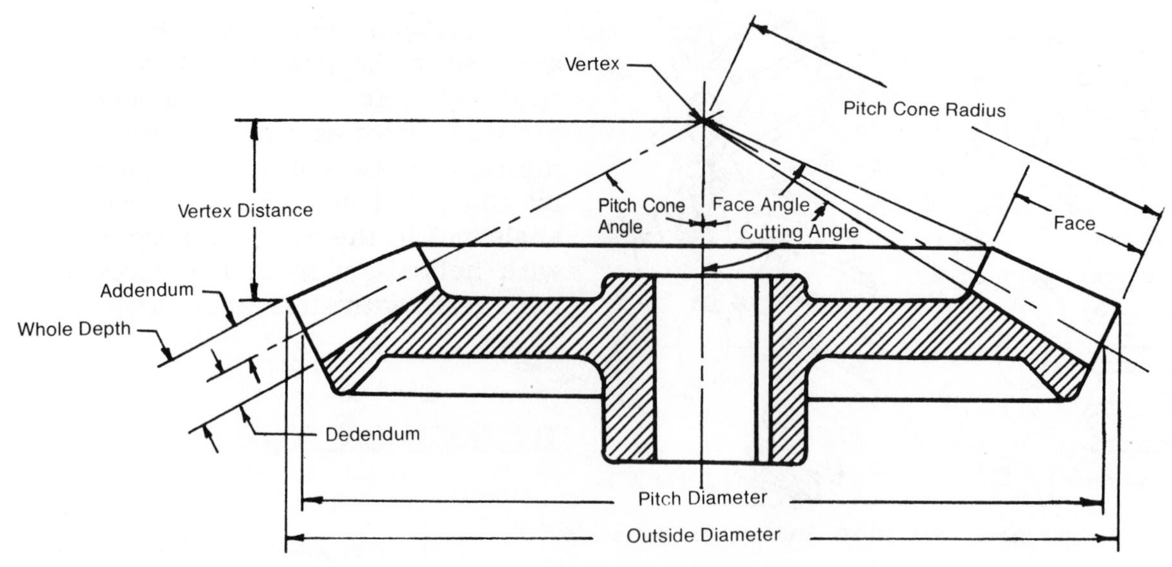

Fig. 10-12 Bevel gear terminology

spiral bevel gears. (See Figure 10-3.) This type of bevel gear runs more smoothly and quietly than the straight bevel gears.

The terminology of bevel gears is similar to that of spur gears, except for the terms relating to conical angles. See Figure 10-12. Standard specification tables for bevel gears are available in machining handbooks. See Figure 10-13.

WORM GEARS

The *worm gear* is another type of gear. The worm gear is especially suited for speed reduction. As shown in Figure 10-14, the gear shafts do not intersect; nor are they parallel.

When the worm (the driver) rotates 1 full turn, it only moves the driven gear 1 tooth. Therefore, if there are 30 teeth on the driven gear, the driven gear shaft will revolve 1 turn for every 30 revolutions of the worm gear.

The terminology of worm gears is similar to that of spur and bevel gears, except for several specialized terms.

GEAR RACKS, SPLINES, AND SERRATIONS

A *gear rack* is used when rotary motion must be changed to reciprocating or straight line motion. The teeth on the rack must have the same profile as its mating gear. The rack terms will also be the same. See Figure 10-15.

Splines are slots or keys machined around the circumference of a shaft and on the inside surface of its mating part. Splines are used to prevent rotation between the shaft and the part it will fit. See Figure 10-16.

The basic shape of spline teeth is either involute (corresponding to gear teeth) or parallel, Figure 10-17. Involute splines are machined to one-half the depth of standard gear teeth. Splines are produced in three classes of fits: *(1) sliding,* (2) *close,* and (3) *press.* The basic dimensions of splines are usually indicated on the production print.

Serrations are similar to splines, but have different tooth proportions. Serrations are intended primarily for permanently assembled parts. Therefore, they are machined for a press fit.

RULES AND FORMULAS FOR BEVEL GEARS WITH SHAFTS AT RIGHT ANGLES*

(When rules or formulas are not given, they are the same as for spur gears)

	TO GET:	HAVING:	RULE:	USE FORMULA:
1	Addendum of small end of tooth	Width of face, pitch cone radius & addendum	Divide the difference of the face width and the pitch cone radius by the pitch cone radius and multiply by the addendum	$A_s = A \times \frac{P_{cr} - F}{P_{cr}}$
2	Addendum angle	Addendum & pitch cone radius	Divide the addendum by the pitch cone radius	$\tan A_a = \frac{A}{P_{cr}}$
3	Angular addendum	Addendum & pitch cone angle	Multiply the addendum by the cosine of the pitch cone angle	$K = A \times \cos P_{ca}$
4	Cutting angle	Dedendum angle & pitch cone angle	Subtract the dedendum angle from the pitch cone angle	$C = P_{ca} - S_a$
5	Dedendum angle	Dedendum & pitch cone radius	Divide the dedendum by the pitch cone radius	$\tan S_a = \frac{S}{P_{cr}}$
6	Face angle	Pitch cone & addendum angles	Subtract the sum of the pitch cone and addendum angles from 90 degrees	$F_a = 90° - (P_{ca} + A_a)$
7	Outside diameter	Angular addendum & pitch diameter	Add twice the angular addendum to the pitch diameter	$O = D + 2K$
8	Pitch cone angle of pinion	Number of teeth in both pinion & gear	Divide the number of teeth in the pinion by the number of teeth in the gear	$\tan P_{ca} = \frac{N_p}{N_g}$
9	Pitch cone angle of gear	Number of teeth in both pinion & gear	Divide the number of teeth in the gear by the number of teeth in the pinion	$\tan P_{ca} = \frac{N_g}{N_p}$
10	Pitch cone radius	Pitch diameter & pitch cone radius	Divide the pitch diameter by twice the sine of the pitch cone angle	$P_{cr} = \frac{D}{2 \times \sin P_{ca}}$
11	Thickness of tooth at pitch line	Circular pitch	Divide the circular pitch by 2	$T = \frac{P}{2}$
12	Thickness of tooth at pitch line at small end	Width of face, pitch cone radius & thickness of the tooth	Divide the difference of the pitch cone radius and width of face by the pitch cone radius and multiply the quotient by the thickness of the tooth	$T_s = T \times \frac{P_{cr} - F}{P_{cr}}$
13	Vertex or apex distance	Outside diameter & face angle	Multiply one-half of the outside diameter by the tangent of the face angle	$V = \frac{O}{2} \times \tan F_a$
14	Vertex distance at small end of tooth	Width of face, pitch cone radius & vertex distance	Divide the difference of the pitch cone radius and width of face by the pitch cone radius and multiply the quotient by the vertex distance	$V_s = V \times \frac{P_{cr} - F}{P_{cr}}$

*Symbols Used in Bevel Gear Calculations

P_d = Diametral Pitch
P = Circular Pitch
P_{ca} = Pitch Cone Angle
D = Pitch Diameter
T = Thickness of Tooth at Pitch Line
A_6 = Addendum at Small End of Tooth
S_a = Dedendum Angle
A_a = Addendum Angle
K = Angular Addendum
V = Vertex Distance
N = Number of Teeth
S = Dedendum
A = Addendum
F = Width of Face
P_{cr} = Pitch Cone Radius
T_s = Thickness of Tooth at Small End
F_a = Face Angle
C = Cutting Angle
O = Outside Diameter
V_s = Vertex Distance at Small End of Tooth

Fig. 10-13 Bevel gear table

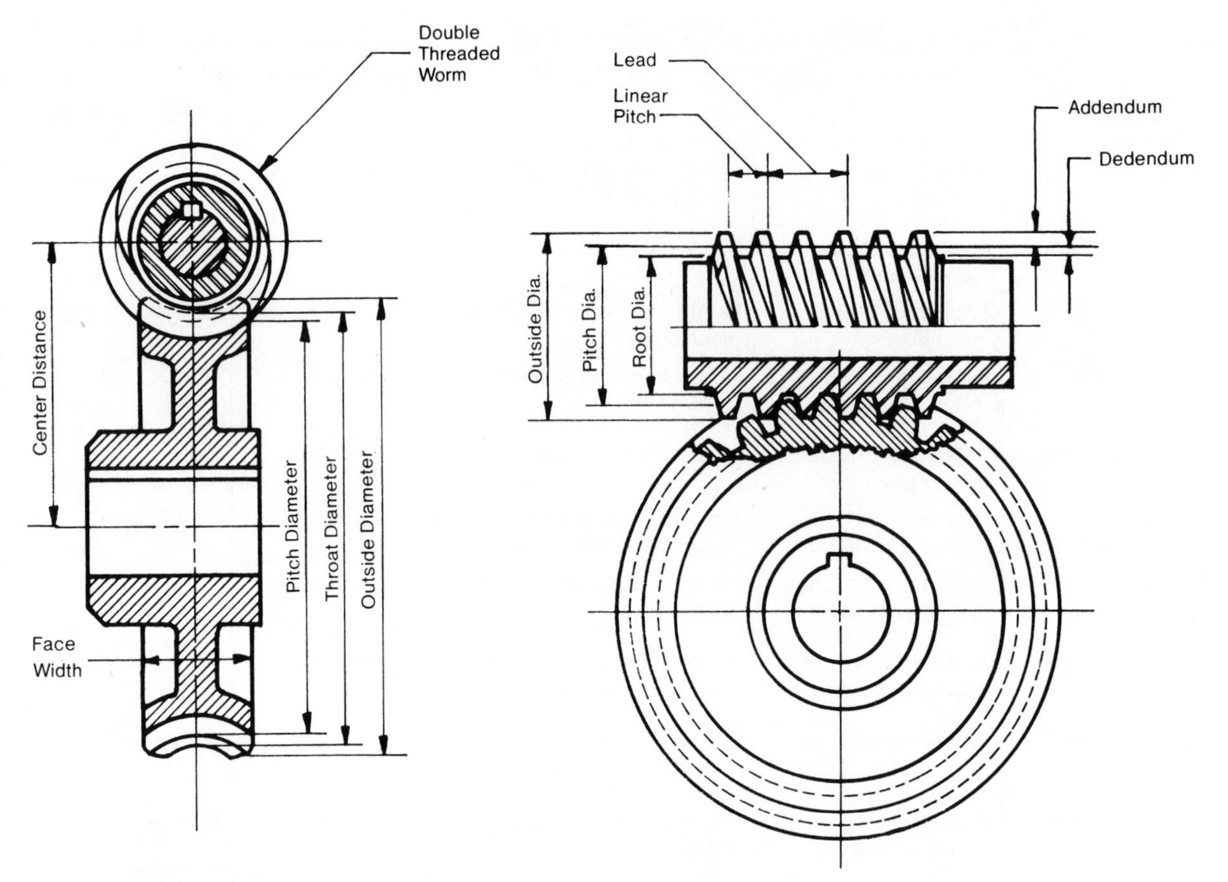

Fig. 10-14 Worm gear terminology

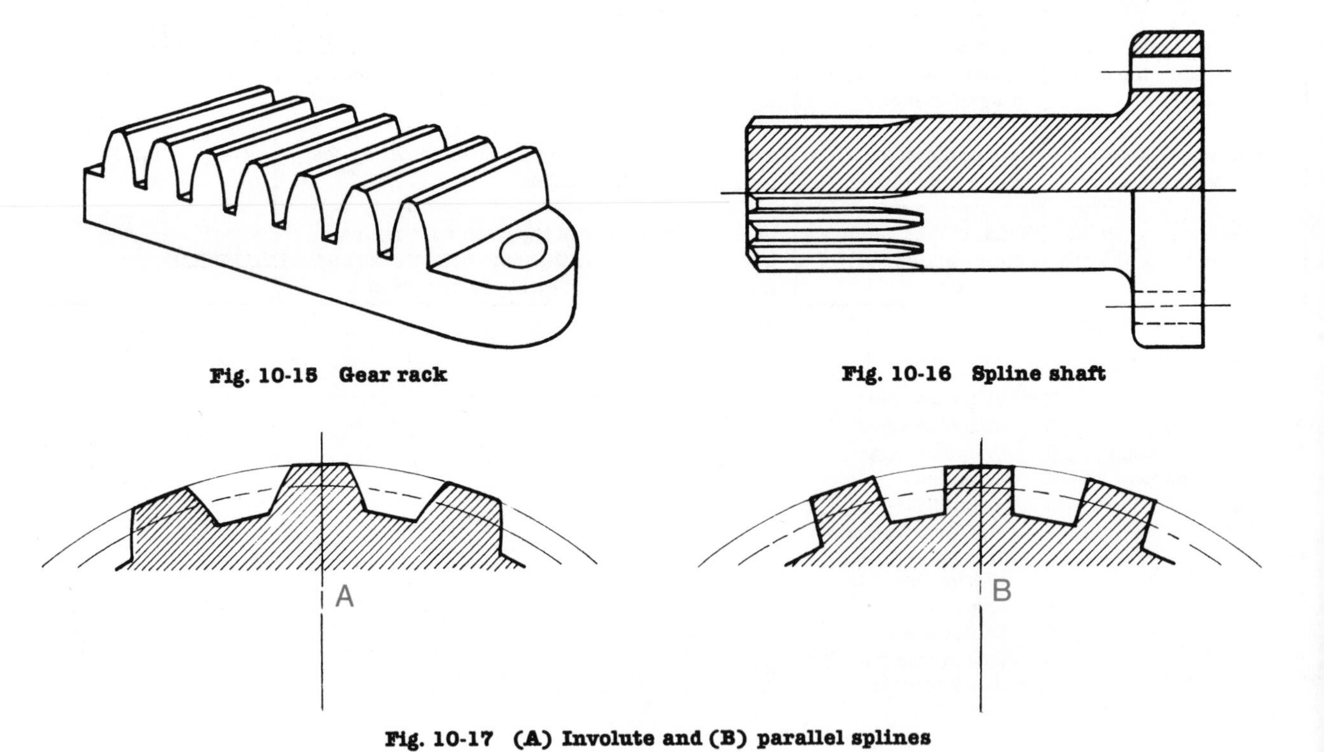

Fig. 10-15 Gear rack

Fig. 10-16 Spline shaft

Fig. 10-17 (A) Involute and (B) parallel splines

REVIEW

This review is provided to serve as reinforcement study material. Fill in the appropriate word(s) to complete the sentences below.

1. Adding teeth to cylinder contact surfaces will eliminate

____________________________.

2. When power is transferred from parallel shafts, ____________ gears are used.

3. When power is transferred from non-parallel shafts, ____________ gears are used.

4. When power is transferred from non-parallel and non-intersecting shafts, ____________________________ gears are used.

5. The number of teeth on a gear per inch of pitch diameter is called

____________________________.

6. The minimum gear information required on a print would be the ____________________________, ____________________________, and

____________________________.

7. The type of gear most often used for speed reduction is

____________________________.

8. When rotary motion must be changed to reciprocating or straight line motion, (a) (an) ____________________________ is used.

9. Splines are produced in three classes of fits: ________________________,

____________________________, and ____________________________.

10. Serrations are usually machined for (a) (an) ____________ fit.

NAME____________________

DATE____________________

SCORE____________________

EXERCISE B10-1

Print #A151309

Answer the following questions after studying Print #A151309 found after this exercise.

This print is a microfilm copy.

QUESTIONS	ANSWERS
1. Of what material is this part made?	**1.** ______
2. Would any other material be allowed?	**2.** ______
3. What kind of gear is shown?	**3.** ______
4. How many changes have been made since 4-18-77?	**4.** ______
5. How would this gear be kept from turning on its shaft?	**5.** ______
6. How many teeth are on the gear?	**6.** ______
7. What is the diametral pitch?	**7.** ______
8. What is the pitch diameter?	**8.** ______
9. What is the tooth clearance?	**9.** ______
10. What is the outside diameter?	**10.** ______
11. Using .2160 inspection pins, what should the outside diameter measure?	**11.** ______
12. What type of heat treatment is indicated?	**12.** ______
13. What type of hardening is indicated?	**13.** ______
14. What should be the final hardness?	**14.** ______
15. How many teeth are on the spline?	**15.** ______
16. At the face run-out, what does the TIR mean?	**16.** ______
17. From what part is this gear machined?	**17.** ______
18. What surfaces have the smoothest finish, and what value is that finish?	**18.** ______ ______

NAME ______

DATE ______

SCORE ______

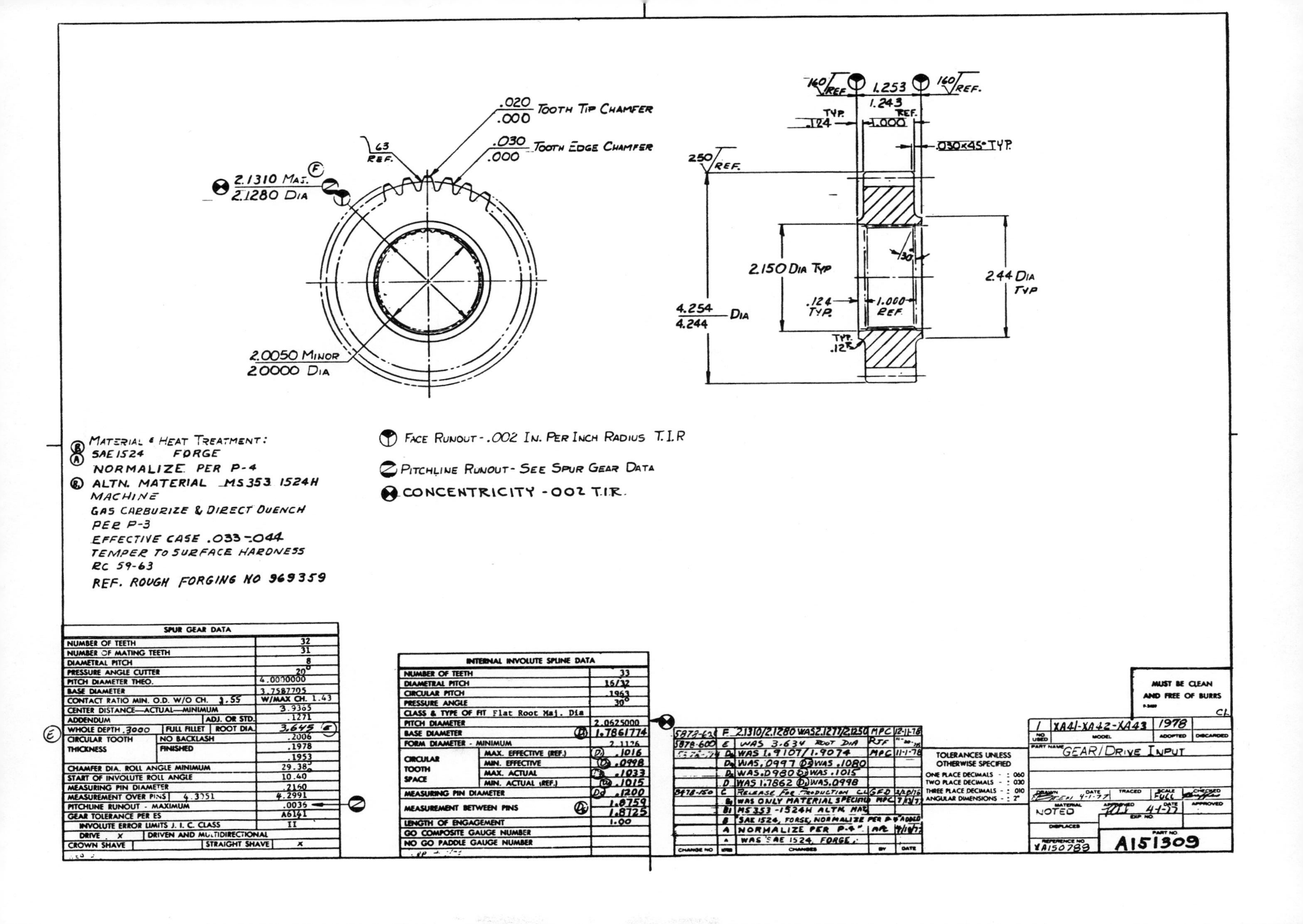

.020 / .000 TOOTH TIP CHAMFER
.030 / .000 TOOTH EDGE CHAMFER
63 REF.
2.1310 MAJ. / 2.1280 DIA
2.0050 MINOR / 2.0000 DIA
160 REF.
1.253 / 1.243
TYP. .124
1.000 REF.
.030x45° TYP.
250 REF.
2.150 DIA TYP
2.44 DIA TYP
.124 TYP.
1.000 REF
4.254 / 4.244 DIA
TYP. .12R
30°
MATERIAL & HEAT TREATMENT:
SAE 1524 FORGE
NORMALIZE PER P-4
ALTN. MATERIAL MS353 1524H
MACHINE
GAS CARBURIZE & DIRECT QUENCH
PER P-3
EFFECTIVE CASE .033-.044
TEMPER TO SURFACE HARDNESS
RC 59-63
REF. ROUGH FORGING NO 969359
FACE RUNOUT - .002 IN. PER INCH RADIUS T.I.R
PITCHLINE RUNOUT - SEE SPUR GEAR DATA
CONCENTRICITY - .002 T.I.R.
SPUR GEAR DATA
NUMBER OF TEETH 32
NUMBER OF MATING TEETH 31
DIAMETRAL PITCH 8
PRESSURE ANGLE CUTTER 20°
PITCH DIAMETER THEO. 4.0000000
BASE DIAMETER 3.7587705
CONTACT RATIO MIN. O.D. W/O CH. 1.55 W/MAX CH. 1.43
CENTER DISTANCE—ACTUAL—MINIMUM 3.9365
ADDENDUM ADJ. OR STD. .1271
WHOLE DEPTH .3000 FULL FILLET ROOT DIA. 3.645
CIRCULAR TOOTH THICKNESS NO BACKLASH .2006
FINISHED .1978 .1953
CHAMFER DIA. ROLL ANGLE MINIMUM 29.38°
START OF INVOLUTE ROLL ANGLE 10.40
MEASURING PIN DIAMETER .2160
MEASUREMENT OVER PINS 4.3051 4.2991
PITCHLINE RUNOUT - MAXIMUM .0036
GEAR TOLERANCE PER ES A6141
INVOLUTE ERROR LIMITS J. I. C. CLASS II
DRIVE X DRIVEN AND MULTIDIRECTIONAL
CROWN SHAVE STRAIGHT SHAVE X
INTERNAL INVOLUTE SPLINE DATA
NUMBER OF TEETH 33
DIAMETRAL PITCH 16/32
CIRCULAR PITCH .1963
PRESSURE ANGLE 30°
CLASS & TYPE OF FIT Flat Root Maj. Dia
PITCH DIAMETER 2.0625000
BASE DIAMETER 1.7861774
FORM DIAMETER - MINIMUM 2.1126
CIRCULAR TOOTH SPACE MAX. EFFECTIVE (REF.) .1016
MIN. EFFECTIVE .0998
MAX. ACTUAL .1033
MIN. ACTUAL (REF.) .1015
MEASURING PIN DIAMETER .1200
MEASUREMENT BETWEEN PINS 1.8759 1.8725
LENGTH OF ENGAGEMENT 1.00
GO COMPOSITE GAUGE NUMBER
NO GO PADDLE GAUGE NUMBER
F 2.1310/2.1280 WAS 2.1277/2.1250 MPC 12-11-78
E WAS 3.634 ROOT DIA RJF
D WAS 1.9107/1.9074 MPC 11-1-78
D WAS .0997 D WAS .1080
D WAS .0980 D WAS .1015
D WAS 1.7862 D WAS .0998
C RELEASE FOR PRODUCTION GFD
B WAS ONLY MATERIAL SPECIFIED MPC
B1 MS353-1524H ALTN. MAT.
B "SAE 1524, FORGE, NORMALIZE PER P-4" ADDED
A NORMALIZE PER P-4" APL
A WAS SAE 1524, FORGE.
CHANGE NO CHANGES BY DATE
TOLERANCES UNLESS OTHERWISE SPECIFIED
ONE PLACE DECIMALS ± .060
TWO PLACE DECIMALS ± .030
THREE PLACE DECIMALS ± .010
ANGULAR DIMENSIONS ± 2°
MUST BE CLEAN AND FREE OF BURRS
XA41-XA42-XA43 1978
MODEL ADOPTED DISCARDED
PART NAME GEAR/DRIVE INPUT
DRAWN DATE 4-1-77 TRACED SCALE FULL CHECKED
MATERIAL NOTED APPROVED DATE 4-1-77
DISPLACES
REFERENCE NO XA150789
PART NO A151309

CHAPTER 11

SECTIONAL VIEWS

OBJECTIVES

After studying this chapter, you will be able to:

- Explain the purpose of sectional views.
- Discuss the importance of the cutting plane location.
- Locate the viewing direction of section views.
- Explain the different types of section views.
- Describe material representation on section views.

Frequently, the three main views can describe an object completely. However, additional views are often needed to thoroughly describe details of complex objects.

The drafter will often redraw one of the main views of an object to better show its inside construction. This is performed by passing a cutting plane through the object, at the place which is most suitable for showing additional details. See Figure 11-1. That portion of the object closest to the observer is "removed." This permits a direct view of the interior of the object. The object is shown on the print as if it were actually cut along this plane. The portion of the object in contact with the cutting plane is crosshatched (closely spaced slanted lines) for better clarity, Figure 11-2. Different types of crosshatch lines are sometimes used to indicate the material of which the part is made.

The position of the cutting plane is indicated on an adjacent view by heavy lines consisting of a long dash and two short dashes. The arrows at the end of the cutting plane line indicate the direction from which the observer will view the object. They also indicate which part of the object is shown in the section view. See Figure 11-3.

SECTIONAL VIEWS

Various types of sectional views are used. The various types are described in this chapter.

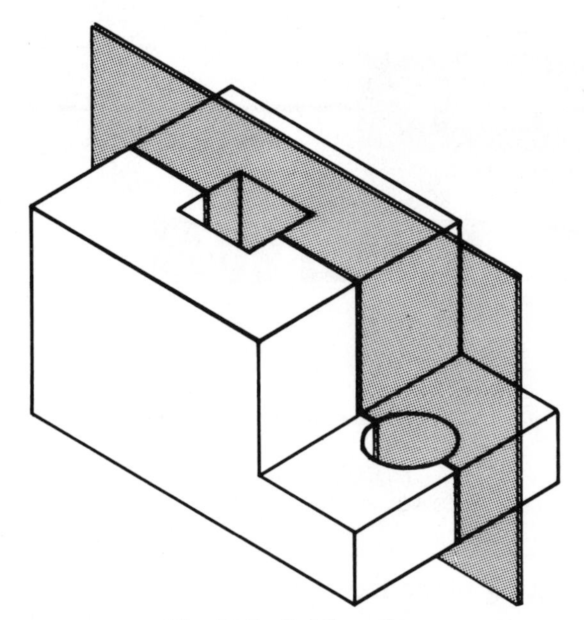

Fig. 11-1 Cutting plane

Full Section

When the cutting plane passes entirely through the object, the view is called a *full section.* A full section is illustrated in Figure 11-3.

In Figure 11-4A, all lines on the section view are visible. Hidden lines are usually omitted. This sectional view of the object is much easier to visualize than the object view in Figure 11-4B, which contains many hidden lines.

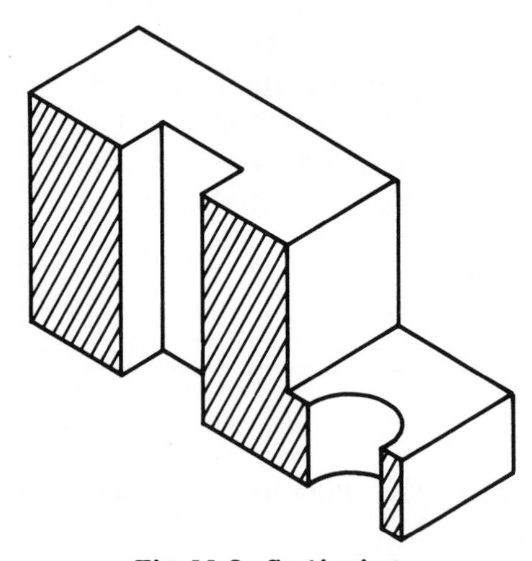

Fig. 11-2 Sectioning

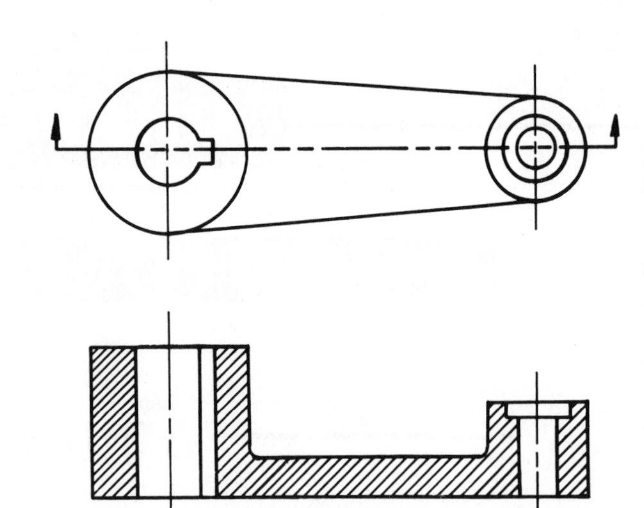

Fig. 11-3 Full section

Half Section

Objects which are uniform or symmetrical are usually sectioned in a different manner. The cutting plane passes only halfway through the object. Such a view is called a *half section.* The other half is shown as a regular external view. See Figure 11-5.

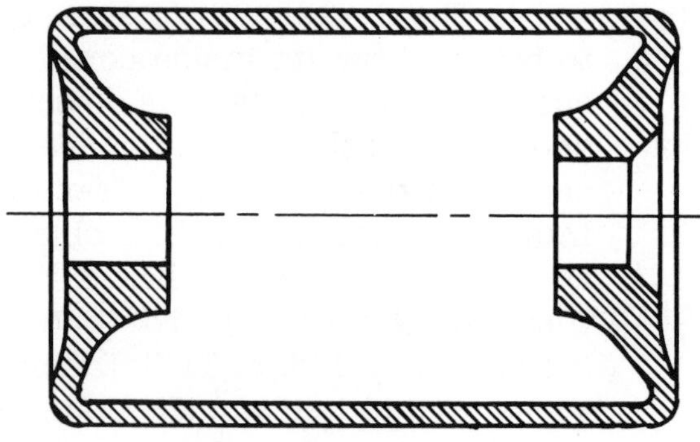

Fig. 11-4A Section view

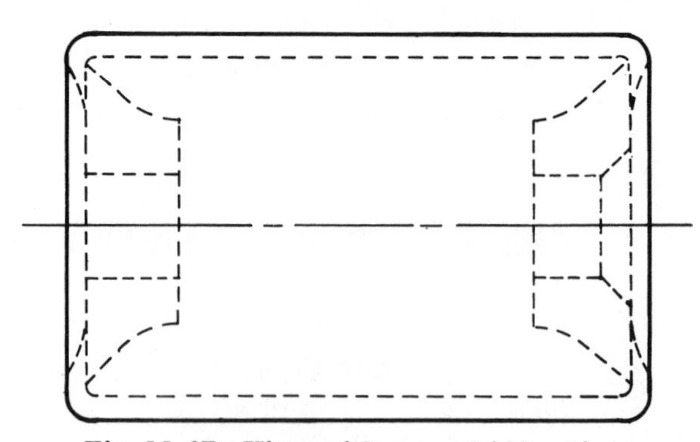

Fig. 11-4B View with many hidden lines

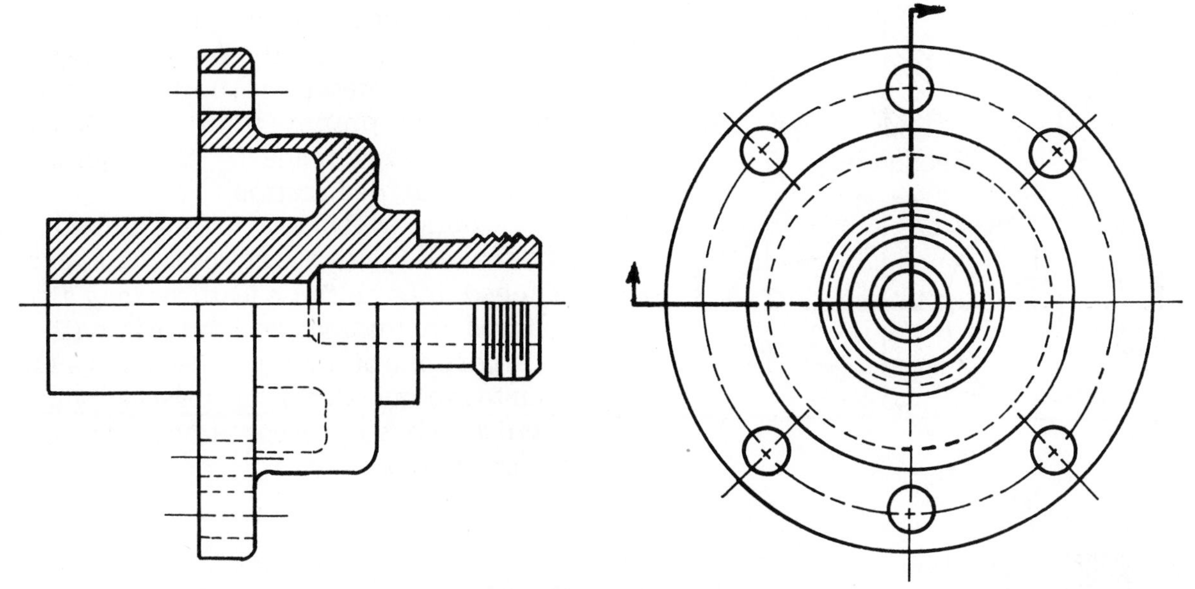

Fig. 11-5 Half section

Offset Section

Details of an object are often not in a straight line. In order to show the desired features, the cutting plane line will be *offset,* or shifted, to indicate the most information possible. See Figure 11-6. Such a view is called an *offset section.*

Sometimes the change in direction of a cutting plane line may be hidden by other construction lines. In such cases the drafter may add small angular lines to indicate a change in direction of the cutting plane line, Figure 11-7. Various methods of showing cutting plane lines are used. The industrial print reader should be familiar with the different methods.

On a drawing of a simple object, the cutting plane line may be completely omitted. Drawings with several section views,

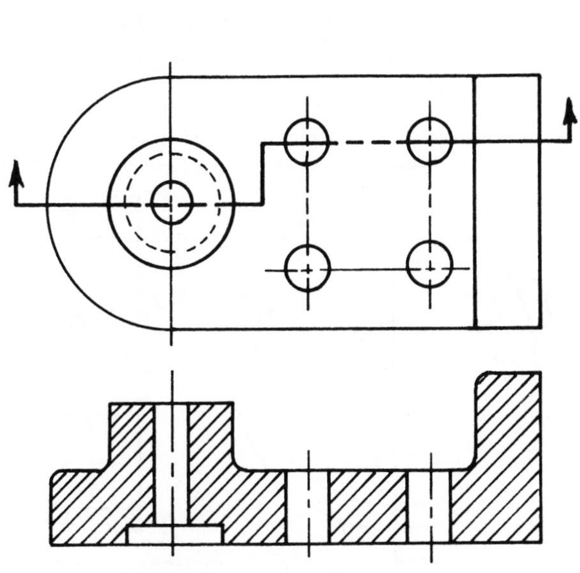

Fig. 11-6 Offset section

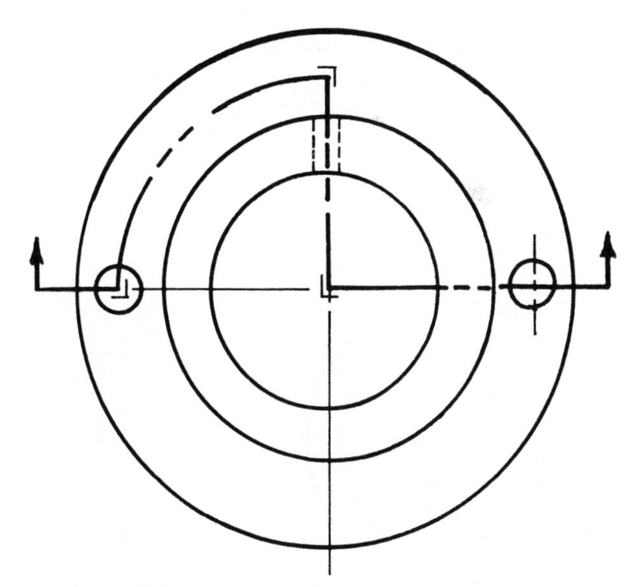

Fig. 11-7 Offset section with angular lines indicating changes in direction of the cutting plane

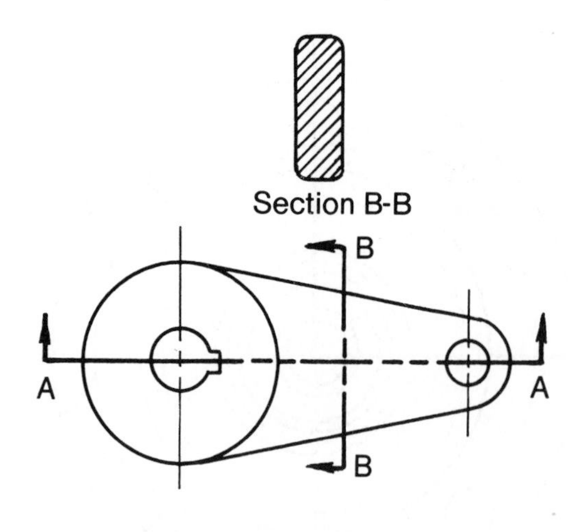

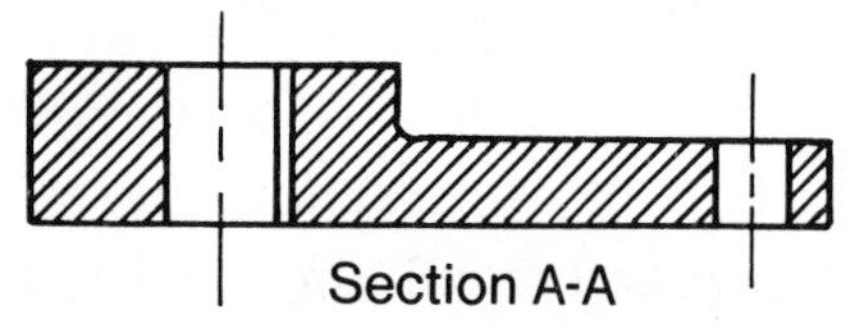

Fig. 11-8 Section view identification

or with section views not in proper relative positions on the drawing, must be identified. The usual method is to identify each as *section A—A, B—B,* etc. See Figure 11-8.

Aligned Section

When drawn as a true projection, sectional views may include distortions of detail locations on an object. Such distortion can usually be eliminated by rotating the detail until it aligns with the cutting plane (Figures 11-9A and 11-9B). A view made by this method is called an *aligned section.*

Solid webs are crosshatched, but spokes and arms are not. The crosshatched lines on these features may also be reduced in number to contrast with the solid portions of the object. See Figure 11-10. This technique alerts the observer that the detail which is not crosshatched is not solid or continuous.

Broken-out Section

A small portion of a part can be exposed to show the interior details by drawing a *broken-out section.* The broken-out portion of the part is outlined by an irregular line, Figure 11-11.

Revolved Section

Sectional views of objects which have rounded edges and corners, such as ribs, spokes, and bars, are shown by revolving the section view at a 90° angle to the cutting plane. These are called *revolved sections,* Figure 11-12. Drafters find this

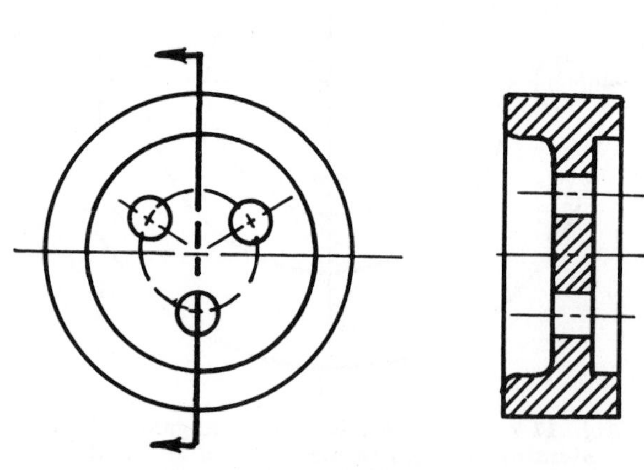

Fig. 11-9A Aligned section

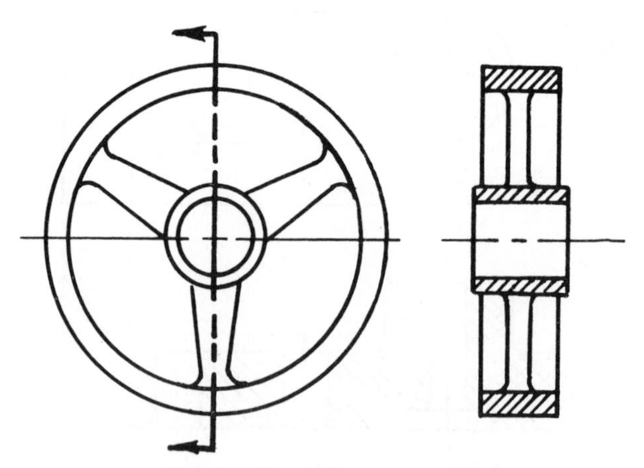

Fig. 11-9B Aligned section

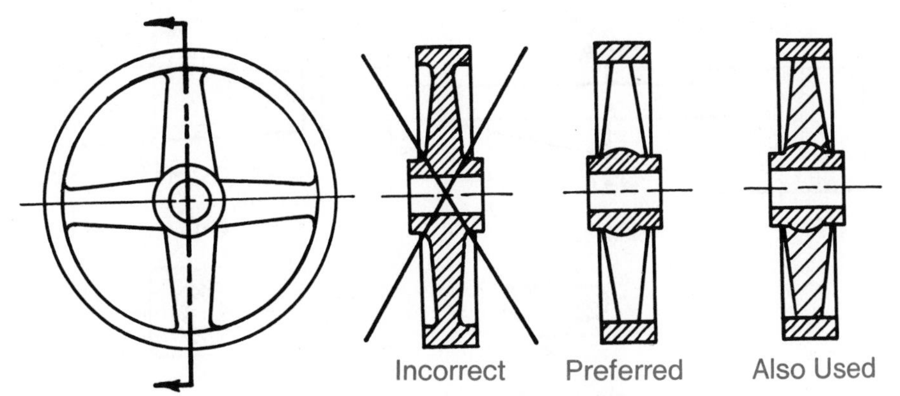

Fig. 11-10 Sectioning spokes

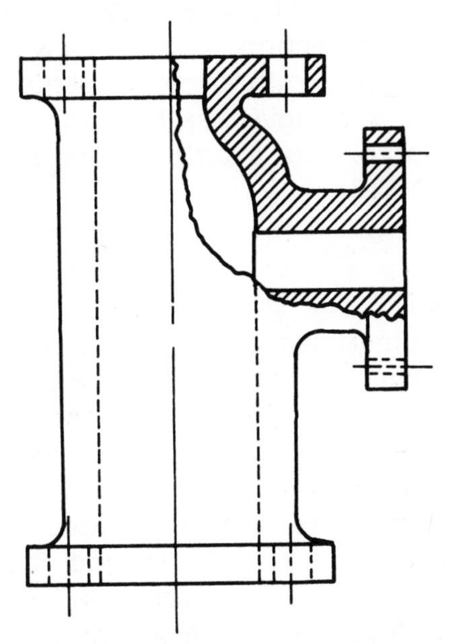
Fig. 11-11 Broken-out section

sectioning method to be most convenient for illustrating casting, forging, or other irregularly shaped objects.

Removed Section

When a revolved section is not on the view but is placed elsewhere on the drawing, it is called a *removed section.* The point of origin is usually identified by letters or by a leader line, Figure 11-13.

Partial Section

A *partial section* is a view that shows only a part of the object as crosshatched. See Figure 11-14.

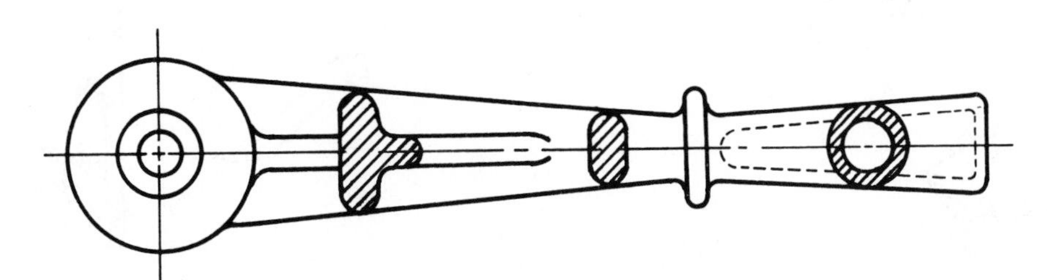
Fig. 11-12 Revolved section

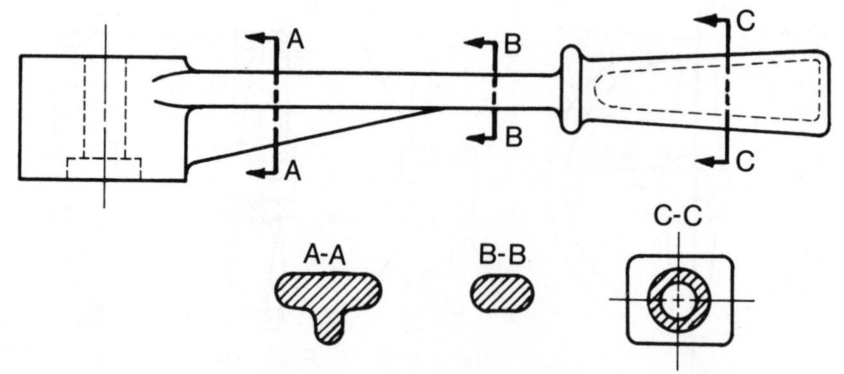

Fig. 11-13 **Removed sections**

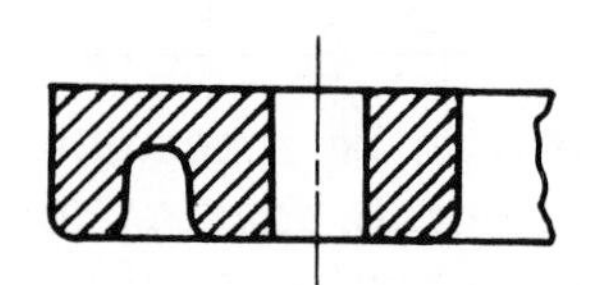

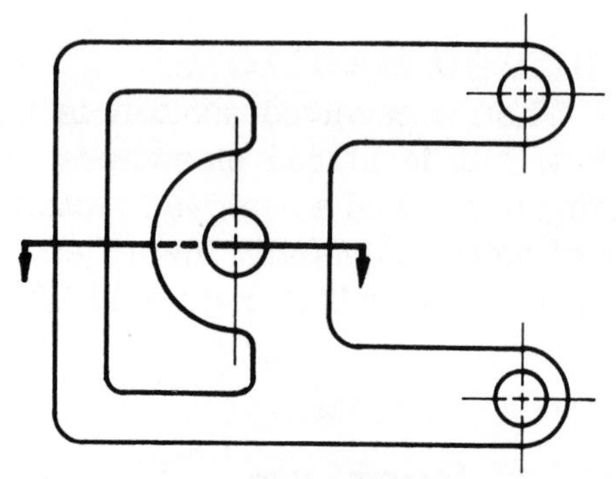
Fig. 11-14 **Partial section**

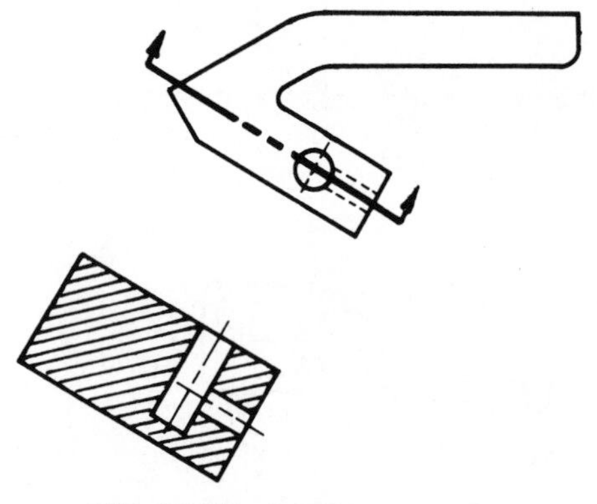
Fig. 11-15 **Auxiliary section**

Auxiliary Section

An *auxiliary view* is an extra view shown to illustrate a portion of an object not parallel to the horizontal or vertical plane of projection. If this view is sectioned to show interior construction, it is called a removed section in auxiliary position, or an *auxiliary section.* See Figure 11-15.

Assembly Section

Sometimes it is useful to show section views on assembly drawings. This technique better illustrates the relative positions of the parts of an assembly, Figure 11-16.

The fasteners illustrated in Figure 11-18 are not sectioned, because they are more easily recognized by their exterior features. The angle of sectioning lines is also reversed on adjoining parts. This angle should not be parallel to a

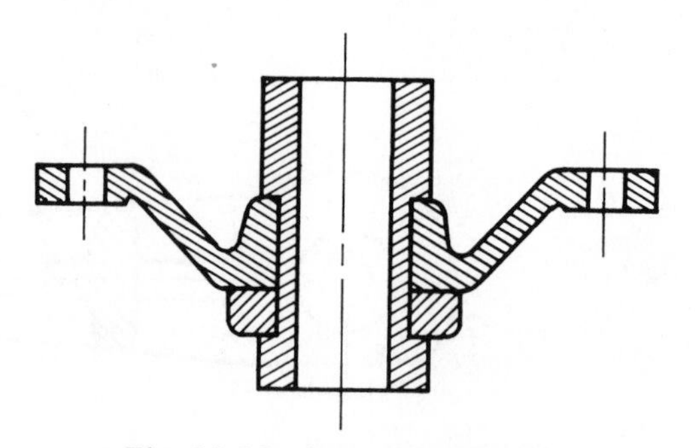
Fig. 11-16 **Assembly section**

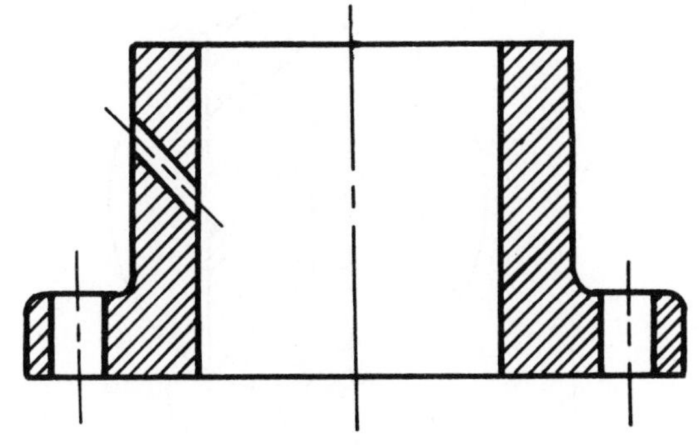

Fig. 11-17A Section lines

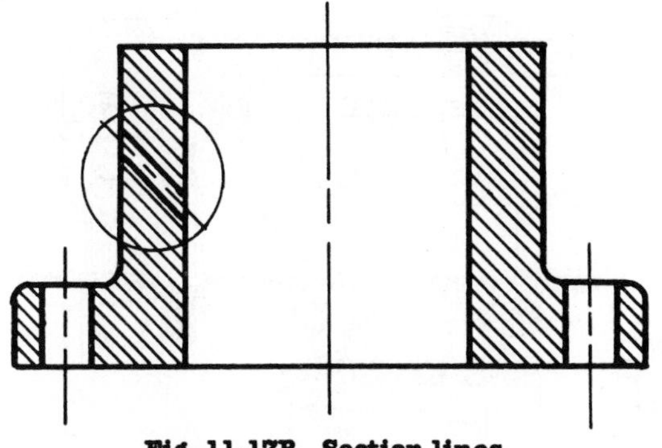

Fig. 11-17B Section lines

hole or other feature of the part, Figures 11-17A and 11-17B. When three or more parts adjoin, the angle of the section lines is changed so as not to align with section lines of the mating parts (Figure 11-16).

A section of a narrow object is usually not crosshatched. It appears as a shaded solid, Figure 11-18.

Outline Section

When large sectioned surfaces are crosshatched, they tend to contrast too much with other views. Therefore, large sectioned surfaces are sometimes crosshatched only along the surface borders. This technique is called *outline sectioning.* An outline section is shown in Figure 11-19.

Shaded Section

The labor costs of making drawings are included in the final cost of producing a part. With this in mind, design departments have developed new methods to reduce the time spent in making drawings. One method is to use ready-made transfers that can be easily attached to areas of the print to illustrate sectioning. These transfers are available in a variety of designs and shades for better contrast of adjoining parts. This system is used extensively in product illustration, and sometimes on industrial prints. A *shaded section* is shown in Figure 11-20.

Enlarged Section

When it is necessary to illustrate a small feature of a part, an enlarged portion

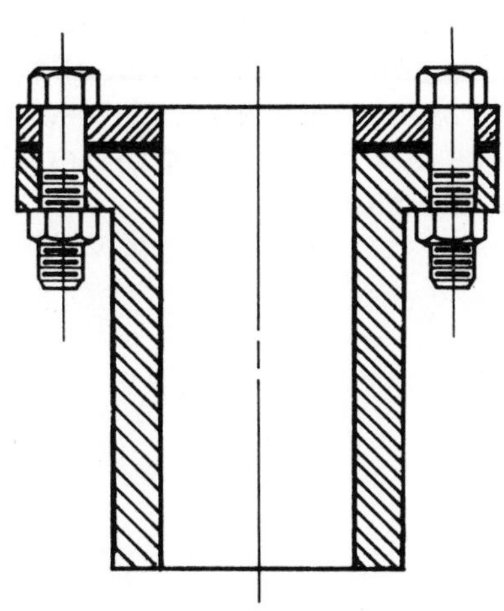

Fig. 11-18 Narrow section

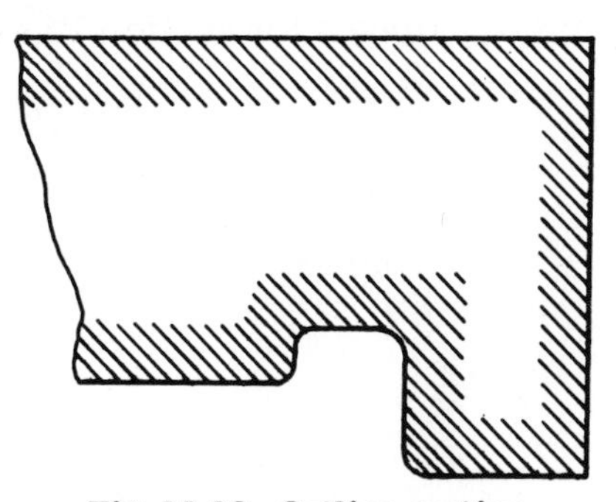

Fig. 11-19 Outline section

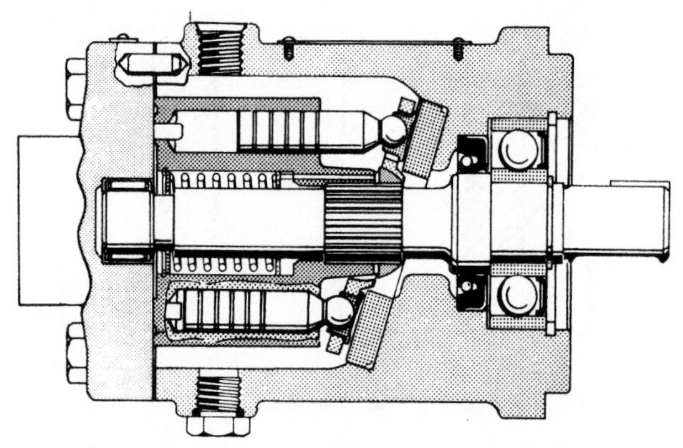

Fig. 11-20 Shaded section (*Courtesy of Sperry Vickers Co.*)

of the part may be sectioned. An *enlarged section* will show interior machining or other important details, Figure 11-21.

MATERIAL REPRESENTATION

When using sectional views, it is a common practice to use a line convention symbol which indicates the material used to make the part. The American National Standards Institute (ANSI) has established a standard for material symbols, Figure 11-22.

The exact specifications of the material cannot be indicated by material symbols. Therefore, exact specifications are indicated elsewhere on the print. Using the standard symbols for section lines, however, does alert the print reader to basic part materials. The use of the symbols also clarifies assembly drawings when the parts are made of different materials.

Some industries do not use all of the material symbols. They usually use the cast-iron symbol for all their section lines, regardless of what kind of material is used to make the part.

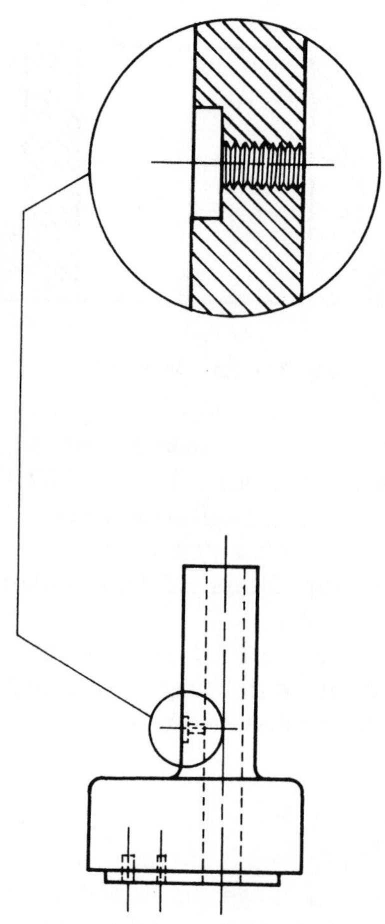

Fig. 11-21 Enlarged section

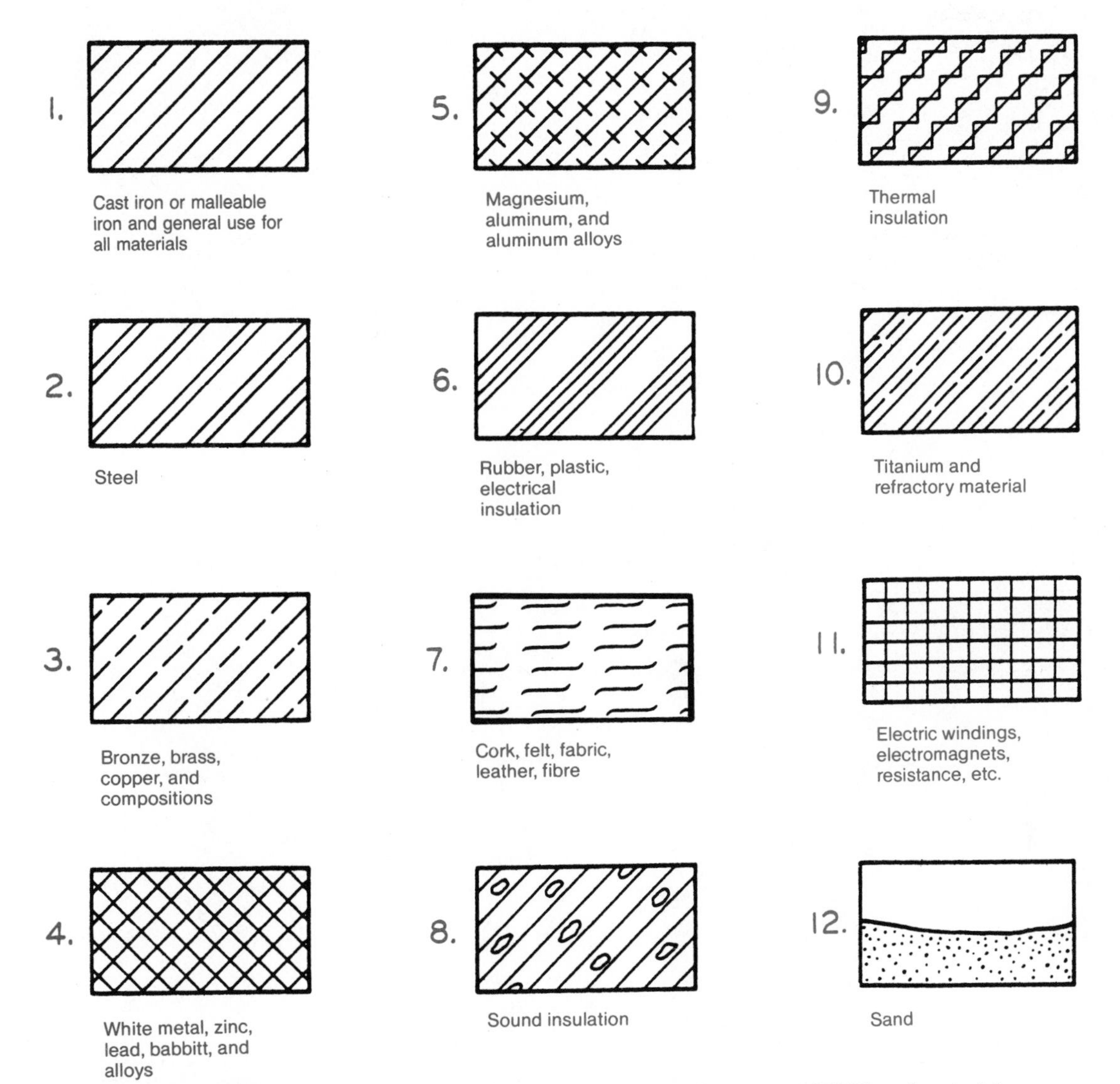

Fig. 11-22 ANSI material symbols for section lining (*Courtesy of ASME, extracted from ANSI Y14.2-1973*)

REVIEW

This review is provided to serve as reinforcement study material. Fill in the appropriate word(s) to complete the sentences below.

1. The viewing direction of a section view is indicated by ______________________ at the end of the ______________________ line.

2. When a cutting plane passes entirely through the object, the view is called (a) (an) ______________________.

3. When a cutting plane passes halfway through the object, the view is called (a) (an) ______________________.

4. Spokes and arms are not ______________________ on the section view.

5. When a small portion of a part is sectioned, the view is called (a) (an) ______________________.

6. Crosshatch lines on adjoining surfaces of an assembly drawing are angled in ______________________.

7. When it is necessary to illustrate small sections of a part, (a) (an) ______________________ section may be used.

8. When the cutting plane changes direction as it passes through the object, the view is called (a) (an) ______________________.

9. When large sectioned surfaces are crosshatched, ____________ section lines may be used.

10. Industries that do not use material symbols on section views usually use the ______________________ symbol.

NAME ______________

DATE ______________

SCORE ______________

EXERCISE A11-1

Sectioning

Complete the section views. Use the cast-iron material symbol.

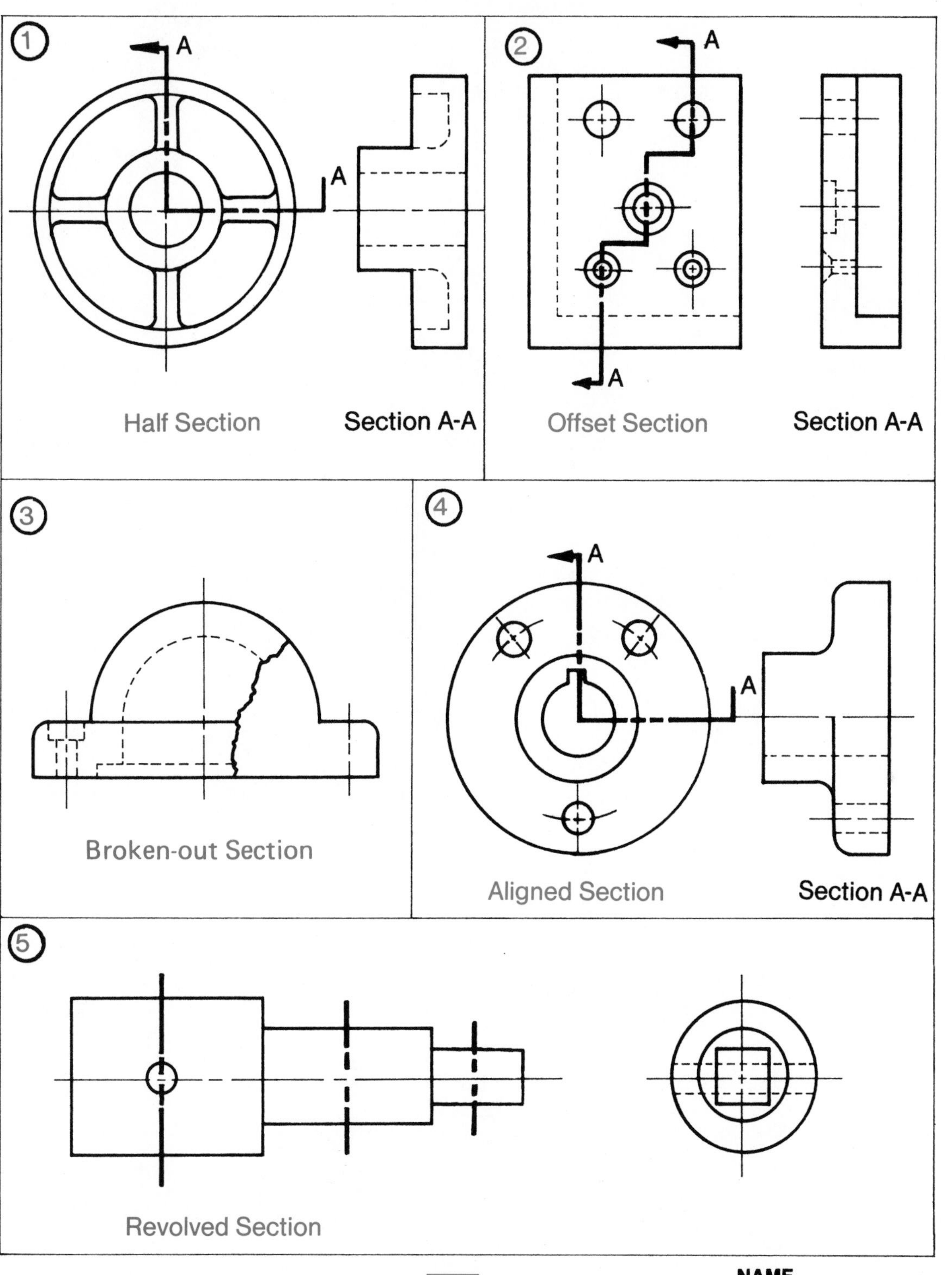

NAME ____________

DATE ____________

SCORE ____________

EXERCISE A11-2

Sectioning

Complete the section views.

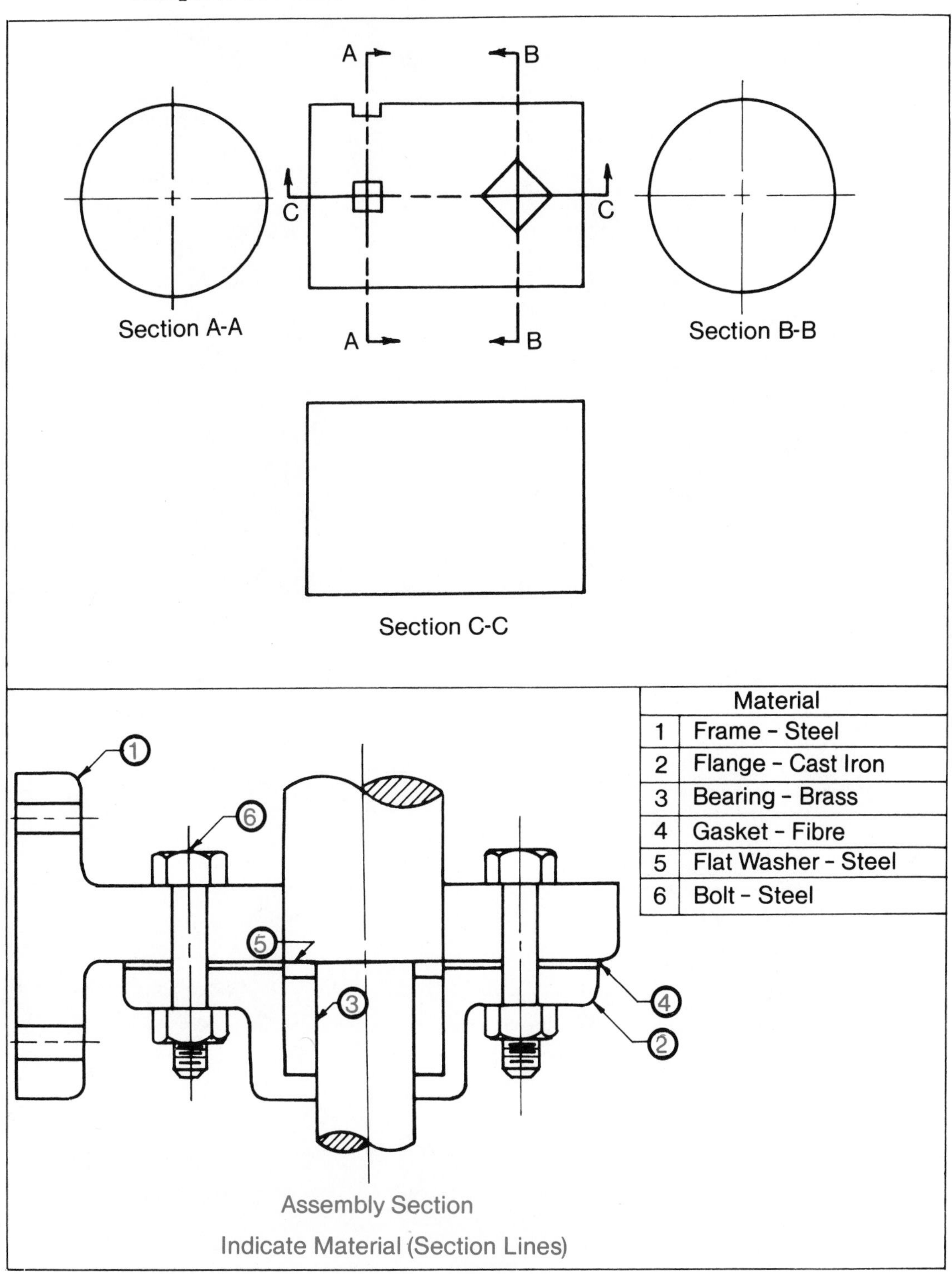

	Material
1	Frame - Steel
2	Flange - Cast Iron
3	Bearing - Brass
4	Gasket - Fibre
5	Flat Washer - Steel
6	Bolt - Steel

Assembly Section

Indicate Material (Section Lines)

NAME ____________

DATE ____________

SCORE ____________

EXERCISE B11-1

Print #317473

Answer the following questions after studying Print #317473 found after this exercise. Note that Print #317473 uses aligned dimensions.

QUESTIONS	ANSWERS
1. Give dimensions *A*, *B*, *C*, *D*, *E*, and *F*. (Indicate correct tolerances.)	1. *A*______ *B*______ *C*______ *D*______ *E*______ *F*______
2. What is this part?	2. ________________
3. On what assembly does this part fit?	3. ________________
4. What scale is used?	4. ________________
5. This part is made of what material?	5. ________________
6. From what size rod is it machined?	6. ________________
7. What is the tolerance on the .083 dimension?	7. ________________
8. What size and kind of thread are used?	8. ________________
9. Prior to 6-23-78, what was .125 ± .005?	9. ________________
10. How is the part identified?	10. ________________
11. Explain what the *TIR* in the note for the .030 DIA hole means.	11. ________________
12. Change *F* was added at what date?	12. ________________
13. What happened to print changes *A* through *F*?	13. ________________

NAME________________

DATE________________

SCORE________________

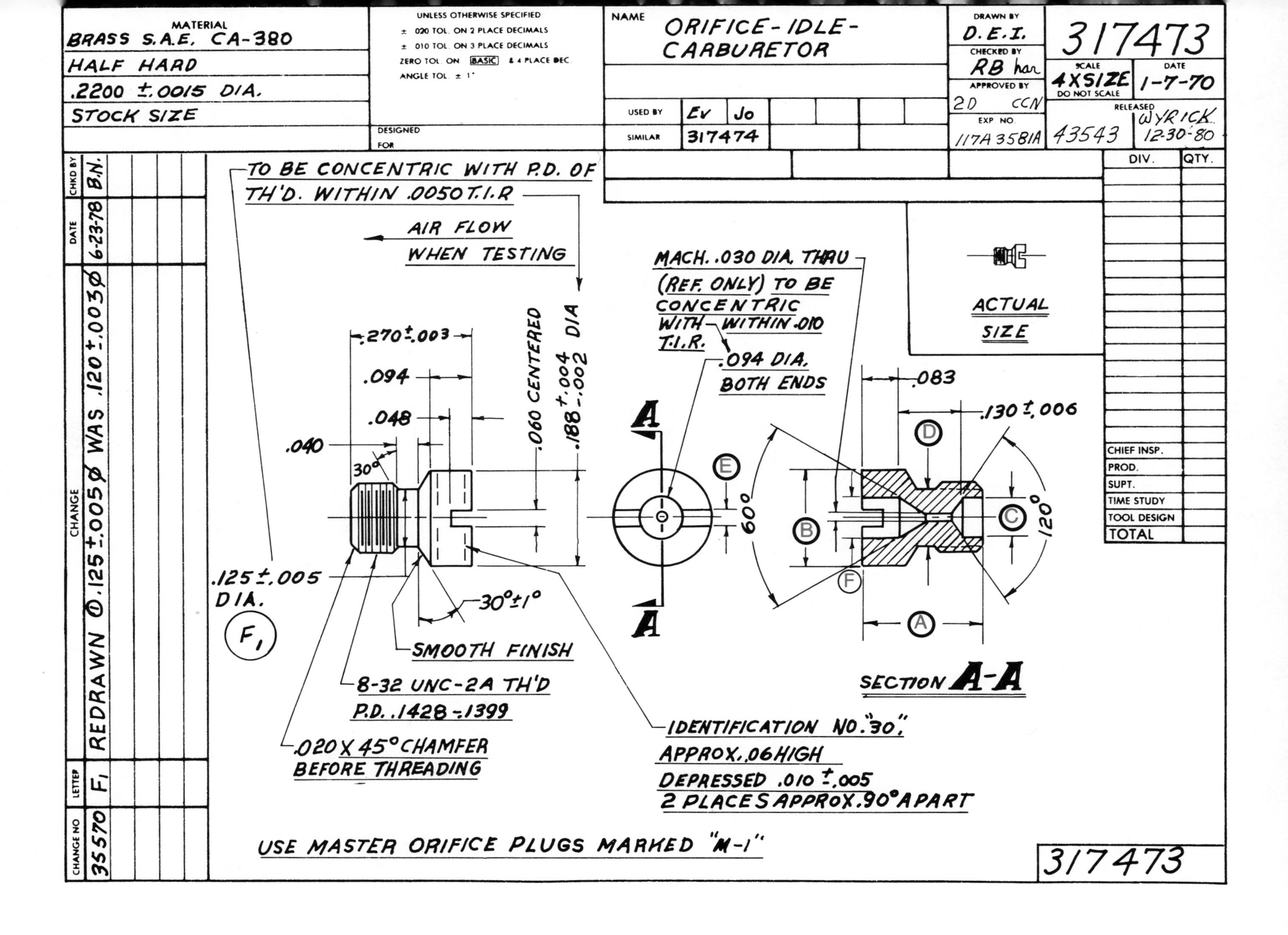
MATERIAL
BRASS S.A.E. CA-380
HALF HARD
.2200 ±.0015 DIA.
STOCK SIZE
UNLESS OTHERWISE SPECIFIED
± .020 TOL. ON 2 PLACE DECIMALS
± .010 TOL. ON 3 PLACE DECIMALS
ZERO TOL. ON BASIC & 4 PLACE DEC.
ANGLE TOL. ± 1°
DESIGNED FOR
NAME
ORIFICE-IDLE-CARBURETOR
USED BY
EV
JO
SIMILAR
317474
DRAWN BY
D.E.I.
CHECKED BY
RB har
APPROVED BY
2D CCN
EXP. NO.
117A3581A
317473
SCALE
4XSIZE
DO NOT SCALE
DATE
1-7-70
RELEASED
43543
WYRICK
12-30-80
DIV.
QTY.
CHIEF INSP.
PROD.
SUPT.
TIME STUDY
TOOL DESIGN
TOTAL
CHANGE NO
LETTER
CHANGE
DATE
CHKD BY
35570
F1
REDRAWN ①.125±.005Ø WAS .120±.003Ø
6-23-78
B.N.
TO BE CONCENTRIC WITH P.D. OF TH'D. WITHIN .0050 T.I.R
AIR FLOW WHEN TESTING
.270±.003
.094
.048
.040
30°
.060 CENTERED
.188 +.004 -.002 DIA
.125±.005 DIA.
F1
30°±1°
SMOOTH FINISH
8-32 UNC-2A TH'D
P.D. .1428-.1399
.020 X 45° CHAMFER BEFORE THREADING
MACH. .030 DIA. THRU (REF. ONLY) TO BE CONCENTRIC WITH — WITHIN .010 T.I.R.
.094 DIA. BOTH ENDS
A
A
E
60°
ACTUAL SIZE
.083
.130±.006
D
B
C
120°
F
A
SECTION A-A
IDENTIFICATION NO. "30"
APPROX. .06 HIGH
DEPRESSED .010±.005
2 PLACES APPROX. 90° APART
USE MASTER ORIFICE PLUGS MARKED "M-1"
317473

CHAPTER 12

TITLE BLOCK INFORMATION

OBJECTIVES

After studying this chapter, you will be able to:

- Discuss the importance of maintaining uniform print sizes.
- List the standard print sizes.
- Identify the standard locations for auxiliary print information.
- Explain the purpose of each type of title block.
- Describe the special types of print notations used by some companies.

The use of part drawings and industrial prints increased greatly as our world became more mechanized. Since drawings were produced in many dimensional sizes, it became more difficult and costly to sort, file, and reproduce them. To avoid these problems, some major industries established a few standard print sizes. The increasing exchange of prints among companies and improved reproduction methods (particularly microfilming) also resulted in a need for standardized print sizes. The standard is outlined by the American National Standards Institute (ANSI) in *ANSI Y14.1—1975.*

As can be seen in Figure 12-1, the most common ANSI print sizes, sizes *A* through *E*, follow a uniform method of size increases. For example, the length of size *A* (11 inches) is the width of size *B*. The length of size *B* (17 inches) is the width of size *C*, and so on.

The standard also provides for uniformity in the location and size for noting supplementary information describing the part or assembly. These guidelines were established because it is an advantage to note similar types of information in the same location in all drawings. It is important for the print reader to easily locate and understand supplementary information.

Each type of information is contained in a block. The net sum of all the blocks is called the *title block.* The title block is usually located in the lower right corner of the print. The print reader should look there first to find basic information about the part. Some companies locate the

Flat Sizes					Roll Sizes					
Size Designation	Width (Vertical)	Length (Horizontal)	Margin Horizontal	Margin Vertical	Size Designation	Width (Vertical)	Length (Horizontal) Min	Length (Horizontal) Max	Margin Horizontal	Margin Vertical
A (Horiz)	8.5	11.0	0.38	0.25	G	11.0	22.5	90.0	0.38	0.50
A (Vert)	11.0	8.5	0.25	0.38	H	28.0	44.0	143.0	0.50	0.50
B	11.0	17.0	0.38	0.62	J	34.0	55.0	176.0	0.50	0.50
C	17.0	22.0	0.75	0.50	K	40.0	55.0	143.0	0.50	0.50
D	22.0	34.0	0.50	1.00						
E	34.0	44.0	1.00	0.50						
F	28.0	40.0	0.50	0.50						

Note: All dimensions are in inches. 1 inch = 25.4 mm.

Fig. 12-1 ANSI standard drawing sheet sizes. (*Courtesy ASME, extracted from ANSI Y14.1—1975*)

title block in the upper right corner of the print, Figure 12-2.

Most illustrations in this chapter conform to the ANSI standards. Other illustrations include samples of title blocks on older prints, and some from industries which have not yet converted to the ANSI standards.

TITLE BLOCK

The following information can usually be found in the title block, with the locations as indicated. The letters in each block, as shown in Figure 12-3, are identified as follows:

- **Block A**—Name and address of the company, division of a corporation, and/or company functional department
- **Block B**—Drawing title
- **Block C**—Drawing number (often the same as the part number)
- **Block D**—Record information, such as date, drafter, checker, approver, project number, and so on
- **Block E**—Approval of designer when different from drafter noted in Block D
- **Block F**—Approval by other department when different from those shown in Blocks D and E
- **Block G**—Main scale of the drawing
- **Block H**—Code identification number of company division or department using the part number
- **Block J**—Drawing size letter (*A, B, C, D,* etc.)
- **Block K**—Actual or estimated weight of the object

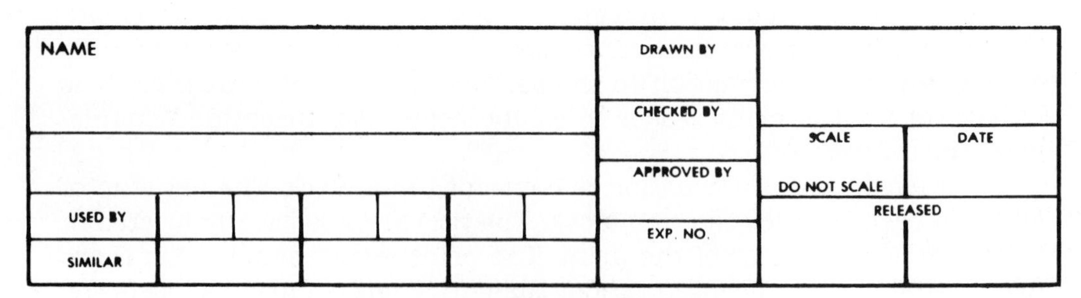

Fig. 12-2 Title block information

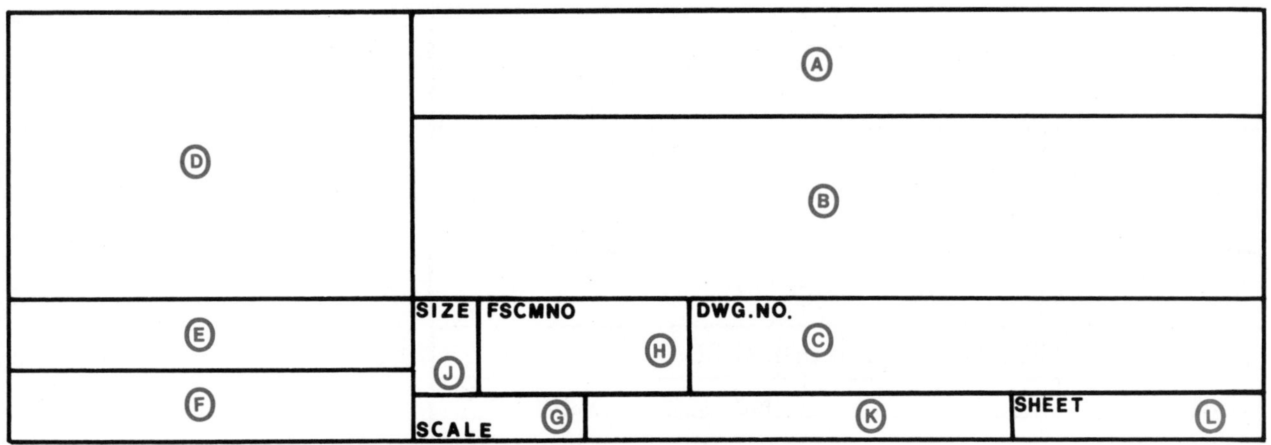

Fig. 12-3 Title block information

SCALE:	APPROVED BY:	DRAWN BY
DATE:		REVISED
		DRAWING NUMBER

Fig. 12-4 Simplified title block

DO NOT SCALE DRAWING	DATE:	SCALE:	FINISH:
	DRAWN BY:	CHECKED:	APPROVED:
ARMSTRONG BROS. TOOL CO. "The Tool People" 5200 W. ARMSTONG AVE. • CHICAGO, ILLINOIS 60646			DRAWING NO:

Fig. 12-5 Custom title block

- **Block L**—Sheet number when more than one drawing sheet is used; for example, "1 of 3"

Smaller companies may use a *simplified title block,* Figure 12-4. Blanks may be added or subdivided as required by the complexity of the part. Companies which require special manufacturing operations such as heat treating or finishing will supply blocks for this information, Figure 12-5.

Often a company will use different title blocks for different departments within the company. These departments may require that specific types of information be listed on the print. For example, the *tooling title block,* Figure 12-6, provides space for indicating the tool name, tool

DET. NO.	SH'T NO.	NO. REQ'D	STOCK SIZE	MAT'L

TOOL DESIGN • MIDTOWN ENGINEERING CO.

PART________ NO.________ MOD.________

TOOL NAME________

TOOL USE________

USE ON MACH.________ DEPT.________

NORMAL TOLERANCES: FRACTIONAL DIMEN. ±.015 DECIMAL DIMEN. ±.003 ANGULAR DIMEN. ±1 DEGREE	DO NOT SCALE DWG. MARK TOOL AND LOOSE PARTS WITH DRAWING NO. BREAK SHARP CORNERS
LAST ENG. PART CHG. ◯ DATE____ BY____ SCALE____ CHK'D____ SHT.____ OF____ APP.____	DWG. NO.

Fig. 12-6 Tooling and parts list tooling block

use, use on machine, etc. The *die design title block,* Figure 12-7, contains blocks for other information.

The *tolerance block* is usually located to the left of the title block. The symbol √ at the bottom of the tolerance block in Figure 12-8 specifies that the part will have a surface finish according to the *ASA Standard B46.1–1962.* The American Standards Association (ASA)

SHT. NO.	DETAIL NO.	NO. REQ'D.	SIZE	MATERIAL	HEAT TREAT
PART NAME					
PART NO.	DRAWN BY:		CHK'D BY:	DATE	
EXP. NO.	SHT. OF SHTS.		LAST ENG. PART CHANGE ◯		
	MACH. USED:	SHRINK PER INCH:	SCALE:		
DIE DESIGN					

Fig. 12-7 Die design title block

UNLESS OTHERWISE SPECIFIED:
±.020 TOL. ON TWO PLACE DECIMALS
±.010 TOL. ON THREE PLACE DECIMALS
ZERO TOL. ON FOUR PLACE DECIMALS
ANGLE TOL. ±1°
DO NOT SCALE
√ ASA B46.1-1962
DESIGNED FOR

Fig. 12-8 Tolerance block

MATERIAL

Fig. 12-9 Material block

has since changed its name to the American National Standards Institute (ANSI). Therefore, *ASA* will appear on older prints; *ANSI* will appear on newer prints.

On assembly drawings, the *parts list* is located just above the title block (Figure 12-6). (Refer to Chapter 14.) On size *A* or size *B* prints, the parts list may be located at another convenient place to save drawing space.

MATERIAL BLOCK

When required, a separate *material block* may be provided. The material block may be located to the left of the tolerance block. It is usually the same height as the title block. See Figure 12-9.

REVISION OR CHANGE BLOCK

On size *C* prints, the *revision*, or *change, block* is located in the upper right corner. On smaller size prints, the revision block is usually located along the left border. The revision block should contain spaces for the change number, designated letter, change information, date, change approval, and print checker. See Figure 12-10.

PRINT DISTRIBUTION BLOCK

Multidivision companies often require that the engineering department distribute prints to all divisions or departments that may need the prints. *Print distribution blocks,* Figure 12-11, are placed on the prints, usually close to the title block. The engineering department will check each division or department to which the print will be sent.

ZONING

On size *B* prints and larger, letters and numbers are placed outside the border

CHANGE NO.	LETTER	CHANGE	DATE	CHKD BY

Fig. 12-10 Change or revision block

DIV.	QTY.
CCP	
AU	
BL	
BT	
CA	
CU	
DS	
EV	
GL	
JO	
ME	
OA	
TW	
CCR	

Fig. 12-11 Print distribution block

lines. These are used as references for locating details on a print. The letters are arranged from the top to the bottom on the print. The numbers are arranged from the right to the left. See Figure 12-12. This zoning system is very similar to the reference system used on road maps.

MICROFILM ALIGNMENT ARROWHEADS

Arrowheads are placed on prints to be used as alignment points when prints are microfilmed. The arrowheads are placed at the horizontal and vertical centers along the print borders, Figure 12-12. Since many industries now microfilm prints for storage, these arrowheads are becoming increasingly important.

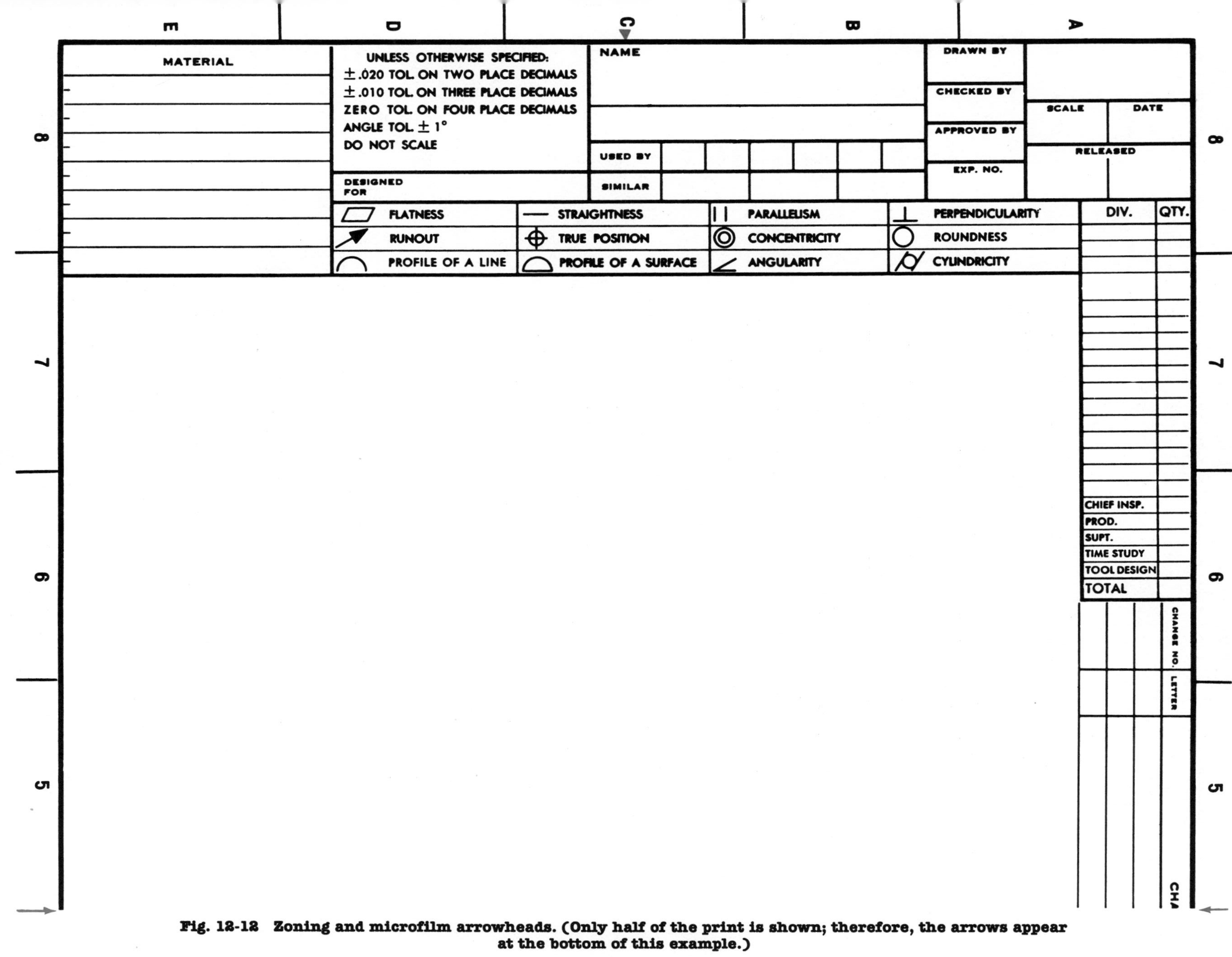

Fig. 12-12 Zoning and microfilm arrowheads. (Only half of the print is shown; therefore, the arrows appear at the bottom of this example.)

REVIEW

This review is provided to serve as reinforcement study material. Fill in the appropriate word(s) to complete the sentences below.

1. Print sizes ____________________ through ____________________ increase in size at a uniform rate.

2. The title block is usually located in the ____________________ corner of the print.

3. The print number is usually the same as the ____________________ number.

4. The symbol ____________________ at the bottom of the tolerance block specifies that the print will have surface finish according to ASA or ANSI standards.

5. The parts list on an assembly print is usually located ____________ the title block.

6. Revision or change blocks are usually located in the ______________ corner of the print.

7. A size *B* print would be ____________________ inches wide and ____________________ inches long.

8. The print distribution block is usually located ________________ the main title block.

9. Outside the print borders, ______________________________ and ____________________ are placed to serve as references for locating specific print details.

10. The arrows placed along print borders are used for ______________ purposes.

NAME ____________________

DATE ____________________

SCORE ____________________

EXERCISE B12-1

Print #E-139

Answer the following questions after studying Print #E-139 found after this exercise.

QUESTIONS	ANSWERS
1. What is this part?	1. ____________
2. It is made of what material?	2. ____________
3. What is the preferred overall length?	3. ____________
4. What is the maximum overall length?	4. ____________
5. What is the minimum overall length?	5. ____________
6. What kind of section view is shown?	6. ____________
7. What is the length of the $^{39}/_{64}$-inch drilled hole?	7. ____________
8. What scale is used?	8. ____________
9. Are there any revisions on this print?	9. ____________
10. What is the tolerance on the .845/.835 dimension?	10. ____________
11. What do you think the note "Automatics Print" means?	11. ____________
12. What are the width and depth of the groove on the left end of the socket?	12. Width ____________ Depth ____________

NAME ____________

DATE ____________

SCORE ____________

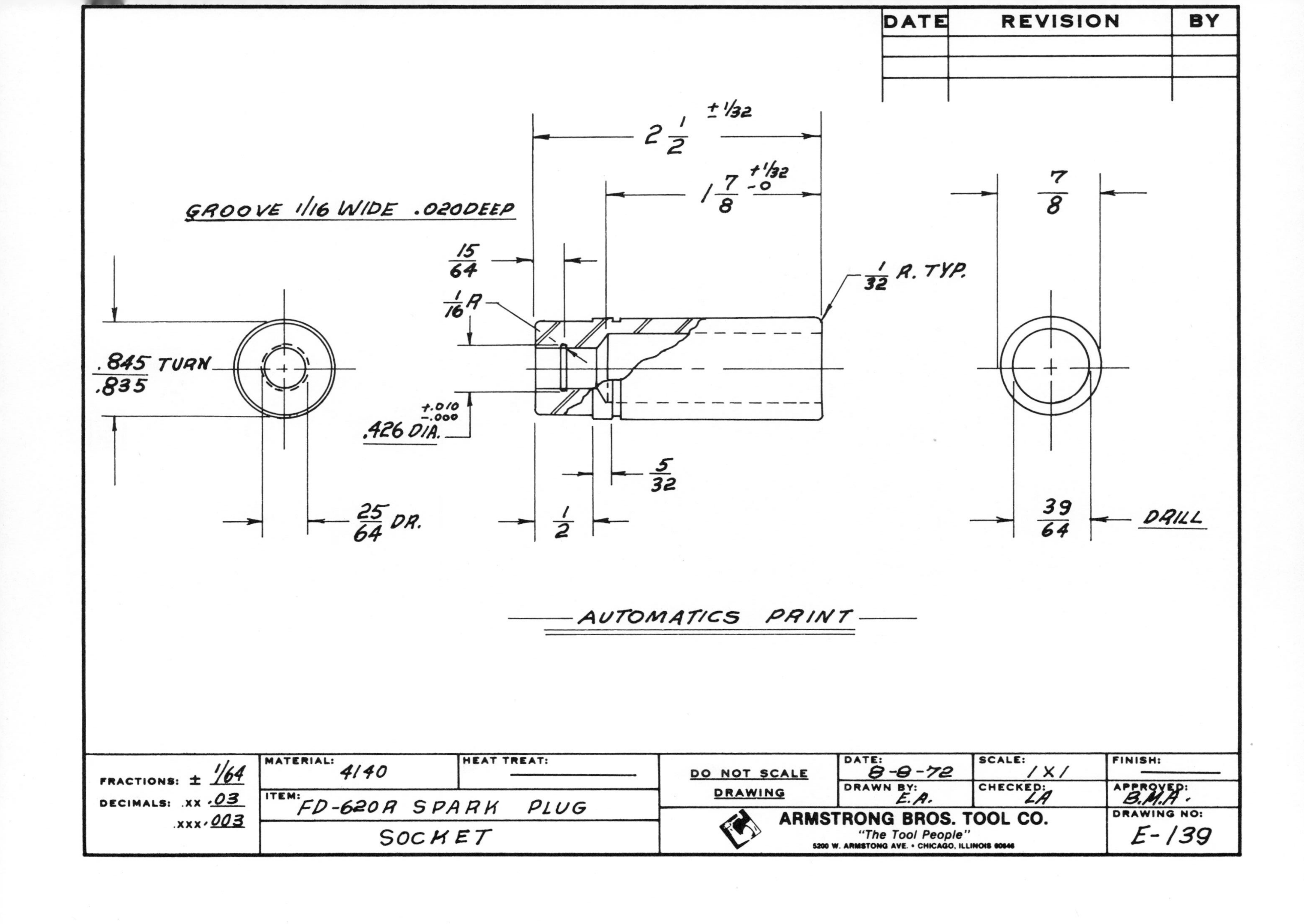
DATE
REVISION
BY
2 1/2 ± 1/32
1 7/8 +1/32 -0
GROOVE 1/16 WIDE .020DEEP
15/64
1/16 R
1/32 R. TYP.
7/8
.845/.835 TURN
.426 DIA. +.010 -.000
5/32
25/64 DR.
1/2
39/64
DRILL
AUTOMATICS PRINT
FRACTIONS: ± 1/64
DECIMALS: .XX .03
.XXX .003
MATERIAL: 4140
HEAT TREAT:
ITEM: FD-620A SPARK PLUG
SOCKET
DO NOT SCALE DRAWING
DATE: 8-8-72
SCALE: 1X1
FINISH:
DRAWN BY: E.A.
CHECKED: LA
APPROVED: B.M.A.
ARMSTRONG BROS. TOOL CO.
"The Tool People"
5200 W. ARMSTONG AVE. • CHICAGO, ILLINOIS 60646
DRAWING NO: E-139

CHAPTER 13

MATERIALS OF THE TRADE

OBJECTIVES

After studying this chapter, you will be able to:

- Explain the factors involved in the selection of materials.
- Name the basic types of materials used in industry.
- Describe the difference between iron and steel.
- List the different types of steels.
- Explain the importance of carbon content in steels.
- Discuss what classifies a steel as an alloy.
- Explain the steel classification numbering system.
- List the common types of non-ferrous metals.
- List the common types of plastics.
- Discuss the various types of material treatments.

The casual print reader need not have a thorough knowledge of the technical aspects of materials specified on industrial prints. It is important, however, to know why a certain type of material is used for a particular application, and how the material is indicated on the print.

People responsible for the technical manufacturing or machining operations required to produce parts illustrated on prints should further their education in the metallurgical field.

SELECTION OF MATERIALS

The designer or engineer must know the materials that can be used to make a part. Many different types of materials may be suitable. To decide which material would be most suitable, the designer or engineer must weigh several factors. Some of the factors considered are strength, weight, corrosive resistance, appearance, machinability, and cost.

The cost factor is important because the finished product must be competitive with similar products on the market. Since costs of basic materials may change from time to time, suitable substitute materials are listed on prints. Some industries employ engineers who have special training in areas of cost reduction. These engineers are concerned with material selection as well as with design and manufacturing.

The factors considered in material selection can be rather involved, since so many different materials are available. In this chapter, therefore, we restrict our discussion to basic materials and how they are indicated on prints.

BASIC MATERIALS

The machining industries use three basic types of material. Listed in the order of their volume usage, these basic types are *ferrous metals* (iron and steel), *nonferrous metals*, and *plastics*.

Even with the increasing use of plastics, ferrous metals still account for about 90% of all materials used in the machine trades. Therefore, iron and steel are the materials most often listed on industrial prints.

Many different types of plastics are now being used. Ceramic materials are also used, in order to satisfy the higher operating temperatures of equipment. In addition, ceramic materials are used in the tooling field.

FERROUS METALS

Steel is a term used for the basic element iron, when it has a low carbon content. Iron, when produced from its natural iron ore, has a carbon content of approximately 5%. When the carbon content of iron is reduced below 1.7%, it is called *steel*. When the carbon content of iron is above 1.7%, the material is called *cast iron*. The molten iron is always "cast" into some type of mold. When cooled, it takes the shape of the mold.

Carbon Steel

Carbon steel is the most widely used material in the machine trades. Carbon steels are classified as low carbon, medium carbon, or high carbon. The percentage of carbon is the most important factor in determining the material's properties and uses.

Low-carbon steels have a range of carbon content from .05% to .30%. These are the most widely used of all steels. Low-carbon steels are tough, ductile, and easily machined. Normally they are not heat treatable, but can be case hardened and then heat treated.

Medium-carbon steels have a carbon content of .30% to .60%. These steels are stronger and harder than low-carbon steels, and can be heat treated. They are not as ductile as the low-carbon types.

High-carbon steels have a range of carbon content from .60% to .75%. Steels with a carbon content of .75% to 1.70% are called *very high-carbon steels*. High-carbon and very high-carbon steels are tough, hard, and can be heat treated. Tools made of these materials will hold a good cutting edge. At the high end of this carbon-content range, however, the steels start to become brittle, similar to cast iron.

Cast Iron

Cast iron is one of the major materials in use today. It is an iron-base material containing a high carbon content (1.7% to 5.0% carbon). Cast iron is used extensively for machine housing, engine blocks, and many automobile parts. Cast iron is made in a variety of types, the most important being gray, white malleable, and nodular cast iron. Each type has special properties and can be used for a variety of applications.

Alloy Steels

An *alloy steel* is a steel which contains an element in addition to carbon in a quantity large enough to alter its mechanical properties. The alloy steels are classified by the major alloying elements. The most common alloying elements are nickel, chromium, manganese, molybdenum, titanium, cobalt, tungsten, and vanadium. The addition of these elements in the proper amounts greatly improves the properties of the steel. Some of these properties are toughness, strength, wear resistance, and corrosive resistance.

STEEL CLASSIFICATION NUMBERING SYSTEM—

A steel classification numbering system, Figure 13-1 and Table 13-1, was adopted by the Society of Automotive Engineers (SAE) and the American Iron and

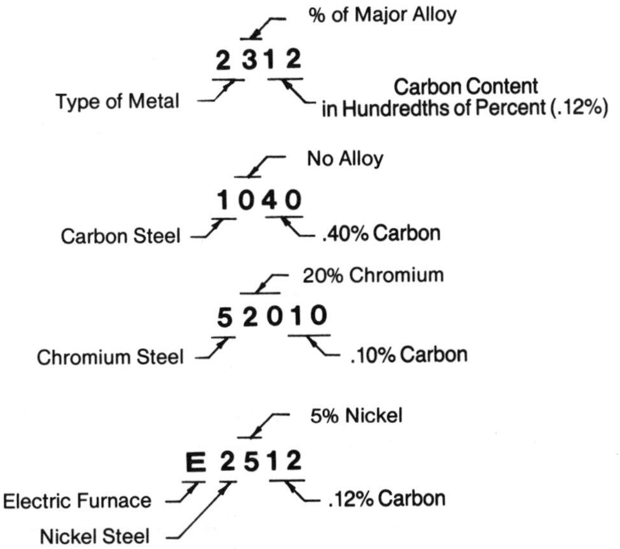

Fig. 13-1 Reading the steel classification numbering system

Table 13–1 Steel Classification Numbering System

Type of Steel	Series Designation
Carbon steels	1XXX
Plain carbon	10XX
Free machining, resulfurized (screw stock)	11XX
Free machining, resulfurized, rephosphorized	12XX
Manganese steels	13XX
High-manganese carburizing steels	15XX
Nickel steels	2XXX
3.50 percent nickel	23XX
5.00 percent nickel	25XX
Nickel-chromium steels	3XXX
1.25 percent nickel, .60 percent chromium	31XX
1.75 percent nickel, 1.00 percent chromium	32XX
3.50 percent nickel, 1.50 percent chromium	33XX
Corrosion and heat resisting steels	30XXX
Molybdenum steels	4XXX
Carbon-molybdenum	40XX
Chromium-molybdenum	41XX
Chromium-nickel-molybdenum	43XX
Nickel-molybdenum	46XX and 48XX
Chromium steels	5XXX
Low chromium	51XX
Medium chromium	52XXX
Corrosion and heat resisting	51XXX
Chromium-vanadium steels	6XXX
Chromium 1.0 percent	61XX
Nickel-chromium-molybdenum	86XX and 87XX
Manganese-silicon	92XX
Nickel-chromium-molybdenum	93XX
Manganese-nickel-chromium-molybdenum	94XX
Nickel-chromium-molybdenum	97XX
Nickel-chromium-molybdenum	98XX
Boron (.0005% boron minimum)	XXBXX

Steel Institute (AISI). The steel identification is based on a code system of four or five numbers. The first digit of the number indicates the type of steel. The second digit indicates the major alloying element. The last two digits indicate the carbon content, in hundredths of 1 percent.

If *SAE 1020* steel is specified, the first digit (1) indicates a carbon steel. The second digit (0) indicates that no alloying element is present. The last two digits (20) indicate .20% carbon. A fifth digit is used when the carbon content is more than .99%.

A specification of *SAE 2540* indicates a nickel steel with 5% nickel and .40% carbon. Further steel chemical composition ranges can be found in technical manuals, in tables such as that shown in Table 13-2.

The typical properties and uses of the different types of carbon and alloy steels are listed in Table 13-3. This table is only a summary of the existing steel alloys. Many special alloys, particularly tool steels of the tungsten, molybdenum, and chromium types, are listed separately, as shown in Tables 13-4, 13-5, and 13-6. Note that the tool steel classification system is quite different from the system illustrated in Tables 13-1 and 13-2.

NON-FERROUS METALS

All other metals are in the non-ferrous groups. The most important non-ferrous metals are aluminum, brass, bronze, copper, lead, magnesium, titanium, and zinc. The other non-ferrous metals are used mainly as alloying agents. These are chromium, cobalt, manganese, molybdenum, and nickel.

Table 13–2 Standard Composition Ranges of Nonresulfurized Carbon Steels**

Steel Designation AISI or SAE	Chemical Composition, Percent					Steel Designation AISI or SAE	Chemical Composition, Percent				
	UNS Number	C	Mn	P Max.	S Max.		UNS Number	C	Mn	P Max.	S Max.
*1005	G10050	0.06 Max.	0.35 Max.	0.040	0.050	1039	G10390	0.37/0.44	0.70/1.00	0.040	0.050
*1006	G10060	0.08 Max.	0.25/0.40	0.040	0.050	1040	G10400	0.37/0.44	0.60/0.90	0.040	0.050
1008	G10080	0.10 Max.	0.30/0.50	0.040	0.050	1042	G10420	0.40/0.47	0.60/0.90	0.040	0.050
1010	G10100	0.08/0.13	0.30/0.60	0.040	0.050	1043	G10430	0.40/0.47	0.70/1.00	0.040	0.050
1012	G10120	0.10/0.15	0.30/0.60	0.040	0.050	1044	G10440	0.43/0.50	0.30/0.60	0.040	0.050
1015	G10150	0.13/0.18	0.30/0.60	0.040	0.050	1045	G10450	0.43/0.50	0.60/0.90	0.040	0.050
1016	G10160	0.13/0.18	0.60/0.90	0.040	0.050	1046	G10460	0.43/0.50	0.70/1.00	0.040	0.050
1017	G10170	0.15/0.20	0.30/0.60	0.040	0.050	1049	G10490	0.46/0.53	0.60/0.90	0.040	0.050
1018	G10180	0.15/0.20	0.60/0.90	0.040	0.050	1050	G10500	0.48/0.55	0.60/0.90	0.040	0.050
1019	G10190	0.15/0.20	0.70/1.00	0.040	0.050	1053	G10530	0.48/0.55	0.70/1.00	0.040	0.050
1020	G10200	0.18/0.23	0.30/0.60	0.040	0.050	1055	G10550	0.50/0.60	0.60/0.90	0.040	0.050
1021	G10210	0.18/0.23	0.60/0.90	0.040	0.050	*1059	G10590	0.55/0.65	0.50/0.80	0.040	0.050
1022	G10220	0.18/0.23	0.70/1.00	0.040	0.050	1060	G10600	0.55/0.65	0.60/0.90	0.040	0.050
1023	G10230	0.20/0.25	0.30/0.60	0.040	0.050	1070	G10700	0.65/0.75	0.60/0.90	0.040	0.050
1025	G10250	0.22/0.28	0.30/0.60	0.040	0.050	1078	G10780	0.72/0.85	0.30/0.60	0.040	0.050
1026	G10260	0.22/0.28	0.60/0.90	0.040	0.050	1080	G10800	0.75/0.88	0.60/0.90	0.040	0.050
1029	G10290	0.25/0.31	0.60/0.90	0.040	0.050	1084	G10840	0.80/0.93	0.60/0.90	0.040	0.050
1030	G10300	0.28/0.34	0.60/0.90	0.040	0.050	*1086	G10860	0.80/0.93	0.30/0.50	0.040	0.050
1035	G10350	0.32/0.38	0.60/0.90	0.040	0.050	1090	G10900	0.85/0.98	0.60/0.90	0.040	0.050
1037	G10370	0.32/0.38	0.70/1.00	0.040	0.050	1095	G10950	0.90/1.03	0.30/0.50	0.040	0.050
1038	G10380	0.35/0.42	0.60/0.90	0.040	0.050						

*Standard Steel Grades for Wire Rods and Wire only.
**By Permission From AISI–"Steel Products Manual–Alloy Carbon and High Strength Low Alloy Steels," August, 1977.

Table 13–3 Properties and Uses of Carbon and Alloy Steels

Type of Steel	SAE No.	Special Characteristics	Common Uses
Low carbon	1010	Low tensile strength, machines rough	Welding steel, sheet iron, tacks, nails, etc.
	1020	Very tough, machines fair	Fan blades, sheet steel, pipe, structural steel
	1030	Heat treats well, stronger than 1020	Seamless tubing, shafting, gears
Medium carbon	1040	Heat treats average, machines fair	Auto axles, bolts, connecting rods
	1045	Thin sections must be carefully quenched	Coil springs, auger bits, screw drivers
	1055	May be oil tempered	Miscellaneous coil springs
	1060	Soft tool steel, does not hold edge	Valve springs, lock washers, nonedged tools
High carbon	1070	Stands severe shocks, very tough and hard	Wrenches, anvils, dies, knives
Very high carbon	1080	Holds edge well, medium hard	Shovels, hammers, chisels, vise jaws
	1085	Hard tool steel	Knife blades, auto bumpers, taps, saws
	1090	Very hard tool steel, thin edges will be brittle	Coil and leaf auto springs, taps, hacksaw blades, cutters
Free cutting	1112	Excellent machining characteristics	Screw machine stock, studs, screws, bolts
	1115	Stronger and tougher but machines slowly	Above items where more strength is required
	X1314	Machines well, case hardens very well	Used where surface hardness and strength are desired
	X1330	Machines very well	Used where better-quality hardness is desired
Manganese	1330	Withstands hard wear, hammering, and shock	Burglar-proof safes, railroad rails (curved)
Nickel	2317	Wear resistant, tough	Structural steels, forgings, and casings
	2515	Easily case hardened, shock resistant	Gears
Nickel-chromium	3115	Very hard and strong	Armor plate
	3130	Very hard and strong	Gears
	3140	Very hard and strong	Springs, axles
	3150	Very hard and strong	Shafts
Molybdenum	4130 4140	Withstands high heat and hard blows	Very fine wire, ball and roller bearings, high-grade auto and machinery parts

(Continued)

Table 13–3 Properties and Uses of Carbon and Alloy Steels (Continued)

Type of steel	SAE No.	Special Characteristics	Common Uses
Molybdenum (cont'd.)	4150	Same as above	Same as above
	4320	Same as above	Same as above
	4615	Same as above	Same as above
	4815	Same as above	Same as above
Chromium	5120	Hard and tough	Burglar-proof safes, springs, cutting tools, bolts and rollers for bearings
	5140	Resists rust, stains and scratches	
	52100	Hard and very tough	Same as above
Stainless-chromium	51210	Not heat treatable	Stainless cooking utensils, sinks
	51310	Can be heat treated	Same as above
	51335	Can be heat treated, does not corrode	Stainless-steel cutlery, tableware, ball bearings
	51710	Same as above	Same as above
Chromium-vanadium	6117	High strength and tough	Crankshafts, axles and springs
	6150	Hard and wear resistant	Bearings, wrenches and tools
Nickel-chromium-molybdenum	8615	Easily carburized	Heavy-duty gears
	8720	Very hard	Aircraft structural parts
Manganese-silicon	9255	Very high strength, good ductility	Coil and leaf springs
	9262	Tough	Chisels and punches
Boron	50B44	Increases hardness of low alloys	Substitute alloy for critical alloy agent shortages
	94B17	Same as above	Same as above

Aluminum

Because of aluminum's light weight (about one-third that of steel), it is an increasingly important metal. It also has an excellent corrosion-resistance quality, a high strength-to-weight ratio, and is easy to fabricate and machine.

Aluminum alloys are classified into two types: casting and wrought. *Casting alloys* are assigned a number such as 43, 112, 357, and so on. This identifying number reveals nothing about the composition of the alloy. *Wrought alloys*, however, are identified by a four-digit numbering system that is very similar to the steel classification system. See Figure 13-2.

As shown in Table 13-7, the first digit, other than 1, indicates the main alloying element. The last two digits are the alloy number. The second digit indicates a modification in the production of this alloy. A letter preceded by a dash after the

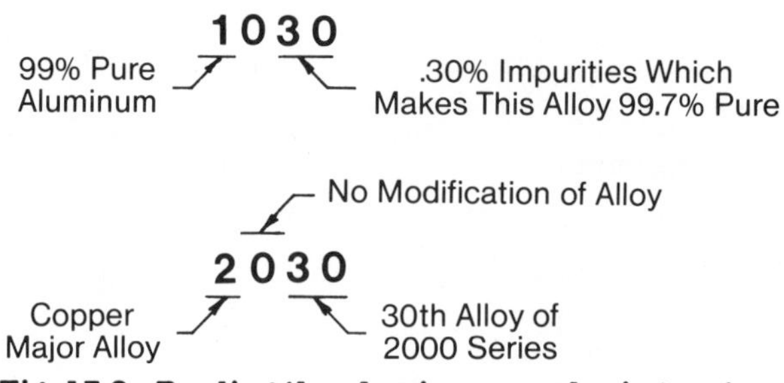

Fig. 13-2 Reading the aluminum numbering system

Table 13–4 Tool Steel Classification Code*

1. High speed:
   - Molybdenum .... M
   - Tungsten .... T
2. Intermediate high speed
   - Molybdenum M50-M59
3. Hot work:
   - Chromium H1-H19
   - Tungsten H20-H39
   - Molybdenum H40-H59
4. Cold work:
   - High carbon, high chromium .... D
   - Medium alloy air hardening .... A
   - Oil hardening .... O
5. Shock resisting .... S
6. Mould steels .... P
7. Special purpose L
8. Water hardening .... W

*By Permission From AISI–"Steel Products Manual-Tool Steels", September 1981

number system indicates a temper or heat treatment.

Most of the remaining non-ferrous metals are used less frequently in the machine trades. No standard numbering system for alloy classification is provided for these materials.

Since more than 2000 metal alloys are now in use, a numbering system was adopted in 1974 by the SAE and the American Society for Testing and Materials (ASTM). This system, the *Unified Numbering System for Metals and Alloys*, is just now being adopted by some industries. The system should begin to appear on industrial prints in the near future. The Unified Numbering System uses a letter to precede the number. In most cases the letter is suggestive of the family of metals identified. A brief description of the system is provided in Table 13-8.

Table 13–5 Tool Steel Composition and Usage*

AISI Type	Composition, Percent								Typical Applications
	C	Mn	Si	W	Mo	Cr	V	Co	
	High Speed, Molybdenum, Types								
M1	.85	—	—	1.50	8.50	4.00	1.00	—	General purpose, low cost drills, taps, lathe tools; popular grade
M2	.85	—	—	6.00	5.00	4.00	2.00	—	Punches, dies; most popular grade
M3-1	1.05	—	—	6.00	5.00	4.00	2.40	—	Universal cutting tool material
M3-2	1.20	—	—	6.00	5.00	4.00	3.00	—	Increased wear resistance for cutting tools
M4	1.30	—	—	5.50	4.50	4.00	4.00	—	Heavy duty cutting tools
M6	.80	—	—	4.00	5.00	4.00	1.50	12.00	
M7	1.00	—	—	1.75	8.75	4.00	2.00	—	Small cutting tools, woodworking tools
M10	.90	—	—	—	8.00	4.00	2.00	—	Similar to M7, popular grade
M15	1.50	—	—	6.50	3.5	4.00	5.00	5.00	
M30	.80	—	—	2.00	8.00	4.00	1.25	5.00	Increased red hardness in cutting tools
M33	.90	—	—	1.50	9.50	4.00	1.15	8.00	Heavy duty high speed steel for machining materials of high hardness

(Continued)

Table 13–5 Tool Steel Composition and Usage (Continued)*

AISI Type	C	Mn	Si	W	Mo	Cr	V	Co	Typical Applications
	Composition, Percent								
	High Speed, Molybdenum, Types								
M34	.90	—	—	2.00	8.00	4.00	2.00	8.00	Similar to M33
M35	.80	—	—	6.00	5.00	4.00	2.00	5.00	
M36	.85	—	—	6.00	5.00	4.00	2.00	8.00	
M45	1.25	—	—	8.00	5.00	4.25	1.60	5.50	
M46	1.25	—	—	2.00	8.25	4.00	3.20	8.25	
	High Speed, Tungsten, Types								
T1	.75	—	—	18.00	—	4.00	1.00	—	General-purpose tungsten high-speed steel for drills, taps, lathe tools
T2	.80	—	—	18.00	—	4.00	2.00	—	Light cuts at high speeds
T4	.75	—	—	18.00	—	4.00	1.00	5.00	For cutting hard, gritty, or tough materials
T5	.80	—	—	18.00	—	4.00	2.00	8.00	High red hardness, heavy-duty cutting
T6	.80	—	—	20.00	—	4.50	1.50	12.00	Very high red hardness
T7	.75	—	—	14.00	—	4.00	2.00	—	Roughing cuts, somewhat erratic in hardening
T8	.75	—	—	14.00	—	4.00	2.00	5.00	
T9	1.20	—	—	18.00	—	4.00	4.00	—	
T15	1.50	—	—	12.00	—	4.00	5.00	5.00	Heavy-duty cutting in abrasive materials

*By permission from AISI "Steel Products Manual — Tool Steels," September, 1981

Table 13–6 Tool Steel Usage*

Application	Suggested Type	Application	Suggested Type
Arbors	L6, L2	Die casting	H13, H22
Automatic screw machine tools	M2, M3-1	Embossing	A2, A11, D2, M48, M61, M62
Beading tools	S1, S5		
Bits, router	M2, M3		
Boring bars	L2, M2	Extrusion, cold	M3, D2
Burnishing tools	W1, D3	Forging, hot	H12, H21
Bushings, ground	W1	Powder metallurgy	A7, D2
Buttons, locating	W1	Plastic mold, machined	P20, H11
Cams	D2, A2	Press brake	L1, L2
Centers, lathe, dead	M2	Stamping	A2, D2
Centers, lathe, live	W1	Wire drawing	D3, M3
Chisels, hand	W1, S5	Drills	M2, W1
Collets, spring	W1, O2, L6	Dummy blocks	M21, H10

(*Continued*)

Table 13–6 Tool Steel Usage (Continued)*

Application	Suggested Type	Application	Suggested Type
Cutting tools:		Finishing tools	F2, M2
Boring	M2, M3	Fingers, feed, cutoff	L6, S2
Broaches	M2, M3	Fixtures	W1, D2
Burrs, metal cutting	M2, M3	Gages, plug, ring, snap, thread	W1, A2, L1
Counterbores	M2, M3	Hammers, ball peen	S2
Cutoff	M3, M33	Hibs (hobs)	A2, D2
End mills	M2, M3	Jaws, chuck	W2, S5
Form tools	M2, M3	Knurling tools	D2, A2
Lathe tools	M2, M3	Knives, wood working	W1, M2
Milling	M2, M3	Liners, mold	A7, D2
Reamers	M2, M3	Pilots, counterbore	F2, W1
Dies:		Pneumatic tools	S1, S2
Bending	A2, D2	Punches	S5, S2
Blanking, cold	A2, D2	Scrapers	F2, W1
Coining	A2, D2	Shear blades, cold	S1, S5
Cold heading	W1, W2	Spindles	L2, L6
Deep drawing	A7, D2	Spinning	M2, M3
		Wear plates	A7, D2

*By permission from AISI–"Steel Products Manual-Tool Steels," September, 1981

NON-METALLIC MATERIALS

The chemical industry has developed plastics with properties that rival the common manufacturing materials of yesterday. Plastics, being synthetic, were first considered to be cheap substitutes for more expensive materials. However, they have earned the right to be classified as separate materials, not merely substitutes.

It would be difficult to pass a day without being exposed to some form of plastic. Many of the clothes we wear, rugs, automobile parts, and even some drafting papers and instruments are some form of plastic. A tremendous number of plastics is now available.

Most plastics are known by their chemical composition or trade name. Plastics are classified into two types: *thermosetting* and *thermoplastic.*

Thermosetting materials will soften only once when exposed to heat and pressure. This heat and pressure is applied during the part-forming operation. No subsequent application of heat or pressure will soften thermosetting materials. The scrap material of this type of plastic cannot be reused. Filler materials of various types are sometimes added to thermosetting plastic to gain desired properties.

Table 13–7 Aluminum Numbering System

Types of Aluminum	Alloying Elements
1XXX	Aluminum (99.00% or greater)
2XXX	Copper
3XXX	Manganese
4XXX	Silicon
5XXX	Magnesium
6XXX	Magnesium and silicon
7XXX	Zinc
8XXX	Other elements

NOTE: A letter following the alloy type, preceded by a hyphen, indicates the basic temper treatment employed in hardening, strengthening, or softening the alloy. Example: O annealed, H strain hardening, etc.

Table 13–8 Unified Numbering System

UNS Series	Metal
Non-Ferrous Metals and Alloys	
A00001-A99999	Aluminum and aluminum alloys
C00001-C99999	Copper and copper alloys
E00001-E99999	Rare earth and rare earth-like metals and alloys
L00001-L99999	Low melting metals and alloys
M00001-M99999	Miscellaneous non-ferrous metals and alloys
N00001-N99999	Nickel and nickel alloys
P00001-P99999	Precious metals and alloys
R00001-R99999	Reactive and refractory metals and alloys
Ferrous Metals and Alloys	
D00001-D99999	Specified mechanical property steels
F00001-F99999	Cast irons and cast steels
G00001-G99999	AISI and SAE carbon and alloy steels
H00001-H99999	AISI H-steels
K00001-K99999	Miscellaneous steels and ferrous alloys
S00001-S99999	Heat and corrosion resistant (stainless) steels
T00001-T99999	Tool steels

Such filler materials are sawdust, shredded fabric, asbestos, mica, and glass fibers.

Thermoplastic materials will soften repeatedly when heat and pressure are applied. This allows the manufacturer to reuse all scrap material. Thermoplastics cannot be used with high temperatures, as the material would simply melt.

Most plastics are available in different colors. This information will usually be specified on the industrial print along with the plastic type. Some of the common plastics and their properties and uses are listed in Tables 13-9 and 13-10.

Most of the other materials, such as rubber and ceramics, that may be specified on prints are usually noted in reference to trade names or supplier types. This determination is usually made by the company's purchasing department and the part designer or engineer.

MATERIAL TREATMENT

Most materials used in manufacturing industries require some additional

Table 13–9 Plastics — Thermosetting

Type	Properties	Typical Uses and Trade Names
Phenolics A. Phenol formaldehyde	Inexpensive, strong, hard, heat resistant, good electrical and heat insulator, only brown and black colors	Appliance parts, ash trays, camera housings, washing machine agitators, electrical and automotive accessories "Bakelite, Durez, Plenco, Resinox, Fiberite, Acetal, Celcon"
B. Phenol furfural	Similar to above but less costly to fabricate	
Urea formaldehyde	More expensive than phenolis, not as heat resistant as phenols, great range of colors, odorless and tasteless	Buttons, clock faces, lighting and electrical appliances, radio cabinets "Plaskon, Urac, Alcylite, Beetle"
Melamine formaldehyde	Good water, chemical and heat resistance, hard, tough, odorless, shock resistant, great range of colors	Dinnerware, hearing aid and shaver housings, buttons "Catalin, Cymel, Permelite, Melmac"

Table 13-10 Plastics—Thermoplastic

Type	Properties	Typical Uses and Trade Names
Cellulosics A. Cellulose acetate	Tough, resistant to chemicals, heat resistant, odorless, tasteless, unlimited colors	Knobs, combs, toys, electrical components
B. Cellulose acetate butyrate	Very resistant to water, resistant to heat and cold	Tool handles, auto steering wheels, pipe, tubing, outdoor uses
C. Cellulose propinate	Similar to cellulose acetate butyrate	Appliance housings
D. Ethyl cellulose	Tough and resilient at subzero temperatures	Flashlights, electrical components, "Bakelite, Chemaco, Kodapak, Tenite I, Tenite II, Tenite Butyrate, Fortical, Pyralin, Ethocel"
Styrene polystyrene	Inexpensive, light, tasteless, odorless, many colors and crystal clear	Wall tile, toys, kitchen items, food and chemical containers "Styrene Polystyrene, Lustron, Styron, Polyco"
A.B.S. (acrylonitrile butadrine styrene)	Strong, crack resistant, many colors	Portable appliance housings (hairdryers, radio, television) "Cycolac, Kralastic"
Vinyl group	Resistant to abrasives and chemicals	Phonograph records, toys, raincoats, packaging, handles, safety glass intelayment, valve seats "Saran, Vinylite A&X, Dow PVC"
Acrylic	Very clear, transmits light, many colors, transmit light around curves	Optical lenses, auto accessories, outdoor signs, T.V. shields "Acrylite, Lucite, Plexiglass, Butaprene"
Polyamide-(nylon)	Very tough, resist abrasion, stable over a wide range of temperatures	Fabric filament, brush bristles bearings, gears, fishing line "Zytel, Dupont Nylon"
Polyethylene	Strong, odorless, tasteless, can be flexible or rigid	Packaging, wire insulation, pipe, tubing, squeeze bottles "Polythene, Marlex"
Polycarbonate	High impact strength, weather and high temperature resistant	Electrical and electronic components "Lexan, Merlon"
Fluorocarbons	Very high heat resistance and good resistance to chemicals	Frying pan coatings, bearing, slides, valve seats, gaskets "Teflon"
Polypropylene	Resistant to chemicals, lightweight, flexible, relatively cheap	Refrigerator parts, pipe and pipe fittings, replacement for more expensive nylon

treatment after machining. It is easier to machine a part when the material is in its soft state. However, the part may be required to perform in a hardened condition. In this case, the part may be case hardened and heat treated after being machined.

When hardness is an important quality of the finished part, this will be specified on the print, usually in the mate-

Table 13–11 Comparison of Hardness Values

Approximate Hardness Conversion					
Rock-well "C"	Rock-well "B"	Brinell BHN	Rock-well "C"	Rock-well "B"	Brinell BHN
65	—	740*	41	—	381
64	—	723*	40	—	370
63	—	705*	39	—	361
62	—	690*	38	—	352
61	—	670*	37	—	344
60	—	655*	36	109	335
59	—	635*	35	—	325
58	—	615*	34	108	318
57	—	596*	33	—	310
56	—	577*	32	107	300
55	—	560*	31	—	294
54	—	544*	30	106	285
53	—	525*	29	105	279
52	—	504	28	104	270
51	—	490	27	103	265
50	—	479	26	—	258
49	—	467	25	102	253
48	—	453	24	101	247
47	—	445	23	100	243
46	—	434	22	99	237
45	—	421	21	98	231
44	—	410	20	97	226
43	—	400	—	96	219
42	—	390	—	95	212

*Standard steel ball replaced by a tungsten carbide ball

rials column. See Table 13-11. The two most common hardness-measuring scales specified on prints are the *Rockwell* and *Brinell*. The Rockwell is the most frequently used system in industry. The *C* scale is used for hard materials. The *B* scale is used for softer materials. In both the Rockwell and Brinell systems, the higher the number, the harder the material.

The part may need to be treated to resist corrosion, or must pass through other operations to satisfy the ultimate operating conditions of the finished product. The specific material treatment will usually be specified on the print. This information is usually located immediately following the part material note.

Eng. Spec. 415
11% Silicon Die Cast Aluminum Casting Specification

Silicon	10–12%
Copper	.6 MAX
Magnesium	.1 MAX
Manganese	.85 MAX
Zinc	.5 MAX
Iron	.4–.8
Nickel	.5 MAX
Chromium	.3 MAX
Titanium	.3 MAX
Others	.2 MAX

Fig. 13-3A Aluminum alloy specifications

Eng. Spec. 31
Treatment For Stainless Steel

A. *Procedure for scale removal*: Applies to scales formed by forging, annealing, or heat treating stainless steels. Forging or annealing scales are very resistant to acids and require the full treatment described below.

A1. *Mechanical loosening or removal:* (Not to be used if part is to have a final polish)
- A. Wheelabrator
- B. Sand blasting
- C. Shot blasting

A2. *Scale removal:*
- A. *Solution composition:*
  Hydrochloric acid
  —5 parts by volume
  Nitric acid
  —1 part by volume
  Water
  —14 parts by volume
- B. *Operation conditions:*
  Maintain solution temperature 125°–160° F.
  Time approximately 10 minutes
  Note: Very dense, resistant scales may be difficult to remove in the above solutions. If necessary, other pickling solutions containing hydrofluoric acid may be substituted. These are more hazardous to use.

A3. *Rinse:*

Fig. 13-3B Stainless-steel treatment

Many industries have established separate engineering specifications which further describe the material and material treatment. These specifications are referred to by a number on the part print. See Figures 13-3A and 13-3B.

Material terms are included in this chapter to familiarize the print reader with material treatment and specifications.

MATERIAL TERMS

Aging—The process of holding metals or alloys at room temperatures for long periods of time after shaping or heat treatment for the purpose of improving hardness, tensile strength, dimensional stability and other properties.

AISI—Abbreviation for American Iron and Steel Institute.

Alloy—A mixture that has metallic properties and is composed of at least two elements, one of which is a metal.

Alloying Elements—Chemical elements added to an alloy to improve the properties of the finished product.

Annealing—Heating a metal to its critical temperature and then allowing it to cool slowly, thus reducing brittleness and increasing ductility.

Anodize—An electrolytic process for transforming the oxide surface coating on aluminum to one that is very resistant to oxidizing agents, thus increasing the life of the base metal.

Arc Weld—To join two metals together by the electric arc process.

Braze—To solder with hard solder made of brass (an alloy of copper and zinc).

Bright Annealing—Annealing in a controlled atmosphere in order to prevent oxidation on the surface of the metal.

Brinell—A scale for measuring metal hardness, usually used for softer metals.

Carbon Steel—A steel with carbon as its main alloying element.

Carburize—To increase the carbon content on the outer surface of a steel, by heating the steel when in contact with a material which has a high carbon content.

Case Harden—A heating and rapid cooling process to harden the surface layer; or in case of a ferrous alloy, making that surface harder than the core of the material.

Chill—The rapid cooling of cast iron to improve its surface hardness.

Corrosion—The deterioration of a metal by exposure to chemicals, liquids (salt water), or atmosphere.

Cyaniding—Case hardening of low-carbon steel by heating in contact with a molten cyanide salt, followed by a rapid cooling.

Deburring—Any of the various operations for removing flash, burrs, sharp edges, and scratches from the finished parts.

Degreasing—A cleaning operation for removing oil, dirt, or other contaminates on the surface of metal.

Descaling—The process of removing oxide scale or corrosion from the surface of metal by acid and mechanical abrasion.

Drop Forging—A metal-forming process in which heated metal is compressed between dies by the use of a drop hammer.

Ductility—The ability of a metal to withstand plastic deformation without fracture.

Electroplating—An electrical process for coating metal parts with a noncorrosive metal such as silver, chromium, copper, tin, etc.

Fatigue Failure—Progressive cracking that takes place in metals that are subjected to repeated loads.

Flame Hardening—A heat treatment process in which only portions of the part exposed to the flame are heat treated.

Forge—To shape hot metal by hammering or by a compressive force.

Galvanize—To coat a metal surface with zinc as a protection against corrosion.

Gray Iron—A cast iron with a 2% to 4% carbon content.

Hard Facing—The fusing of a natural hard-metal alloy to a metal part requiring improved surface qualities.

Hardening—The operation of fast cooling a heated metal to improve hardness.

Heat Treatment—Any operation involving the heating and cooling of a metal or alloy.

High-carbon Steel—Steel with a carbon range between .60% and .75% and valued for its hardness.

High-speed Steel—A special alloy steel used for high-speed metal cutting tools.

Induction Hardening—A heat treatment process in which electrical induction coils are used to heat treat controlled portions of a part.

Malleable Iron—A cast iron which is heat treated to improve its toughness.

Nitriding—A case hardening treatment in which nitrogen gas is used instead of carbon. It has certain property advantages over carburizing.

Normalizing—A heat treatment and slow cooling of cast steel to improve its grain structure.

Rockwell—A measurement for metal hardness. The "B" scale is used for softer metal (cast iron) and the "C" scale is used for harder metal (tool steels).

SAE—Abbreviation for Society of Automotive Engineers.

Scale—A dark oxide coating which forms on the surface of hot steel.

Stainless Steel—A name given to a steel alloy that is resistant to corrosion and heat.

Stress Relieving—A heating and cooling process to reduce internal stresses induced by machining, casting, cold working, or welding.

Temper—A heating and cooling process of improving the hardness of a high-carbon alloy steel.

Tensile Strength—The most important factor in determining the strength of a metal. This value is determined by the maximum load on a metal sample before it breaks.

Work Hardening—The increase in hardness and strength due to the cold working of metal.

REVIEW

This review is provided to serve as reinforcement study material. Fill in the appropriate word(s) to complete the sentences below.

1. The three basic types of materials used in industry are ________, ____________________, and ____________________.

2. The materials listed most often on industrial prints are ____________________ and ____________________.

3. Steel has a carbon content under ____________________ percent.

4. Tool steels usually have a carbon content between ____________________ percent and ____________________ percent.

5. The first digit in the steel classification numbering system indicates the ____________________ of steel.

6. The second digit in the steel classification number system indicates the ____________________ of the major alloy.

7. The most important types of cast iron are ____________________, ____________________, and ____________________.

8. The two main types of aluminum alloys are ____________________ and ____________________.

9. The main types of plastics are ____________________ and ____________________.

10. The most frequently used hardness-measuring system is the ____________________ system.

NAME ____________________

DATE ____________________

SCORE ____________________

EXERCISE B13-1

Print #8-2025

Answer the following questions after studying Print #8-2025 found after this exercise.

QUESTIONS	ANSWERS
1. What is this part?	1. ______
2. What was it machined from?	2. ______
3. What is the tolerance on a three-place dimension?	3. ______
4. What is the chamfer angle around the .859 hole?	4. ______
5. What is the drill size of the four tapped holes?	5. ______
6. What is the finish on the back surface?	6. ______
7. What was the .010 dimension prior to 6-23-67?	7. ______
8. What is the depth of the .859 hole?	8. ______
9. On what date was this part released for production?	9. ______
10. What are the maximum and the minimum allowed sizes of the .859 holes?	10. Maximum ______ Minimum ______
11. What are the maximum overall height, width, and thickness dimensions?	11. Height ______ Width ______ Thickness ______
12. What are the limits on the removal of burrs and sharp edges?	12. ______

NAME ______

DATE ______

SCORE ______

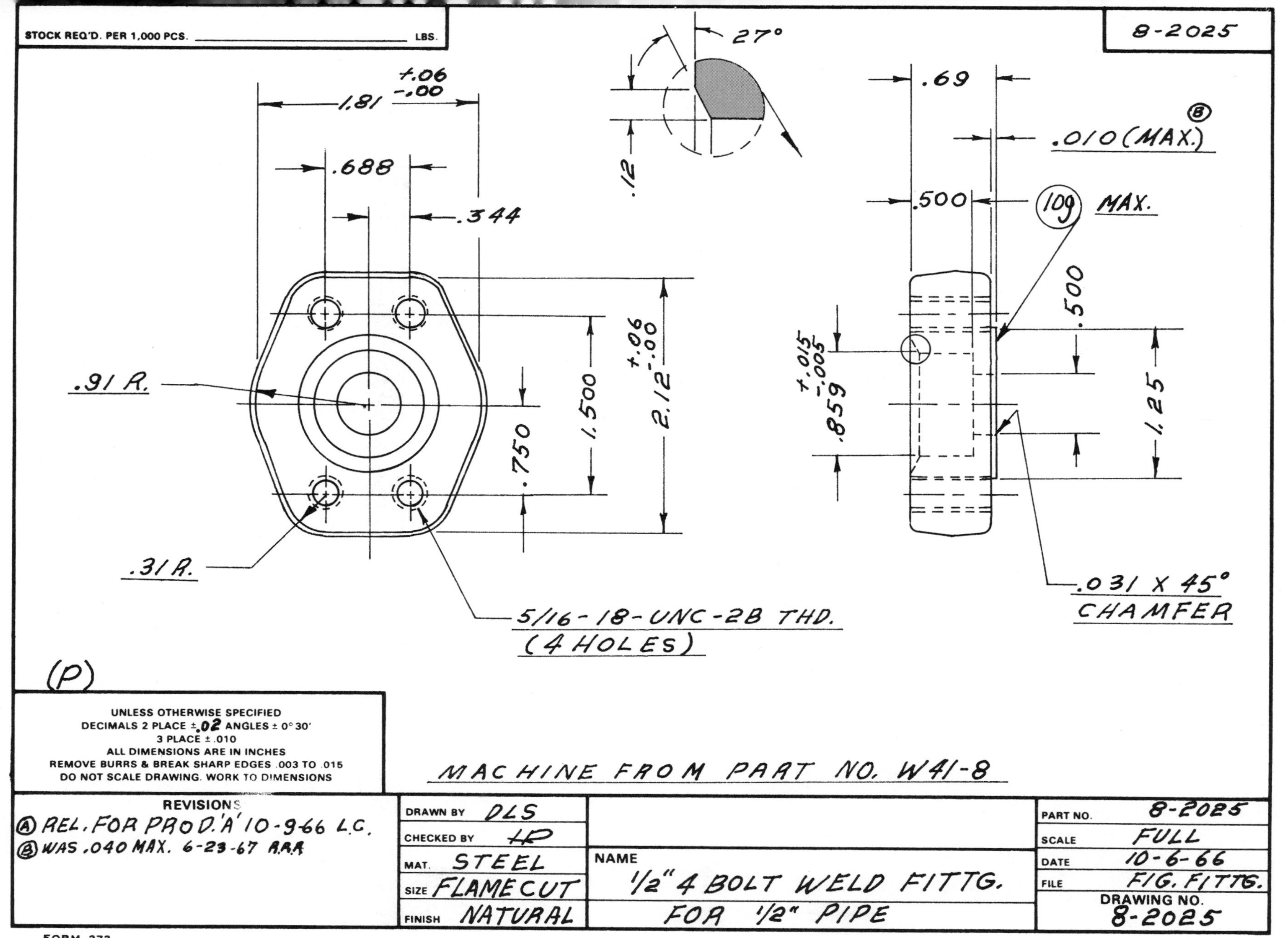

STOCK REQ'D. PER 1,000 PCS. ____ LBS.
8-2025
27°
.12
1.81 +.06 -.00
.688
.344
.91 R.
.31 R.
.750
1.500
2.12 +.06 -.00
5/16-18-UNC-2B THD. (4 HOLES)
.69
.010 (MAX.) (B)
.500
(109) MAX.
.500
.859 +.015 -.005
1.25
.031 X 45° CHAMFER
(P)
UNLESS OTHERWISE SPECIFIED
DECIMALS 2 PLACE ±.02 ANGLES ± 0° 30'
3 PLACE ±.010
ALL DIMENSIONS ARE IN INCHES
REMOVE BURRS & BREAK SHARP EDGES .003 TO .015
DO NOT SCALE DRAWING. WORK TO DIMENSIONS
MACHINE FROM PART NO. W41-8
REVISIONS
(A) REL. FOR PROD.'A' 10-9-66 L.C.
(B) WAS .040 MAX. 6-23-67 R.R.R.
DRAWN BY DLS
CHECKED BY HP
MAT. STEEL
SIZE FLAMECUT
FINISH NATURAL
NAME
1/2" 4 BOLT WELD FITTG.
FOR 1/2" PIPE
PART NO. 8-2025
SCALE FULL
DATE 10-6-66
FILE FIG. FITTG.
DRAWING NO. 8-2025
FORM 273

CHAPTER 14

TYPES OF DRAWINGS

OBJECTIVES

After studying this chapter, you will be able to:

- Explain why some older industrial prints may be in use for many more years.
- Recognize that the title block contains most of the print's basic information.
- Discuss the new trend of providing separate prints for each manufacturing department.
- Describe the combined foundry and machining prints.
- Describe how to locate drawings for individual parts of an assembly.
- Discuss the different kinds of hydraulic prints.
- Describe air logic prints.
- Describe basic electrical control prints.
- Identify the basic welding symbols.

A strong effort has been made in recent years to standardize industrial prints. However, we still must work with older, unstandardized prints since it would not be economical to redraw all prints. Also, many industries have not yet changed their drafting methods to conform to the standards.

Many different types of industrial prints are shown in this chapter. The title block is the part of the print to read first. Usually all of the basic information is included in or near the title block. If required information cannot be found easily and quickly, ask an instructor or supervisor for assistance.

PART PRINTS

A new trend in the machine trades is to make more than one drawing of the same part. This allows a company to simplify the reading of prints in the various departments. Each manufacturing department which will work on the part will receive prints of its own specialized drawing. For example, a print may be provided to the foundry department where the part casting will be made. That department may need the dimensions necessary for the rough casting, but not the machining tolerance dimensions. Therefore, the print provided to the foundry department will not include the machining tolerance dimensions. Another drawing may be made for the machining departments, which require the more critical and closely toleranced machining dimensions.

This system makes for a less confusing print. Only the drawings and dimensions needed by a particular department will be on the prints used by that department. When such a print is used, a notation in the title block will indicate the print as being a foundry, tooling, machining, inspection, or assembly print.

PRINT #1116—MINIATURE WRENCH FORGING PRINT

Study Figure 14-1, print #1116. This print would be used by the workers who

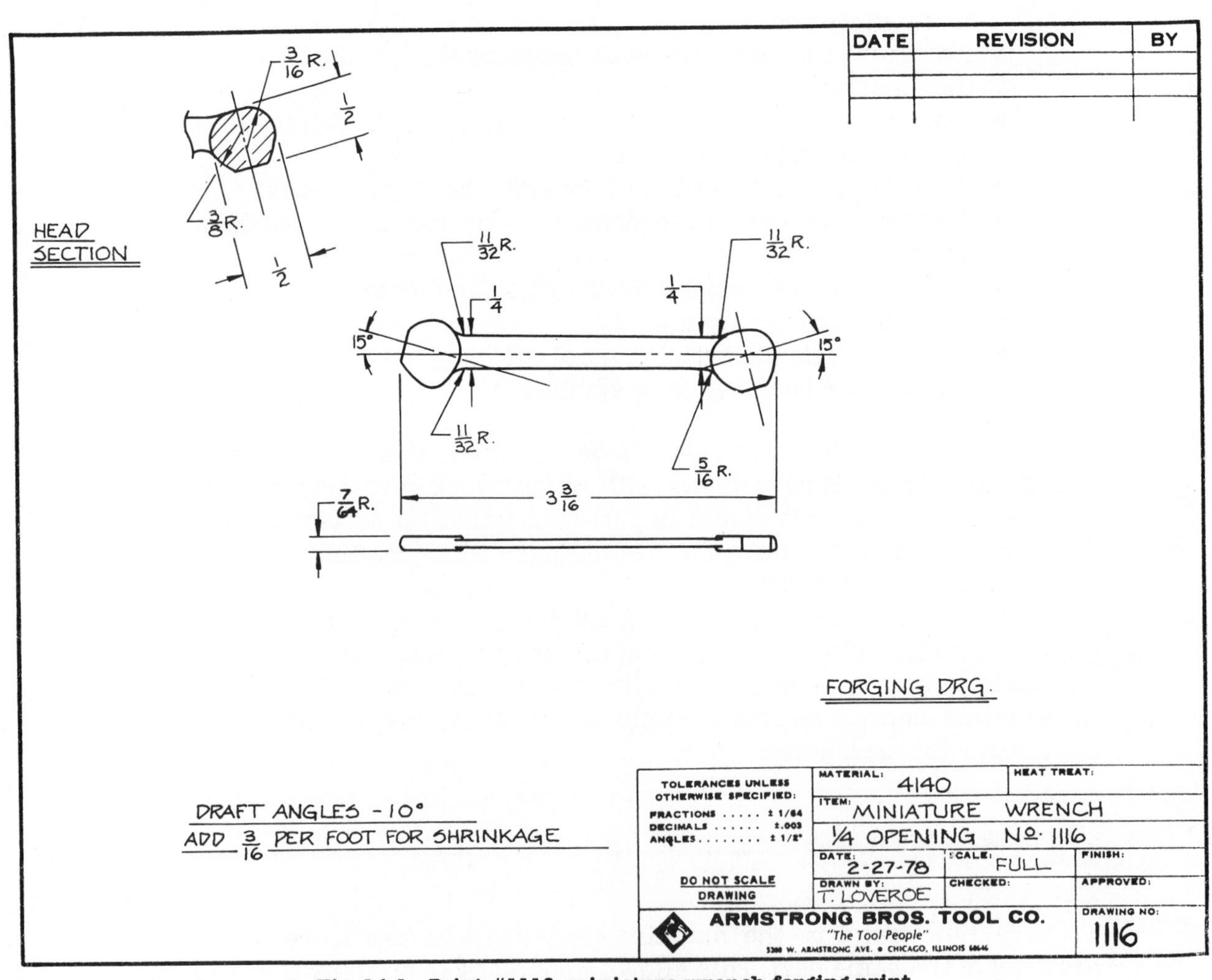

Fig. 14-1 Print #1116, miniature wrench forging print

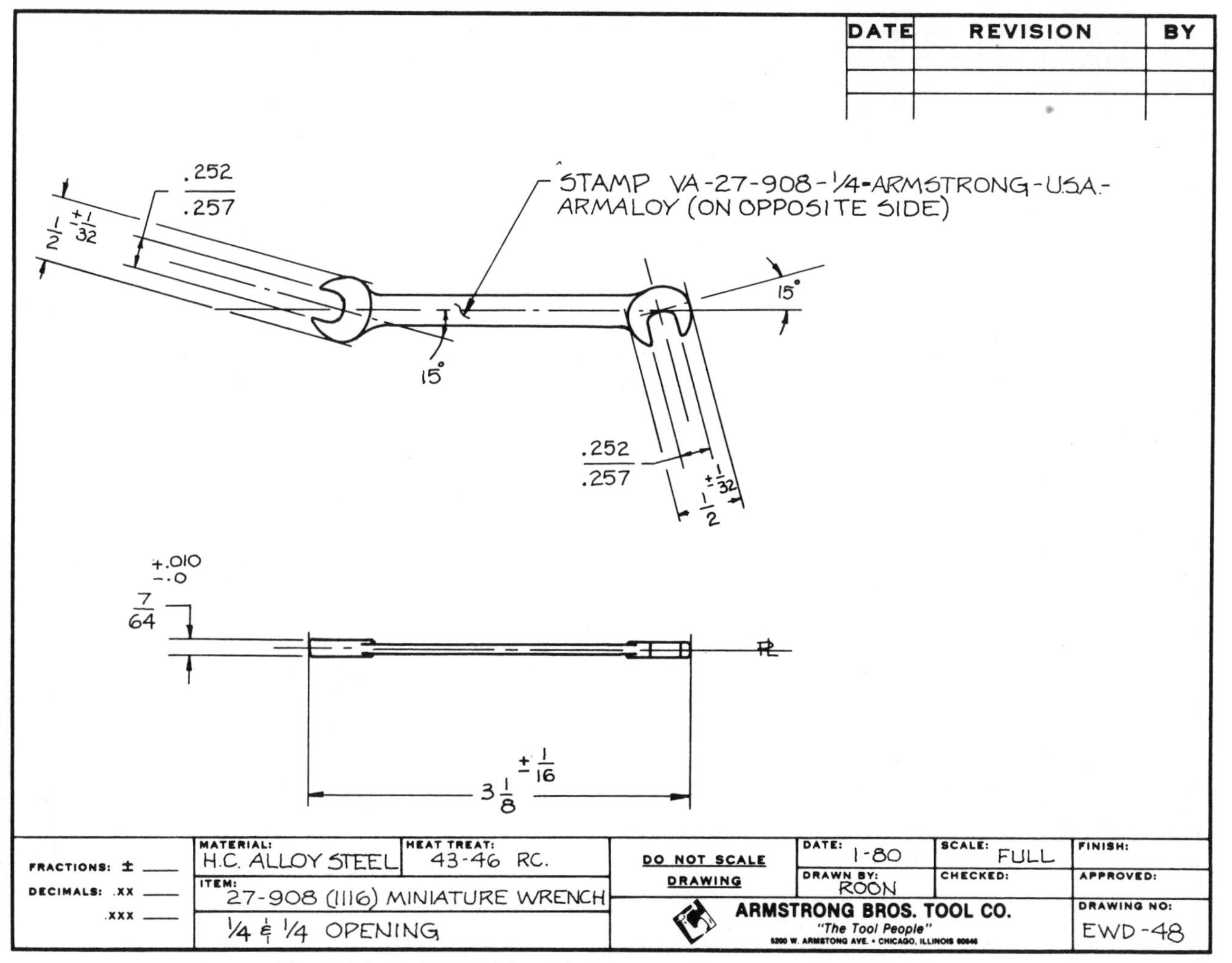

Fig. 14-2 Print #EWD-48, miniature wrench machining print

make the forging dies, and by the employees in the department producing the part.

The draft angle, which is stated as 10°, is the amount allowed for easy removal of the forging from the dies. This part is formed when the metal is heated to about 2200° Fahrenheit (F). Therefore, the die cavity dimensions must be increased to the specified shrinkage allowance. The shrinkage allowance is stated as 3/16 inch per foot which guarantees that all dimensions of the part will be as specified when the part has cooled to room temperature.

The forging alloy is 4140, a chromium-molybdenum alloy with a .40% carbon content. This particular forging is used for the 1/4-inch miniature wrench or for wrenches close to that size.

PRINT #EWD-48 — MINIATURE WRENCH MACHINING PRINT

Study Figure 14-2, print #EWD-48. The item column indicates #27-908 as the part number; #1116 is the number that is listed in the sales catalog. The drawing number (#EWD-48) is a code number used in the company's print filing system.

The term *Armaloy*, as stamped on the finished product, is a company trade name which indicates an alloy steel. This term should not be confused with the surface treatment trade name as illustrated on print #982102 (Figure 14-16) later in this chapter.

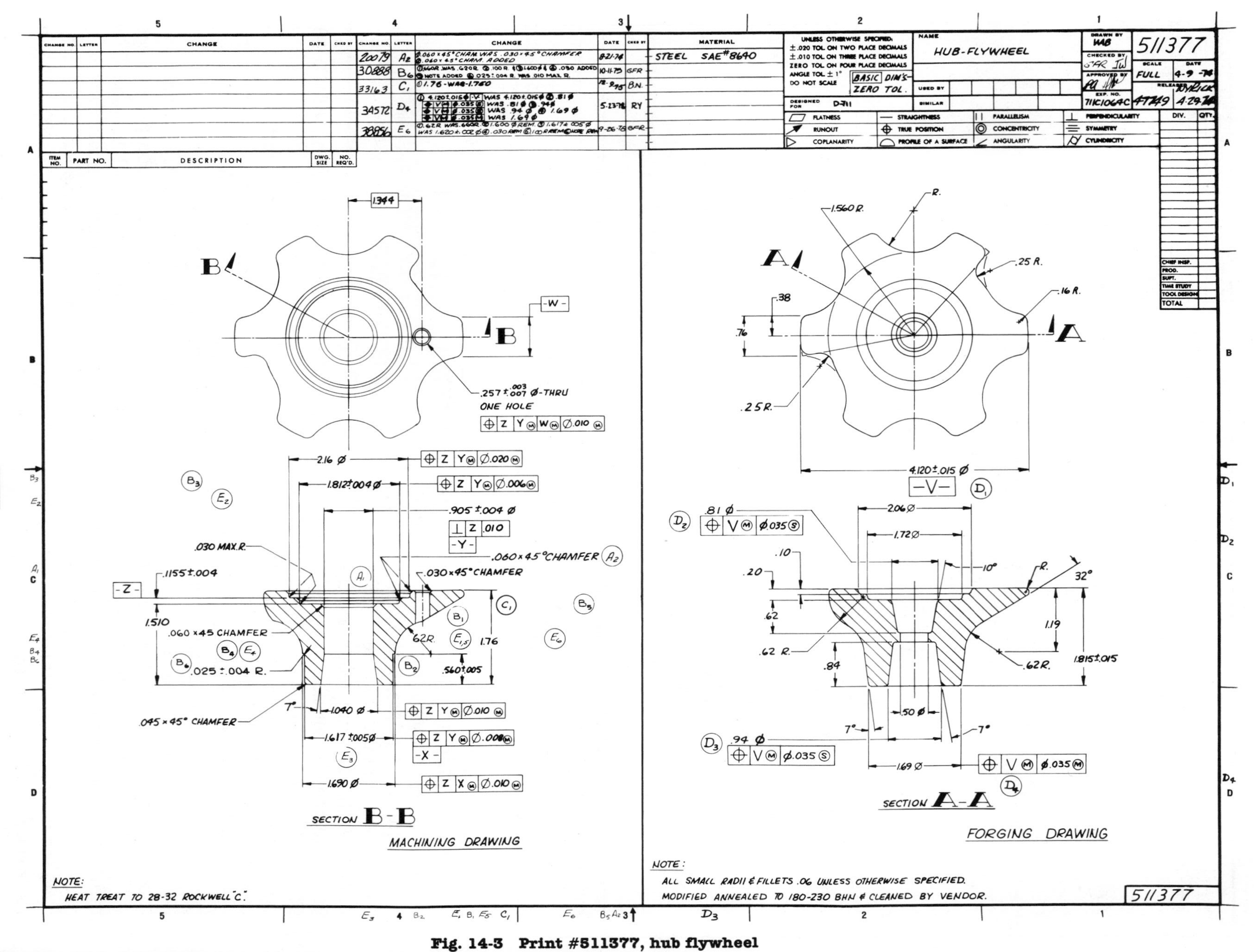

Fig. 14-3 Print #511377, hub flywheel

Print #EWD-48 is the machining or finish print for the part drawn in the forging print #1116 (Figure 14-1).

The dimensional tolerances are not stated in the tolerance block, because all dimensions on the print include tolerance.

PRINT #511377—HUB FLYWHEEL

Some companies prefer to place the casting or forging drawing and the machining drawing on the same print, as illustrated in Figure 14-3, print #511377. The personnel in the forge shop receive the machining dimensions as well as the dimensions necessary to produce the forging. Therefore, the forge shop workers can determine forging irregularities that may cause machining problems.

This arrangement also allows the forge shop personnel to keep abreast of any machining changes, since both drawings are on the same print.

PRINT #981859—SHIFT ASSEMBLY PRINT

Study Figure 14-4, print #981859. The company uses a standard title block which is located at the top of the print rather than at the bottom. The part name, *shift assembly*, indicates that it is an assembly drawing. The word *gearcase* tells the location of the assembly; the term *hydraulic assist* tells its function. The rest of the information in the title block is standard. However, the symbols shown below the title block indicate that this company uses the newer geometric dimensioning system.

The assembly shown in the print is drawn at twice its actual size. However, the assembly is also shown on the print at its actual size.

The note in the material block states *see details*. The details are described in the bill of materials; the details indicate the material used for each component of this assembly. Where appropriate, the prints for the details are shown in this chapter to illustrate some of the techniques used in industry. The details are shown in the same order as listed in the bill of materials, and a few important facts about each detail are discussed in the accompanying text.

The change block, located to the left of the material block in print #981859 (Figure 14-4), lists the change numbers and the dates the drawing and changes were made. The notation A_6, change #14947, shows that six changes were made on 5/30/78.

The bill of materials is listed to the left of the change block, Figure 14-4. The bill of materials includes 12 items (these are the details discussed above). Detail #9 is left blank; it was probably omitted due to redesign of the original part. The information about each detail is shown on separate prints. In the bill of materials, each detail is identified by a part number, description, drawing size, and how many of each item this assembly requires.

The notation (A_4) is indicated to the right of detail #3. This tells the print reader that four changes will be shown on print #320176. This type of information is also indicated after detail #10. The ($A_{2/3}$) indicates that two changes will be shown on one subassembly print, and three on another subassembly print. (A_1) was added at a later date to one of the subassembly prints.

Important notes that can only be shown on the assembly print are also indicated. The triangular symbols at details #5 and #7 are safety-related symbols. (Safety-related symbols are discussed in more detail in Chapter 16.)

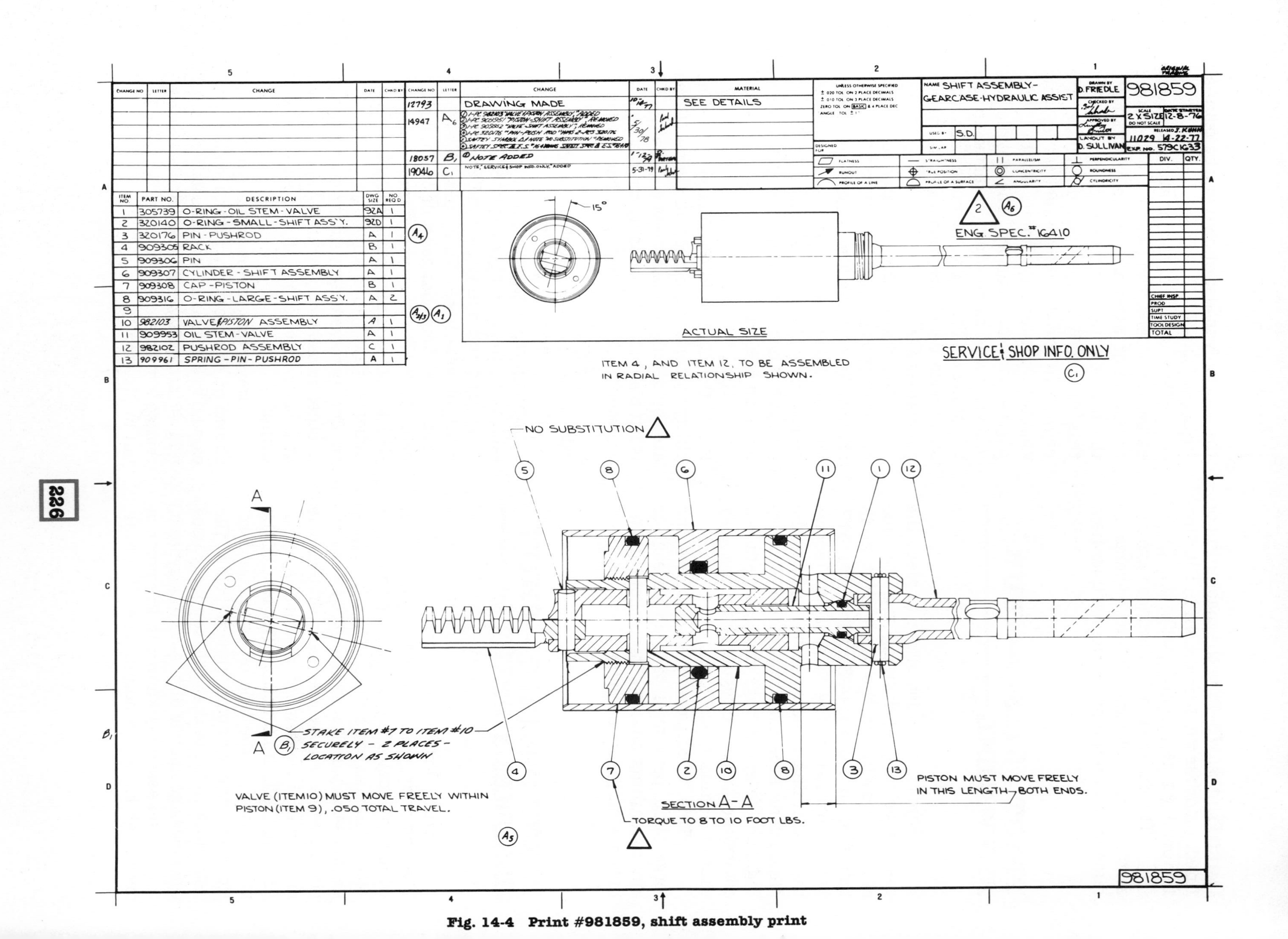

ITEM NO	PART NO.	DESCRIPTION	DWG SIZE	NO REQ'D
1	305739	O-RING-OIL STEM-VALVE	92A	1
2	320140	O-RING-SMALL-SHIFT ASS'Y.	92D	1
3	320176	PIN-PUSHROD	A	1
4	909305	RACK	B	1
5	909306	PIN	A	1
6	909307	CYLINDER-SHIFT ASSEMBLY	A	1
7	909308	CAP-PISTON	B	1
8	909316	O-RING-LARGE-SHIFT ASS'Y.	A	2
9				
10	982103	VALVE & PISTON ASSEMBLY	A	1
11	909953	OIL STEM-VALVE	A	1
12	982102	PUSHROD ASSEMBLY	C	1
13	909961	SPRING-PIN-PUSHROD	A	1

Fig. 14-4 Print #981859, shift assembly print

Detail #1, O-ring (Part #305739)

O-rings are specially purchased hardware items. No prints are made for these items. Information on the O-rings is found on separate hardware lists similar to that shown in Figure 14-5 (hardware list #92A).

O-RINGS
SPECIAL HARDWARE

92 A

(F) FLASH FROM PARTING SEAM NOT TO EXCEED .003 x .005 THICK

A

B

PART NO.	A	B	RUBBER MAT'L TO CONFORM TO ENG. SPEC. #312 AS NOTED (*) OR AS OTHERWISE SPECIFIED.	F	REMARKS
305739	.244 .234	.073 .067	* SB515 Z_3 (V.S. 10) Z_4 Z_{12}	X	
305242	.306 .296	.143 .135	* SB515 Z_3 (V.S. 19) Z_4 Z_{12}	X	
202893	.369 .359	.073 .067	* SB720 B Z_3 (V.S. 19) Z_4 Z_{12}	X	
303347	.429 .419	.106 .100	* SB715 Z_3 (V.S. 19) Z_4 Z_{12}	X	
307450	.431 .421	.073 .067	* SB715 Z_3 (V.S. 19) Z_4 Z_{12}	X	
308797	.868 .856	.106 .100	* SB715 Z_3 (V.S. 10) Z_4 Z_{12}	X	
305123	3.749 3.719	.143 .135	* SB715 Z_3 (V.S. 19) Z_4 Z_{12}	X	
308624	.930 .918	.106 .100	* SB615 Z_3 (V.S. 26) Z_4 Z_{12}	X	
302540	.990 .978	.143 .135	* SB715 Z_3 (V.S. 19) Z_4 Z_{12}	X	
303360	1.052 1.040	.143 .135	* SB515 Z_3 (V.S. 19) Z_4 Z_{12} (H) 40 to 50	X	
313340	1.684 1.664	.106 .100	* SB515 Z_3 (V.S. 19) Z_4 Z_{12}		
202591	1.240 1.228	.143 .135	* SB720 B Z_3 (V.S. 19) Z_4 Z_{12}	X	
308626	1.430 1.418	.106 .100	* SB615 Z_3 (V.S. 26) Z_4 Z_{12}	X	
318372	.306 .296	.179 .169	* SB515 Z_3 (V.S. 19) Z_4 Z_{12}	X	
308458	1.622 1.602	.106 .100	* SB615 Z_3 (V.S. 19) Z_4 Z_{12} (H) 60 to 70	X	
318717	1.485 1.465	.215 .205	* SB715 B Z_3 (V.S. 20) Z_4 Z_{12} Z_{18}	X	
302588	1.744 1.724	.143 .135	* SB515 Z_3 (V.S. 19) Z_4 Z_{12}	X	
305276	2.137 2.117	.143 .135	* SB515 Z_3 (V.S. 10) Z_4 Z_{12}	X	

Fig. 14-5 Hardware list #92A, for O-rings (detail #1)

Detail #2, O-ring (Part #320140)

The description for detail #2 is found on another hardware list, #92D, which is similar to #92A.

Detail #3, Pin-push Rod — Shift Assembly (Part #320176)

Only one pin is required. It is drawn on a size *A* print, Figure 14-6. From this print it can be determined that this pin, similar to pin #317864, was used in a shift assembly produced in 1973.

This part is made of SAE 4037, 50100, or 52100 steel, and is heat treated to 42-50 Rockwell "C" hardness. From the change block it can be seen that the safety-related symbol was added to the print on 7/13/77.

Detail #4, Rack — Valve — Shift Assembly (Part #909305)

Study Print #909305, Figure 14-7. This part is made of SAE 1040-1050 cold drawn steel. According to the change block, the part was redrawn and redesigned on 1/7/76.

A *rack* is defined as a type of gear. This print (Figure 14-7) shows rack data information that is quite similar to the information shown on a gear print. The

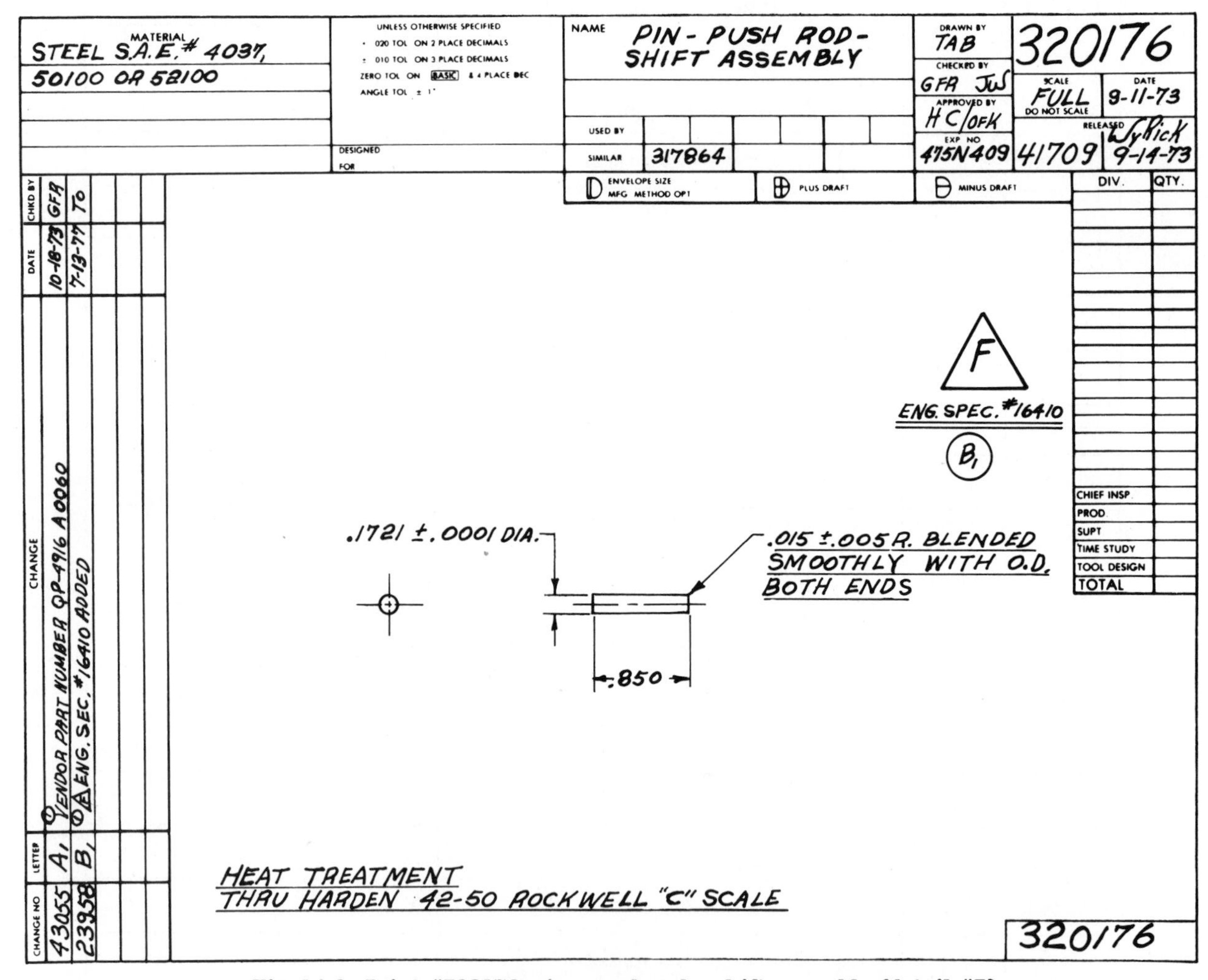

Fig. 14-6 Print #320176, pin — push rod — shift assembly (detail #3)

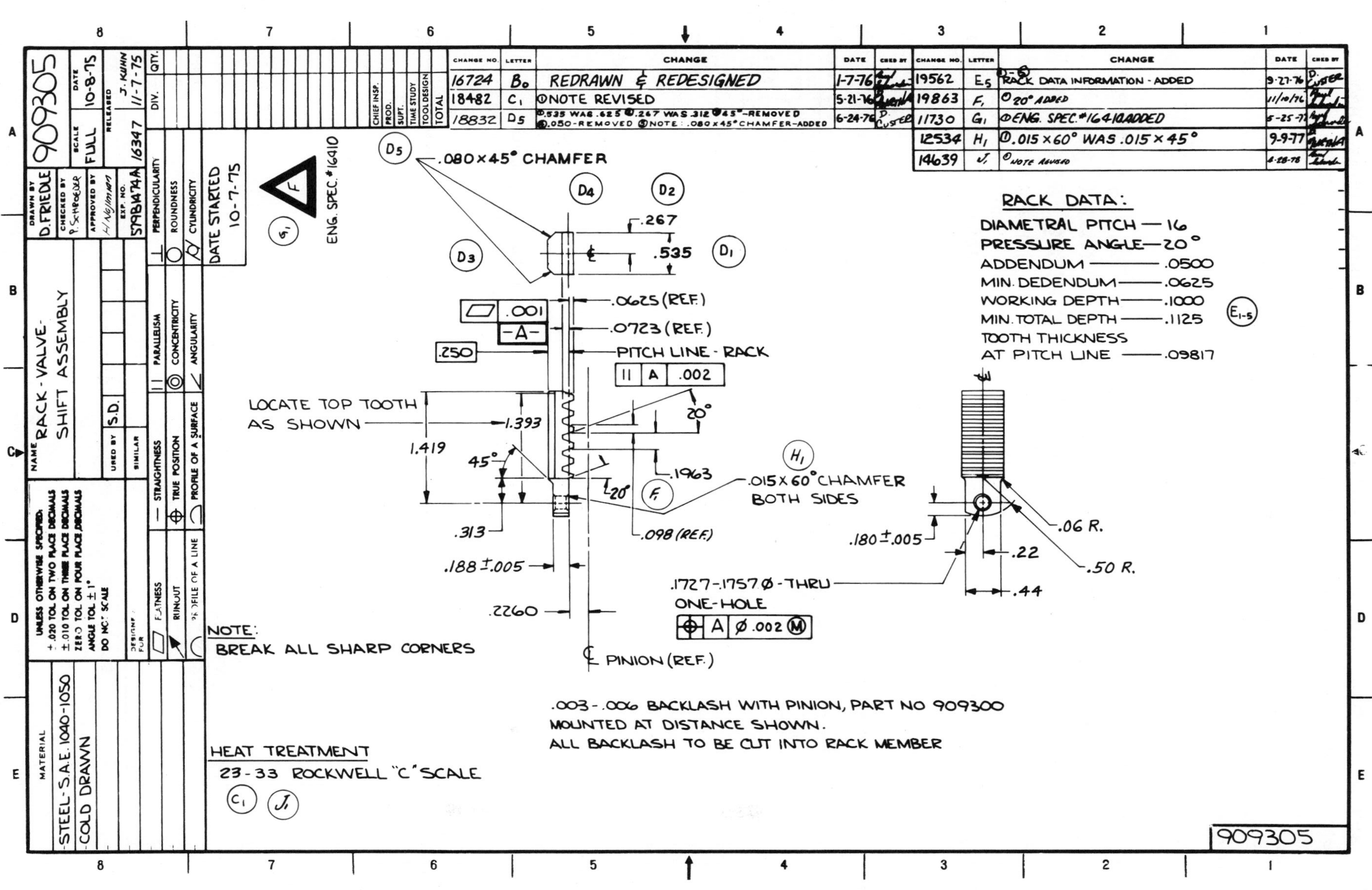

Fig. 14-7 Print #909305, rack—valve—shift assembly (detail #4)

company which made this print changed to the newer geometric dimensioning system in about 1974. Prints drawn by the company after that date would incorporate the geometric dimensioning system, as shown by this print.

purchased part and is heat treated to 42-50 Rockwell "C" scale.

The pin is similar to part #320176 (Figure 14-6), detail #3, in that it has the same diameter. However, this pin is shorter in length.

Detail #5, Pin — Rack to Valve (Part #909306)

This detail, Figure 14-8, is on a size *A* print (#909306). This pin keeps the rack attached to the valve. It is made of steel SAE 4037, 50100, or 52100. It is a purchased part and is heat treated to 42-50 Rockwell "C" scale.

Detail #6, Cylinder — Shift Assembly — Gearcase (Part #909307)

Detail #6, Figure 14-9, is made of 2024 wrought aluminum of a copper alloy family. The notation *T351* on the print in-

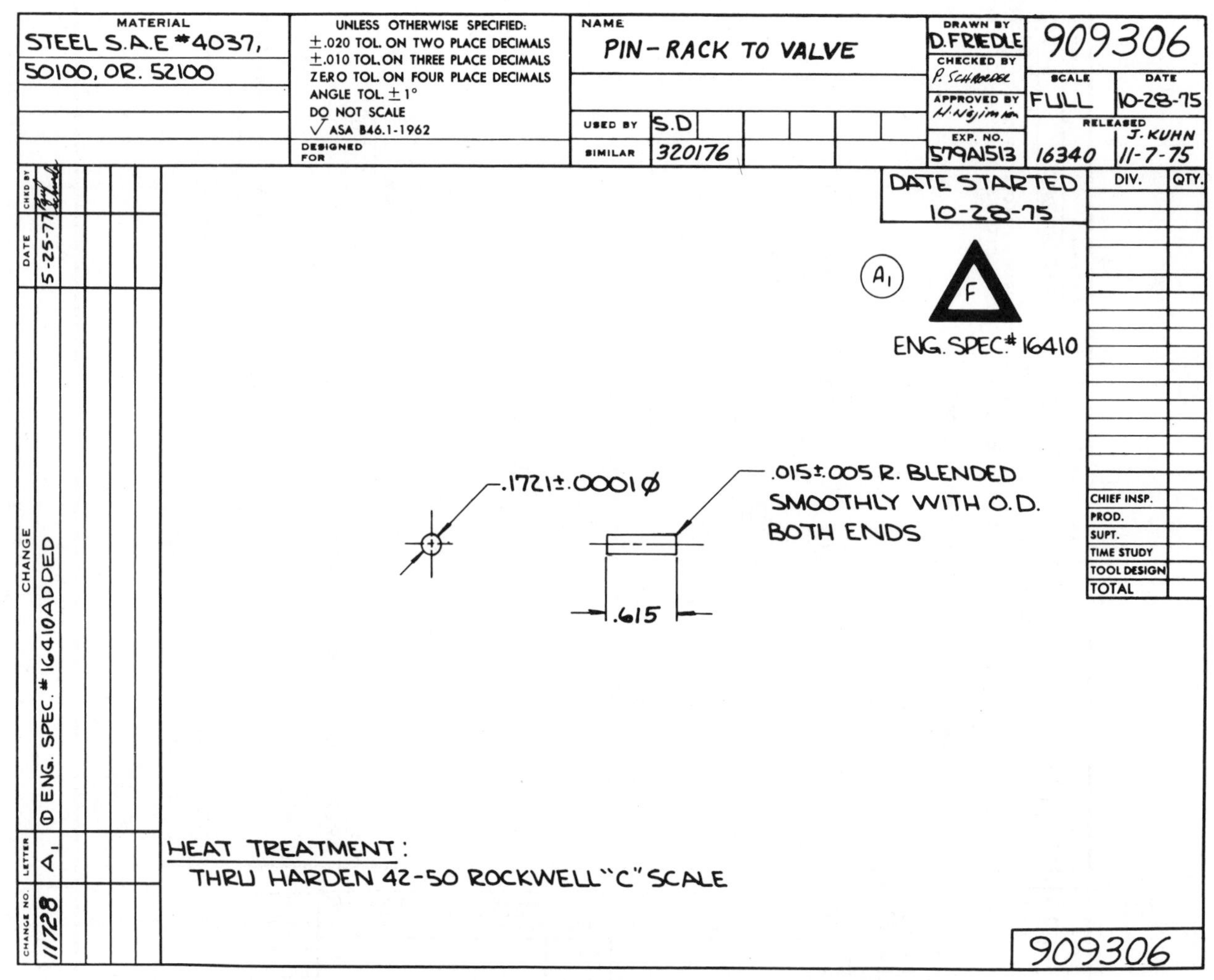

Fig. 14-8 Print #909306, pin — rack to valve (detail #5)

dicates the type of heat treatment required. The engineering specification #8 denotes the company's anodizing treatment for improving the corrosion resistance of unpainted aluminum.

Detail #7, Cap—Piston—Shift Assembly (Part #909308)

As shown in print #909308, Figure 14-10, the part is made of the same material as part #909307 (Figure 14-9). However, detail #7 is given the engineering specification treatment #4, which is the company's special Lyfanite treatment for aluminum. This special chemical surface treatment improves the bonding power of paint. The treatment also moderately improves the corrosion resistance of unpainted aluminum surfaces.

Detail #8, O-ring—Piston—Hydraulic Assist (Part #909316)

This part, Figure 14-11, is a large O-ring of a special size, not listed on hard-

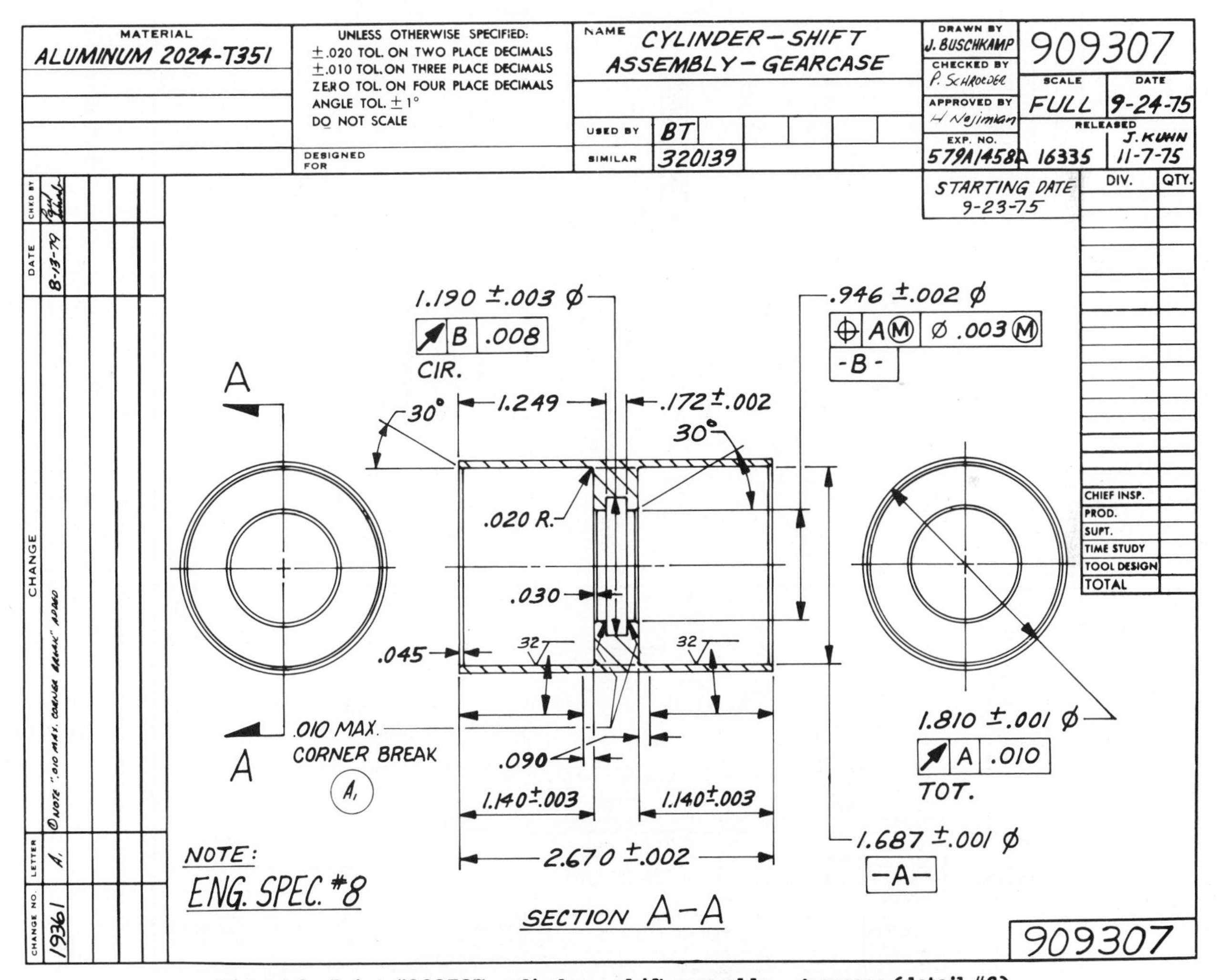

Fig. 14-9 Print #909307, cylinder—shift assembly—gearcase (detail #6)

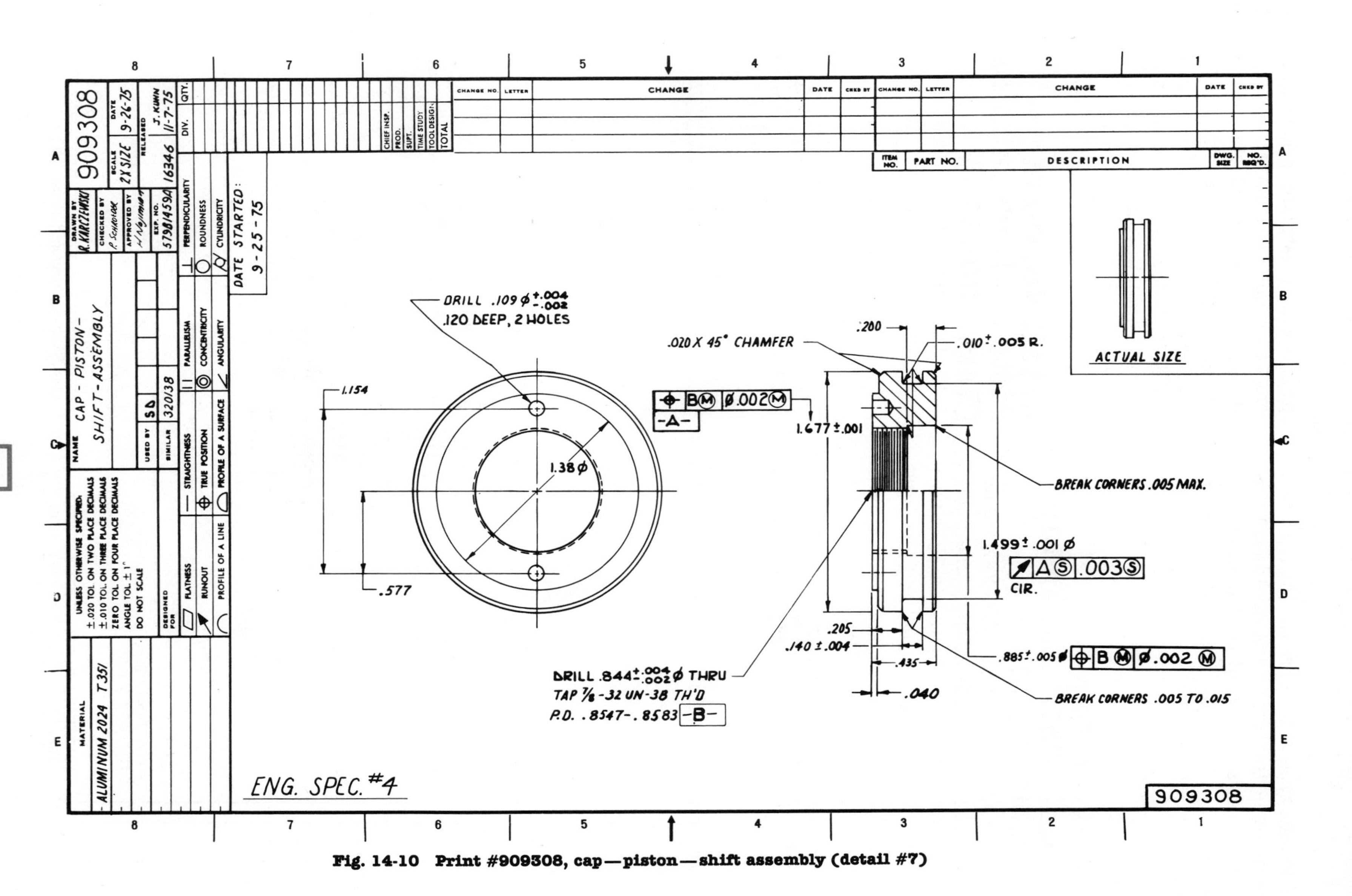

Fig. 14-10 Print #909308, cap—piston—shift assembly (detail #7)

ware list #92A. This special O-ring must be made to satisfy the requirements of print #909316 (Figure 14-11).

Detail #9

Detail #9, as mentioned earlier, is not listed on the material list.

Detail #10, Valve and Piston Assembly (Part #982103)

Study print #982103, Figure 14-12. This shows a subassembly of the original shift assembly. (The original shift assembly is on print #981859, Figure 14-4.) The details on the print in Figure 14-12 are as follows:

- Detail #1 (part and print #320176) on this print is the same as detail #3 on the original assembly print. Therefore, print #320176 is not shown here.

- Detail #2 (part and print #909951, piston—shift assembly—gearcase) is on a size *D* print. See Figure 14-13. It is made of 2024 aluminum, and is similar to part #909309 (not shown).

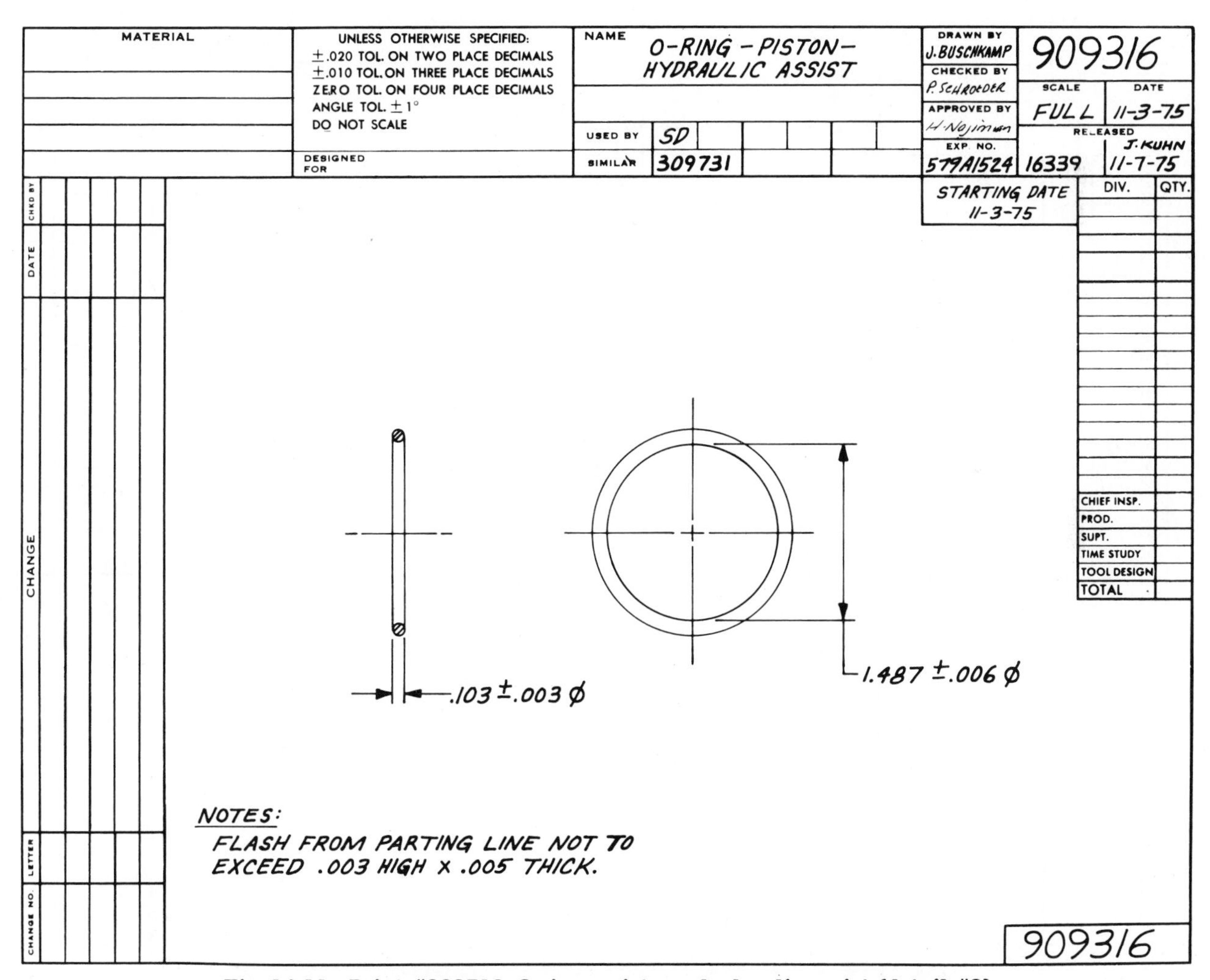

Fig. 14-11 Print #909316, O-ring—piston—hydraulic assist (detail #8)

• Detail #3 (part and print #909952, valve—shift assembly—gearcase) is on a size *C* print. See Figure 14-14.

On Figure 14-14 note that there are several basic dimensions and angles, such as [.500], [.660], [90°], and [15°]. The outside diameter is machined to four decimal places, .6253 ±.0003, and is machined to a smooth 32$\sqrt{\ }$ micro-finish.

The notes at the lower left corner of the print indicate the heat and rust-inhibitor treatments.

Detail #11, Oil Stem—Valve (Part #909953)

From section A-A of assembly print #981859 (Figure 14-4), it can be seen that detail #11 is the innermost component of the assembly. See print #909953, Figure 14-15. The part is made of an aluminum copper alloy. This part is a critical component of the assembly because it has a safety-related notation after a four-place dimension, .3100 ± .0005.

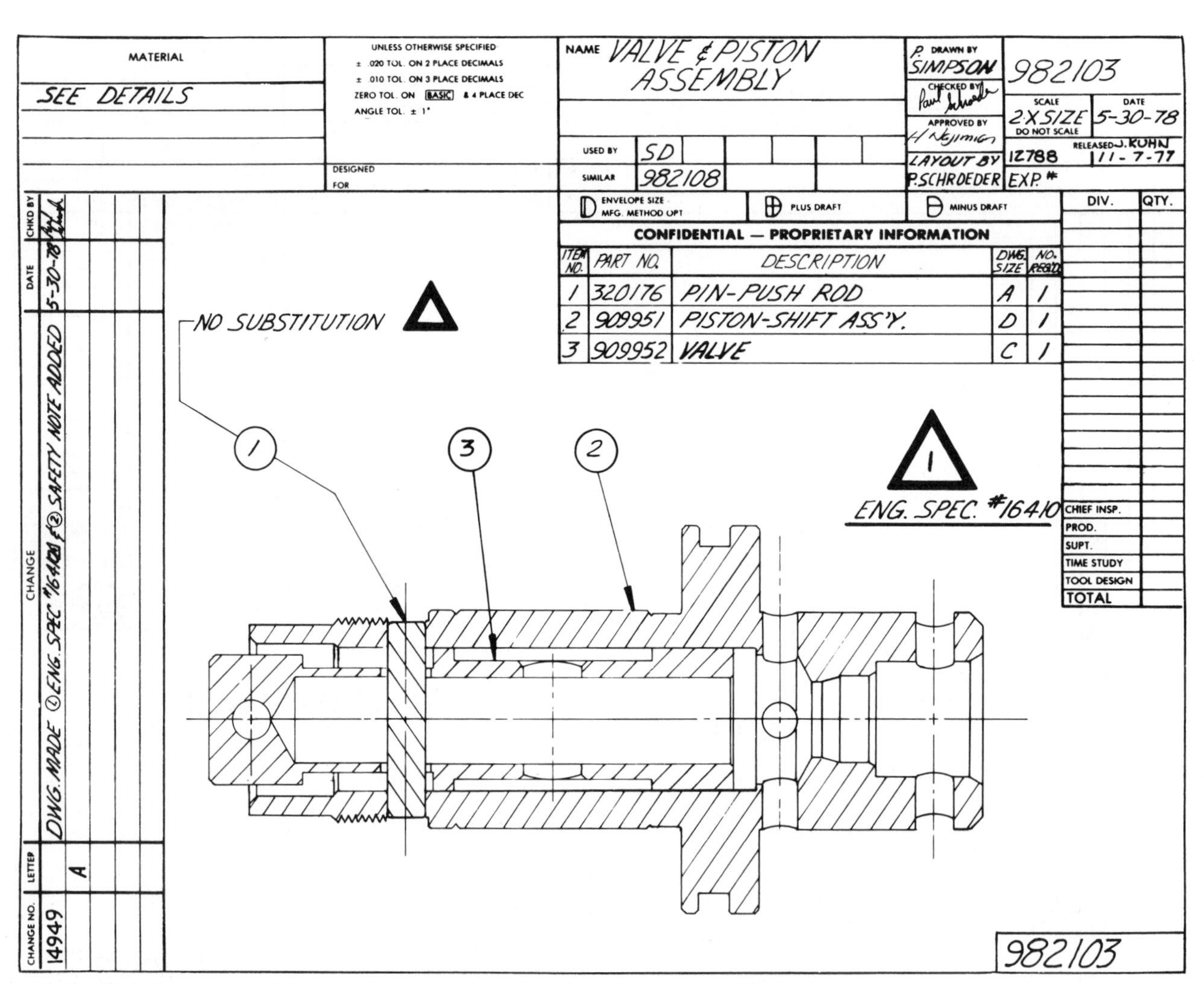

Fig. 14-12 Print #982103, valve and piston assembly (detail #10)

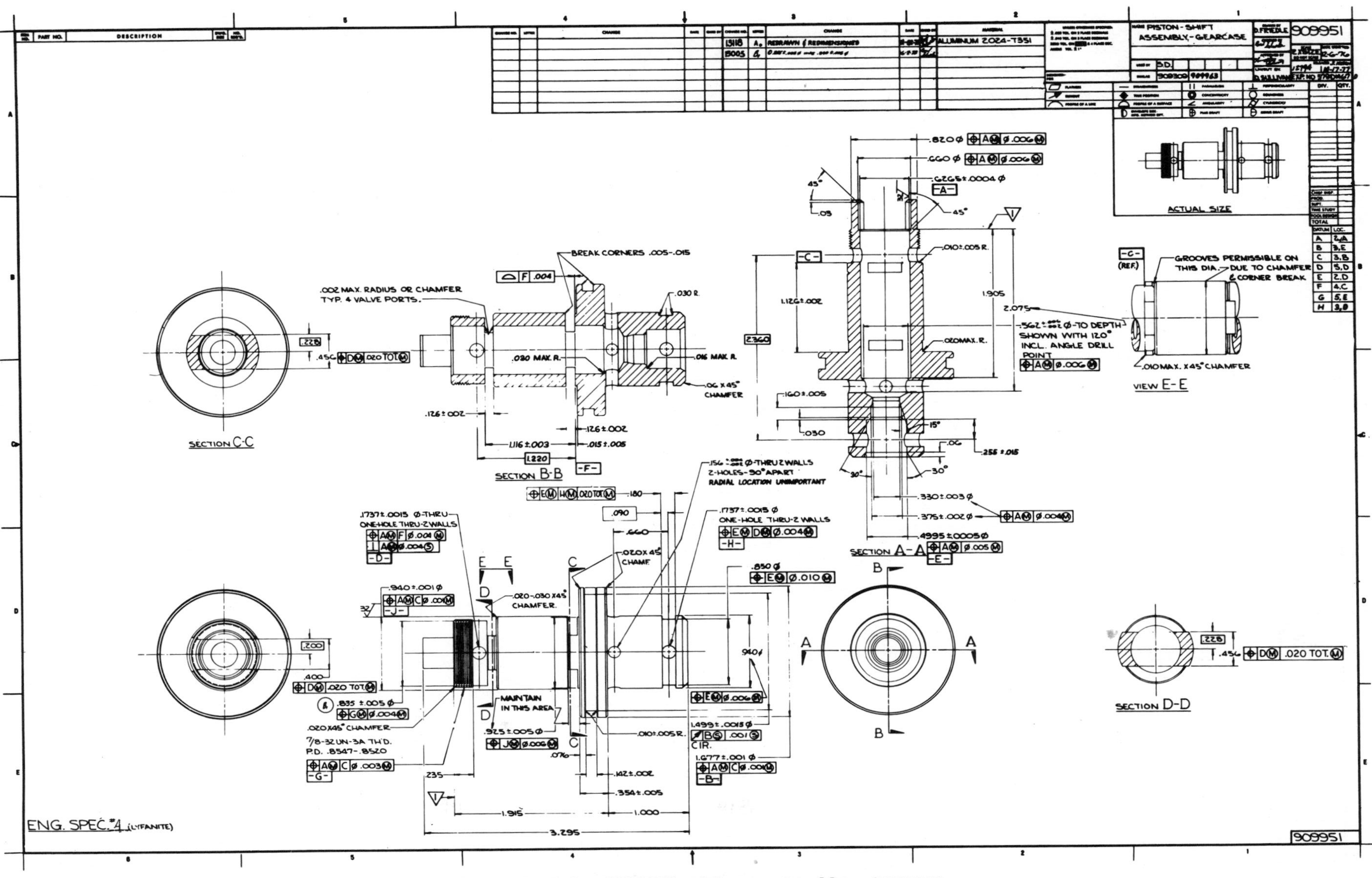

Fig. 14-13 Print #909951, piston—assembly—gearcase

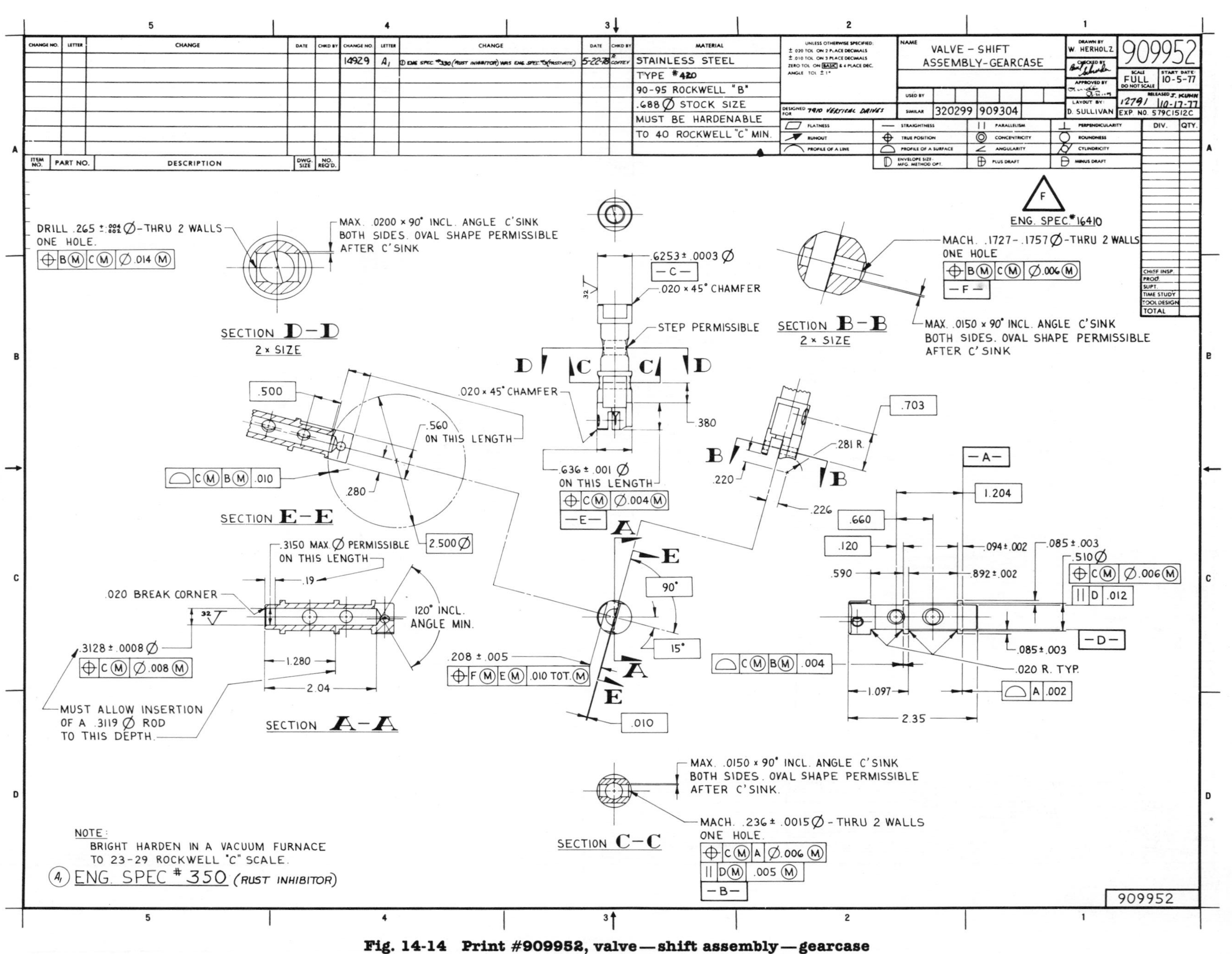

Fig. 14-14 Print #909952, valve—shift assembly—gearcase

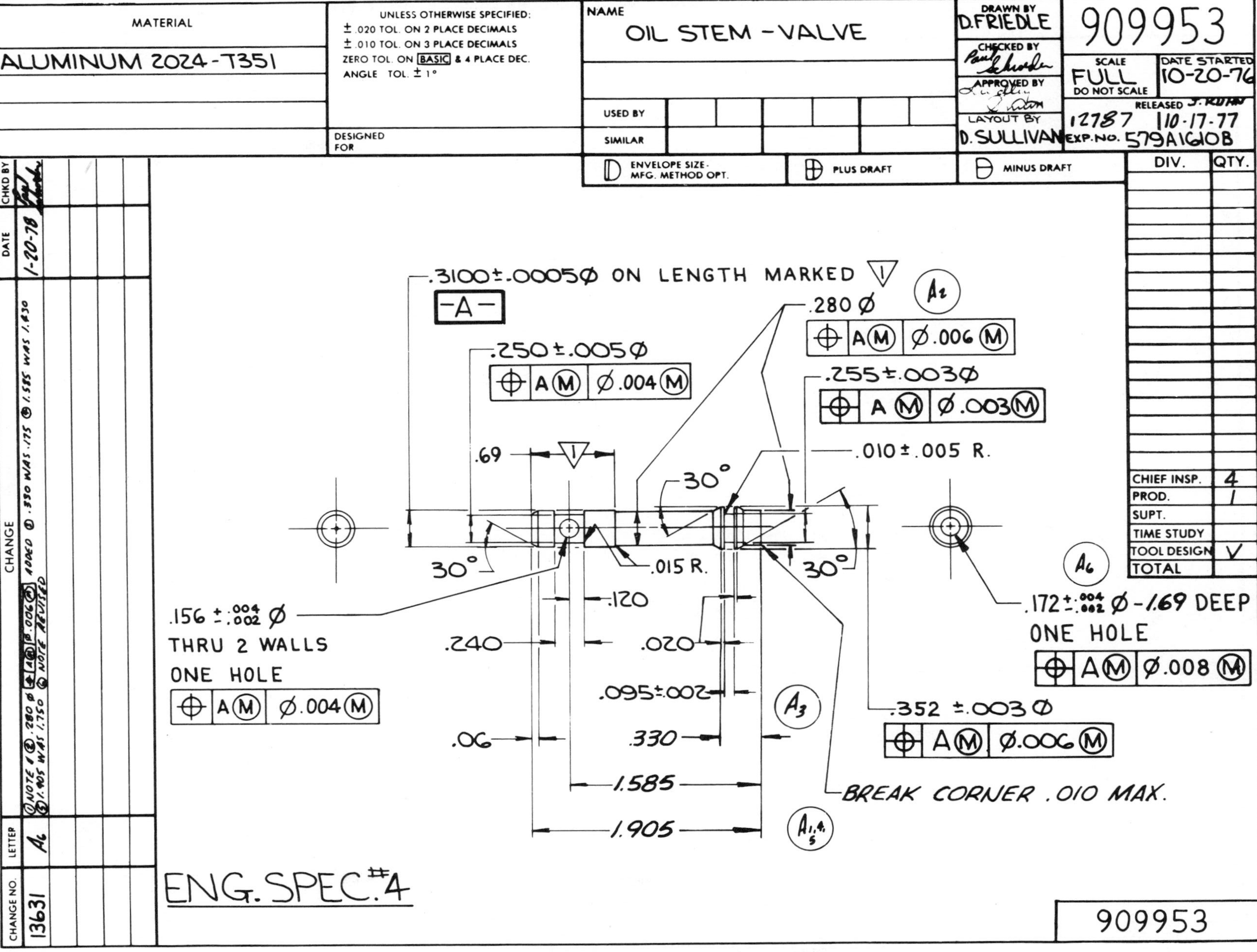

Fig. 14-15 Print #909953, oil stem—valve (detail #11)

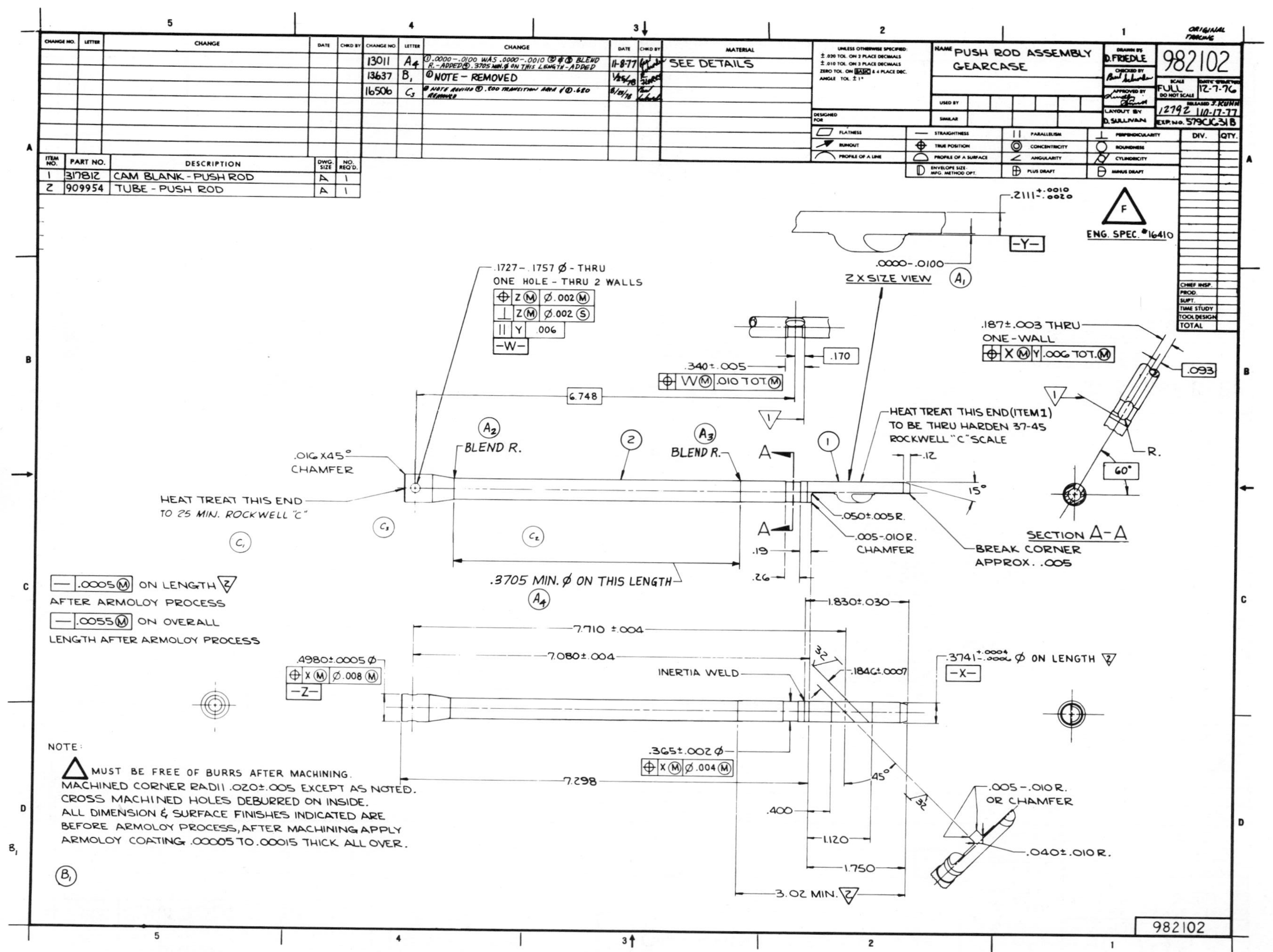

Fig. 14-16 Print #982102, push rod assembly gearcase (detail #12)

Detail #12, Push Rod Assembly Gearcase (Part #982102)

Print #982102, Figure 14-16, is an unusual assembly drawing because all of the machining dimensions are shown. The detail prints #909954 (Figure 14-17) and #317812 (Figure 14-18) show only the material and rough stock size. These two parts, one hollow and one solid, are inertia-welded together, and are then machined as indicated on the assembly print. Several different or unusual kinds of machining, heat treating, or other operations are performed on this assembly. Armaloy is a patented surface treatment process that greatly increases surface hardness and reduces the sliding friction of mating parts.

Detail #13, Spring — Pin — Push Rod (Part #909961)

Study print #909961, Figure 14-19. This is the last detail listed on assembly print #981859 (Figure 14-4). From section A-A of the assembly print, it can be seen that this part, Figure 14-19, is used to keep the pin (detail #3) from loosening in the assembly.

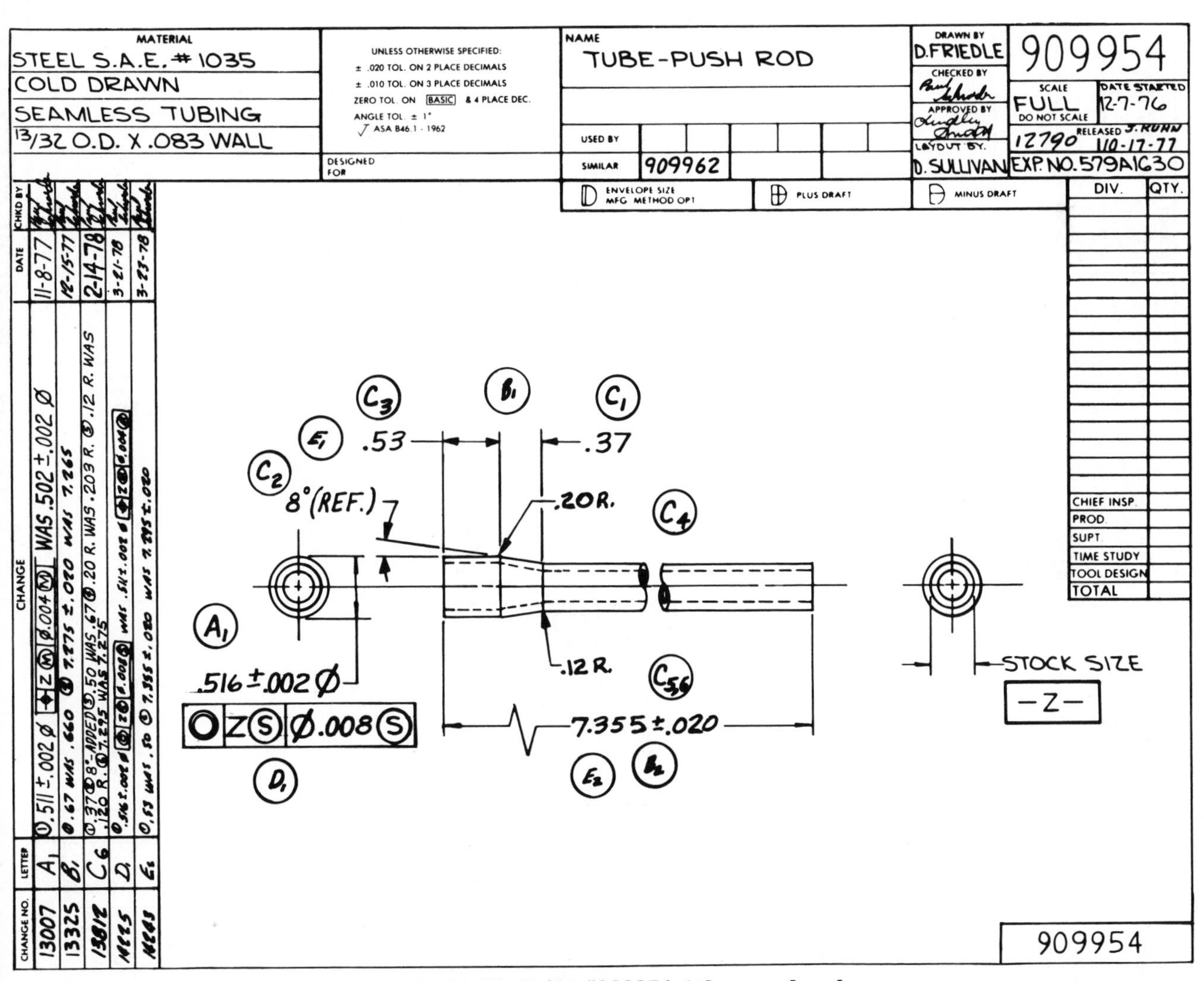

Fig. 14-17 Print #909954, tube — push rod

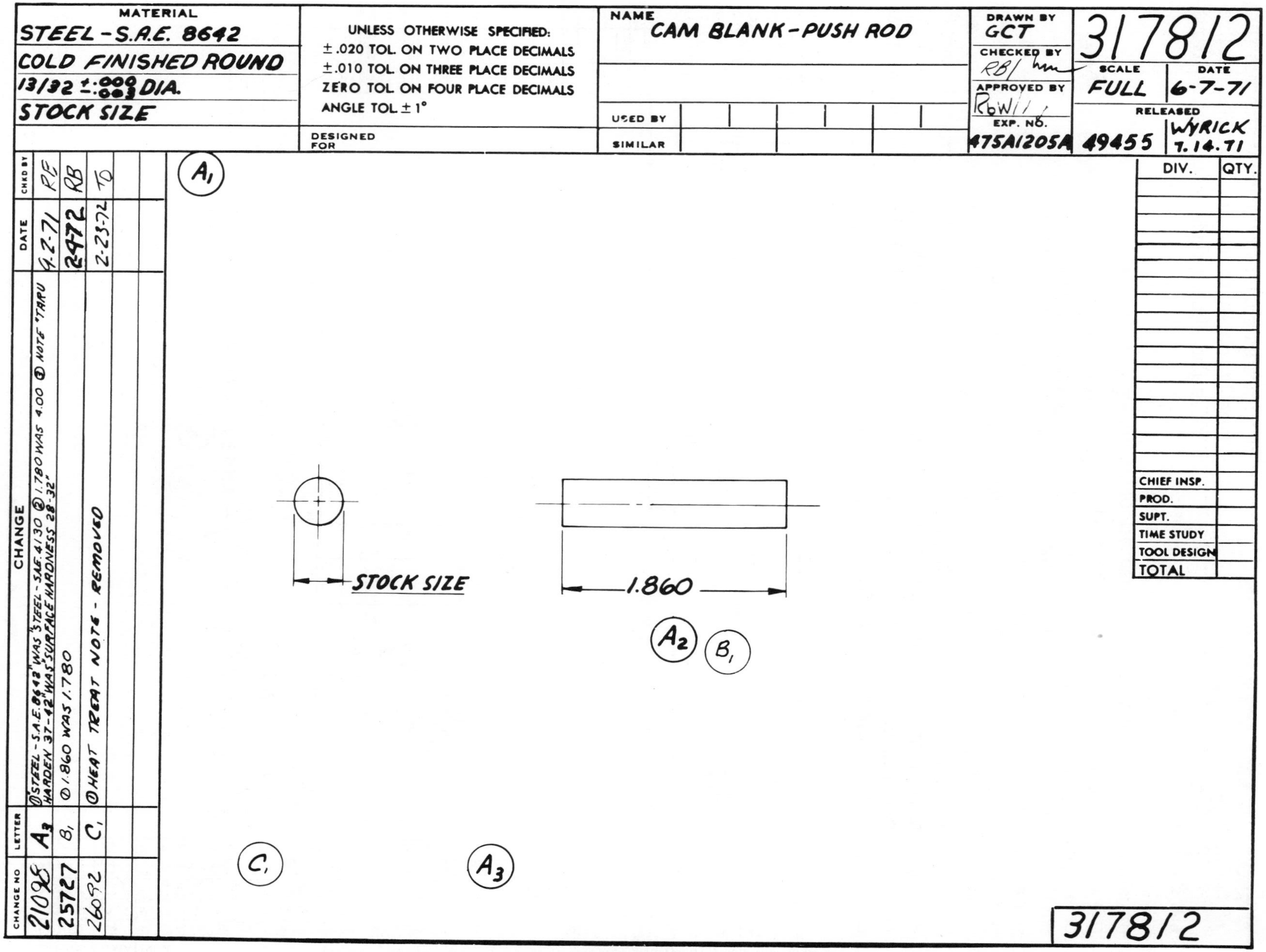

MATERIAL	UNLESS OTHERWISE SPECIFIED:	NAME	DRAWN BY	317812	
STEEL - S.A.E. 8642	±.020 TOL. ON TWO PLACE DECIMALS	CAM BLANK - PUSH ROD	GCT		
COLD FINISHED ROUND	±.010 TOL. ON THREE PLACE DECIMALS		CHECKED BY: RB/ [illegible]	SCALE: FULL	DATE: 6-7-71
13/32 $^{+.000}_{-.003}$ DIA.	ZERO TOL. ON FOUR PLACE DECIMALS		APPROVED BY: RbW[illegible]	RELEASED	
STOCK SIZE	ANGLE TOL ±1°	USED BY	EXP. NO.		
	DESIGNED FOR	SIMILAR	475A1205A	49455	WYRICK 7.14.71

CHANGE NO	LETTER	CHANGE	DATE	CHKD BY
21098	A_3	① STEEL - S.A.E. 8642" WAS "STEEL - SAE 4130 ② 1.780 WAS 4.00 ③ NOTE "THRU HARDEN 37-42" WAS "SURFACE HARDNESS 28-32"	4-2-71	RE
25727	B_1	① 1.860 WAS 1.780	2-4-72	RB
26092	C_1	① HEAT TREAT NOTE - REMOVED	2-23-72	TO

DIV.	QTY.
CHIEF INSP.	
PROD.	
SUPT.	
TIME STUDY	
TOOL DESIGN	
TOTAL	

A_1

STOCK SIZE

1.860

A_2 B_1

C_1 A_3

317812

Fig. 14-18 Print #317812, cam blank—push rod

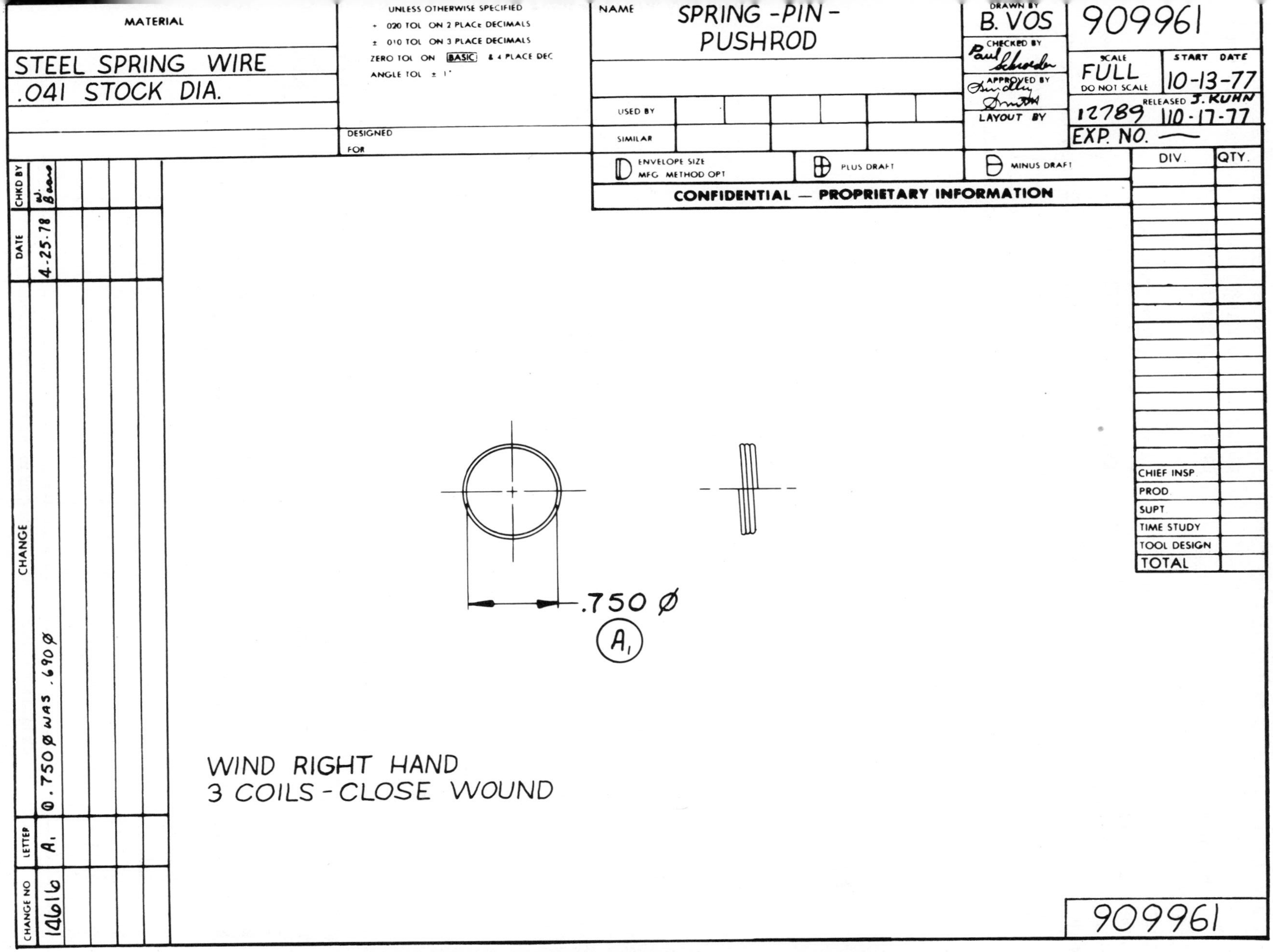

MATERIAL	UNLESS OTHERWISE SPECIFIED	NAME	DRAWN BY	
STEEL SPRING WIRE .041 STOCK DIA.	± .020 TOL ON 2 PLACE DECIMALS ± .010 TOL ON 3 PLACE DECIMALS ZERO TOL ON BASIC & 4 PLACE DEC ANGLE TOL ± 1°	SPRING - PIN - PUSHROD	B. VOS	909961
	DESIGNED FOR	USED BY / SIMILAR	CHECKED BY / APPROVED BY / LAYOUT BY	SCALE FULL (DO NOT SCALE) · START DATE 10-13-77 · RELEASED J. KUHN 12789 10-17-77 · EXP. NO. —

ENVELOPE SIZE MFG METHOD OPT · PLUS DRAFT · MINUS DRAFT

CONFIDENTIAL — PROPRIETARY INFORMATION

CHANGE NO	LETTER	CHANGE	DATE	CHKD BY
14616	A₁	0.750 Ø WAS .690 Ø	4-25-78	W. Boone

DIV.	QTY.
CHIEF INSP	
PROD	
SUPT	
TIME STUDY	
TOOL DESIGN	
TOTAL	

.750 Ø (A₁)

WIND RIGHT HAND
3 COILS - CLOSE WOUND

909961

Fig. 14-19 Print #909961, spring—pin—push rod (detail #13)

PICTORIAL DRAWINGS

Pictorial drawings are normally used for assembly or maintenance of equipment. See Figure 14-20. The table included in this figure lists the part numbers (which are the same as the manufacturing drawing numbers) related to the assembly print #981859 (Figure 14-4). Also listed are the name of the part and how many of the part are required in each assembly.

Dimensions are rarely shown on pictorial drawings, unless they are needed in the assembly operation.

HYDRAULIC DRAWINGS

A hydraulic drawing may be one of three different types: *cutaway diagram*, *graphical diagram*, or *pictorial diagram*. These types are illustrated in Figure 14-21.

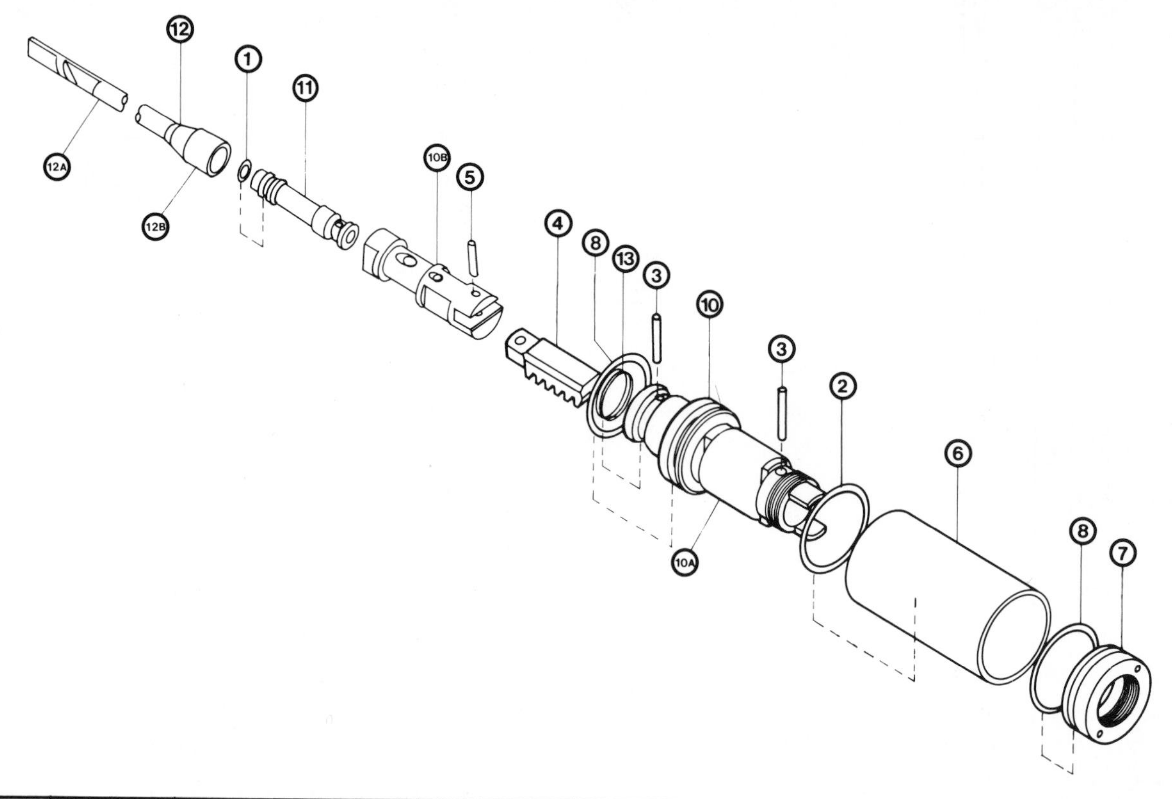

Ref. no.	Part no.	Name of part	QTY. Per assy.	Ref. no.	Part no.	Name of part	QTY. Per assy.
1	305739	O-Ring-Oil Stem Valve	1	10	982103	Valve and Piston Assy.	1
2	320140	O-Ring-Small-Shift Assy.	1	10A	909951	Piston-Shift Assy.	1
3	320176	Pin-Push Rod	2	10B	909952	Valve	1
4	909305	Rack	1	11	909953	Oil Stem-Valve	1
5	909306	Pin	1	12	982102	Push Rod Assy.	1
6	909307	Cylinder-Shift Assy.	1	12A	317812	Cam Blank-Push Rod	1
7	909308	Cap-Piston	1	12B	909954	Tube-Push Rod	1
8	909316	O-Ring-Large-Shift Assy.	2	13	909961	Spring-Push Rod	1
9	—	—	—				

Fig. 14-20 Pictorial drawing of print #981859 assembly

APPLICATION DATA VICKERS INCORPORATED DETROIT, MICH.		PAGE 1 OF 1	CLASSIFICATION NUMBER SC-5.2.4.2.2.1
DATE ISSUED	BY ORDER OF	REVISED	SUPERSEDES

SUBJECT

SYSTEM ILLUSTRATING TYPICAL TYPES OF COMPONENTS AND CIRCUIT DIAGRAMS

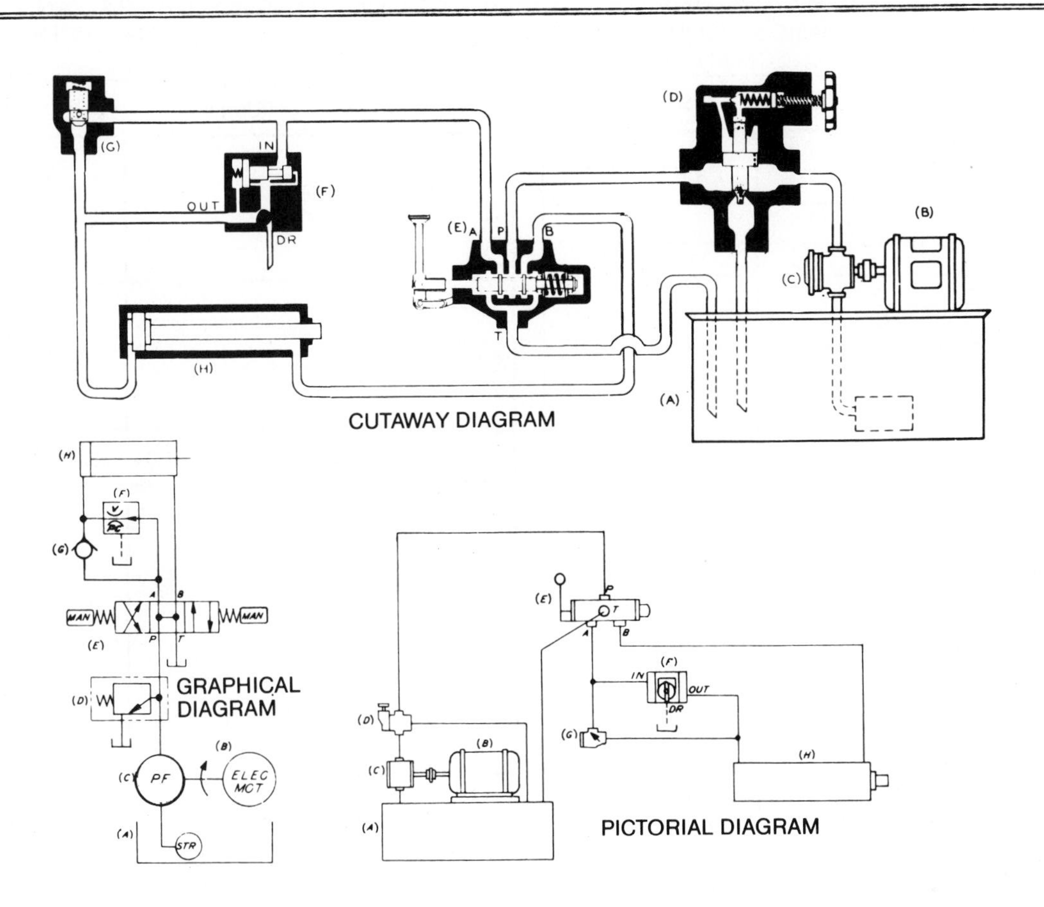

A Hydraulic Circuit may be divided into four principal parts: 1) the "A-end" (pump), 2) the "B-end" (motor), 3) the control valves, and 4) the interconnecting pipe or tubing. The control valves may be subdivided into three parts; 1) pressure controls, 2) directional controls, and 3) volume controls. Pressure controls limit maximum or assure minimum force or torque. Directional controls determine direction of motion. Volume controls limit maximum rate of motion.

A cutaway diagram emphasizes the principal internal working parts of the components.

A graphical diagram emphasizes purpose of each component in the system by indicating flow paths and method of operation.

A pictorial diagram emphasizes the piping arrangement between components.

Fig. 14-21 Hydraulic drawing ***(Courtesy of Sperry Vickers Co.)***

The graphical diagram is the type used most commonly in industry.

Standard hydraulic symbols are shown in Table 14-1.

Table 14–1 Standard Hydraulic Graphic Symbols

THE SYMBOLS SHOWN CONFORM TO THE AMERICAN NATIONAL STANDARDS INSTITUTE (ANSI) SPECIFICIATIONS. BASIC SYMBOLS CAN BE COMBINED IN ANY COMBINATION. NO ATTEMPT IS MADE TO SHOW ALL COMBINATIONS.

LINES AND LINE FUNCTIONS	
LINE, WORKING	
LINE, PILOT (L>20W)	
LINE, DRAIN (L<5W)	
CONNECTOR	
LINE, FLEXIBLE	
LINE, JOINING	
LINE, PASSING	
DIRECTION OF FLOW, HYDRAULIC PNEUMATIC	
LINE TO RESERVOIR ABOVE FLUID LEVEL BELOW FLUID LEVEL	
LINE TO VENTED MANIFOLD	
PLUG OR PLUGGED CONNECTION	
RESTRICTION, FIXED	
RESTRICITION, VARIABLE	

PUMPS	
PUMP, SINGLE FIXED DISPLACEMENT	
PUMP, SINGLE VARIABLE DISPLACEMENT	
MOTORS AND CYLINDERS	
MOTOR, ROTARY, FIXED DISPLACEMENT	
MOTOR, ROTARY VARIABLE DISPLACEMENT	
MOTOR, OSCILLATING	
CYLINDER, SINGLE ACTING	
CYLINDER, DOUBLE ACTING	
CYLINDER, DIFFERENTIAL ROD	
CYLINDER, DOUBLE END ROD	
CYLINDER, CUSHIONS BOTH ENDS	

Sperry Vickers Co.

Table 14-1 Standard Hydraulic Graphic Symbols (*continued*)

MISCELLANEOUS UNITS	
DIRECTION OF ROTATION (ARROW IN FRONT OF SHAFT)	
COMPONENT ENCLOSURE	
RESERVOIR, VENTED	
RESERVOIR, PRESSURIZED	
PRESSURE GAGE	
TEMPERATURE GAGE	
FLOW METER (FLOW RATE)	
ELECTRIC MOTOR	M
ACCUMULATOR, SPRING LOADED	
ACCUMULATOR, GAS CHARGED	
FILTER OR STRAINER	
HEATER	
COOLER	
TEMPERATURE CONTROLLER	
INTENSIFIER	
PRESSURE SWITCH	
BASIC VALVE SYMBOLS	
CHECK VALVE	
MANUAL SHUT OFF VALVE	
BASIC VALVE ENVELOPE	
VALVE, SINGLE FLOW PATH, NORMALLY CLOSED	

BASIC VALVE SYMBOLS (CONT.)	
VALVE, SINGLE FLOW PATH, NORMALLY OPEN	
VALVE, MAXIMUM PRESSURE (RELIEF)	
BASIC VALVE SYMBOL, MULTIPLE FLOW PATHS	
FLOW PATHS BLOCKED IN CENTER POSITION	
MULTIPLE FLOW PATHS (ARROW SHOWS FLOW DIRECTION)	
VALVE EXAMPLES	
UNLOADING VALVE, INTERNAL DRAIN, REMOTELY OPERATED	
DECELERATION VALVE, NORMALLY OPEN	
SEQUENCE VALVE, DIRECTLY OPERATED, EXTERNALLY DRAINED	
PRESSURE REDUCING VALVE	
COUNTER BALANCE VALVE WITH INTEGRAL CHECK	
TEMPERATURE AND PRESSURE COMPENSATED FLOW CONTROL WITH INTEGRAL CHECK	
DIRECTIONAL VALVE, TWO POSITION, THREE CONNECTION	
DIRECTIONAL VALVE, THREE POSITION, FOUR CONNECTION	
VALVE, INFINITE POSITIONING (INDICATED BY HORIZONTAL BARS)	

WELDING PRINTS

There are many uses of welding in the fabrication of metal products. Therefore, it is important to understand welding prints.

On prints, the welding information is indicated around a base weld symbol, Figure 14-22. The arrow portion of the symbol points to the weld joint. The basic joint types are *butt*, *corner*, *T*, *lap*, and *edge*, Figure 14-23. The actual types of welds are *fillet*, *plug*, *spot*, *seam*, and *groove*. See Figure 14-24.

Additional weld symbol information is indicated in Figure 14-25. Figure 14-26 shows some common abbreviations used in welding. A complete listing of welding symbols can be found in pamphlets published by the American Welding Society (AWS) or in textbooks specializing in the subject.

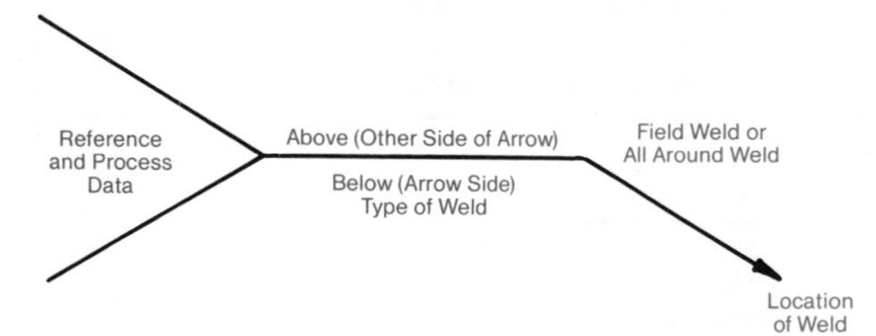

Fig. 14-22 Basic weld symbol data locations

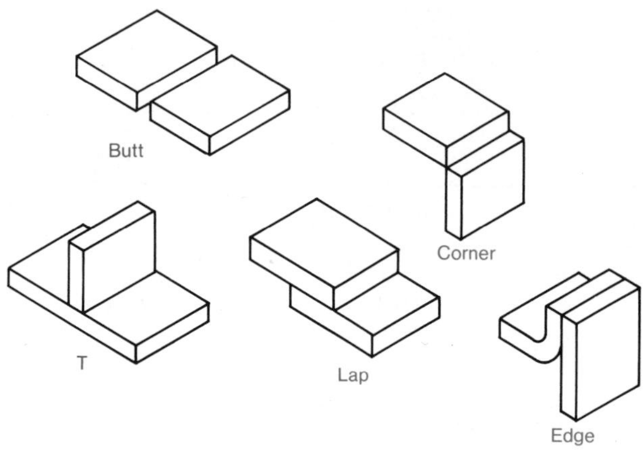

Fig. 14-23 Types of weld joints

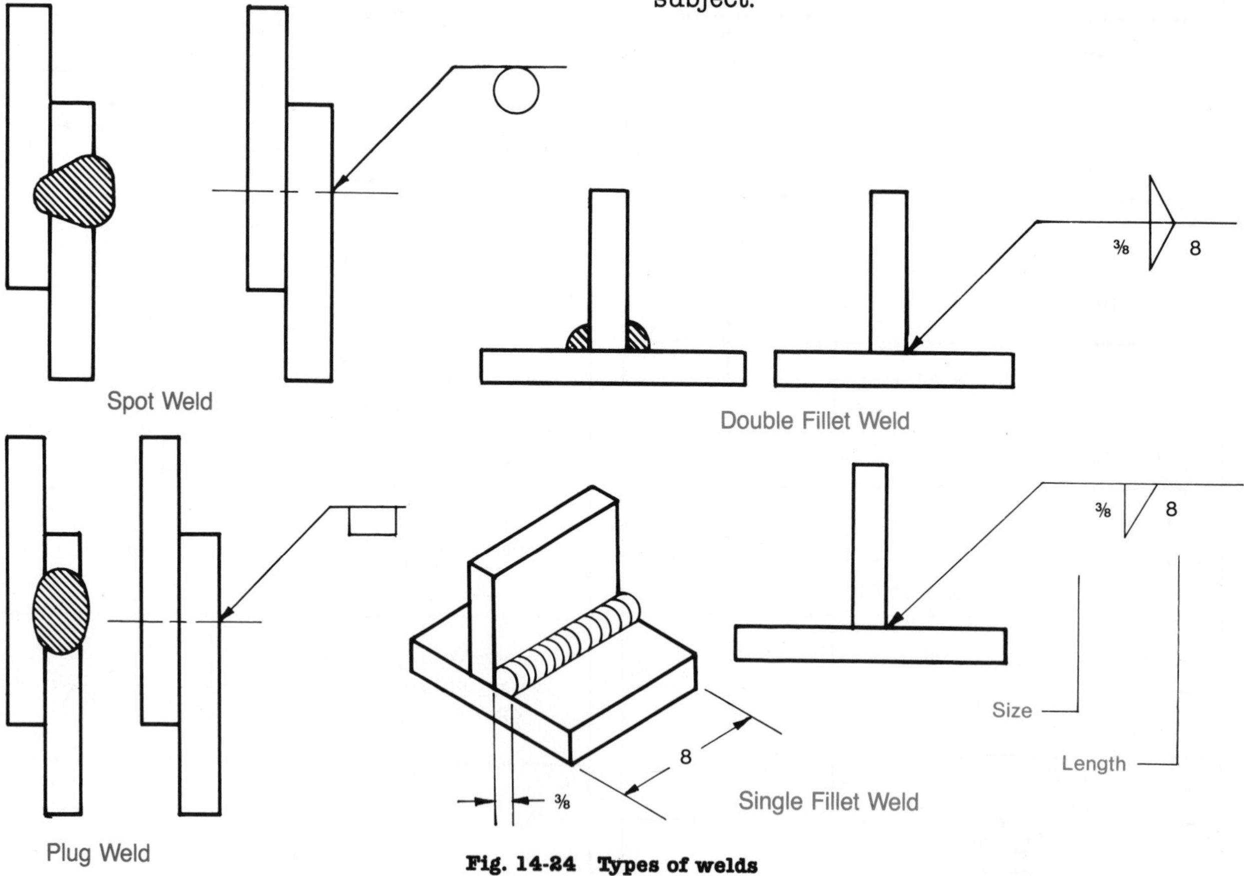

Fig. 14-24 Types of welds

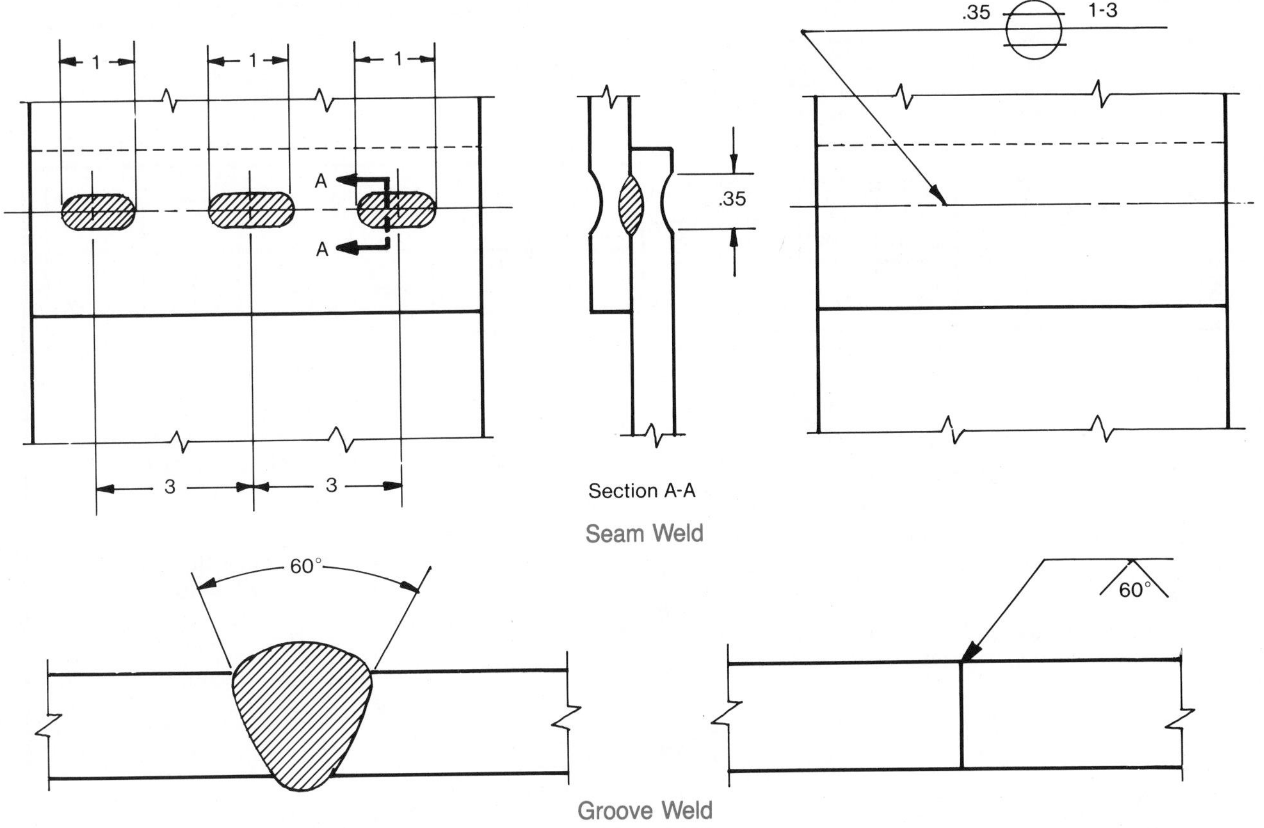

Fig. 14-24 ***(continued)***

AIR LOGIC PRINTS

Air logic is a relatively new system of machine control. Air logic is now used by many manufacturing industries. The air logic system uses compressed air to regulate machine controls. The devices in this system operate in a manner similar to those used in hydraulic systems, but are much smaller in size. The control devices of an air logic system can duplicate many of the electrical functions in a hydraulic circuit.

Air logic control prints resemble electrical prints more than hydraulic prints.

The circuit print in Figure 14-27 is used in a drilling operation. The first operation is to clamp the part in a fixture, P1A. The second operation is to drill, P2. The next operation is to unclamp, P1B. The final operation is to eject the part from the fixture, P3A.

ELECTRICAL PRINTS

The electrical industry uses some of the most complicated prints, particularly when large drive motors must be controlled. The print shown in Figure 14-28 is used for a vacuum pump system. The system has two complete pump and motor drives; this allows the operator to easily switch from one pump to the other. This type of equipment may be used in hospitals, which usually require a secondary system in case of mechanical failure.

The wiring connection diagram for the starter switches is shown just above the title block in Figure 14-28. The wiring connection diagram for the starter relays

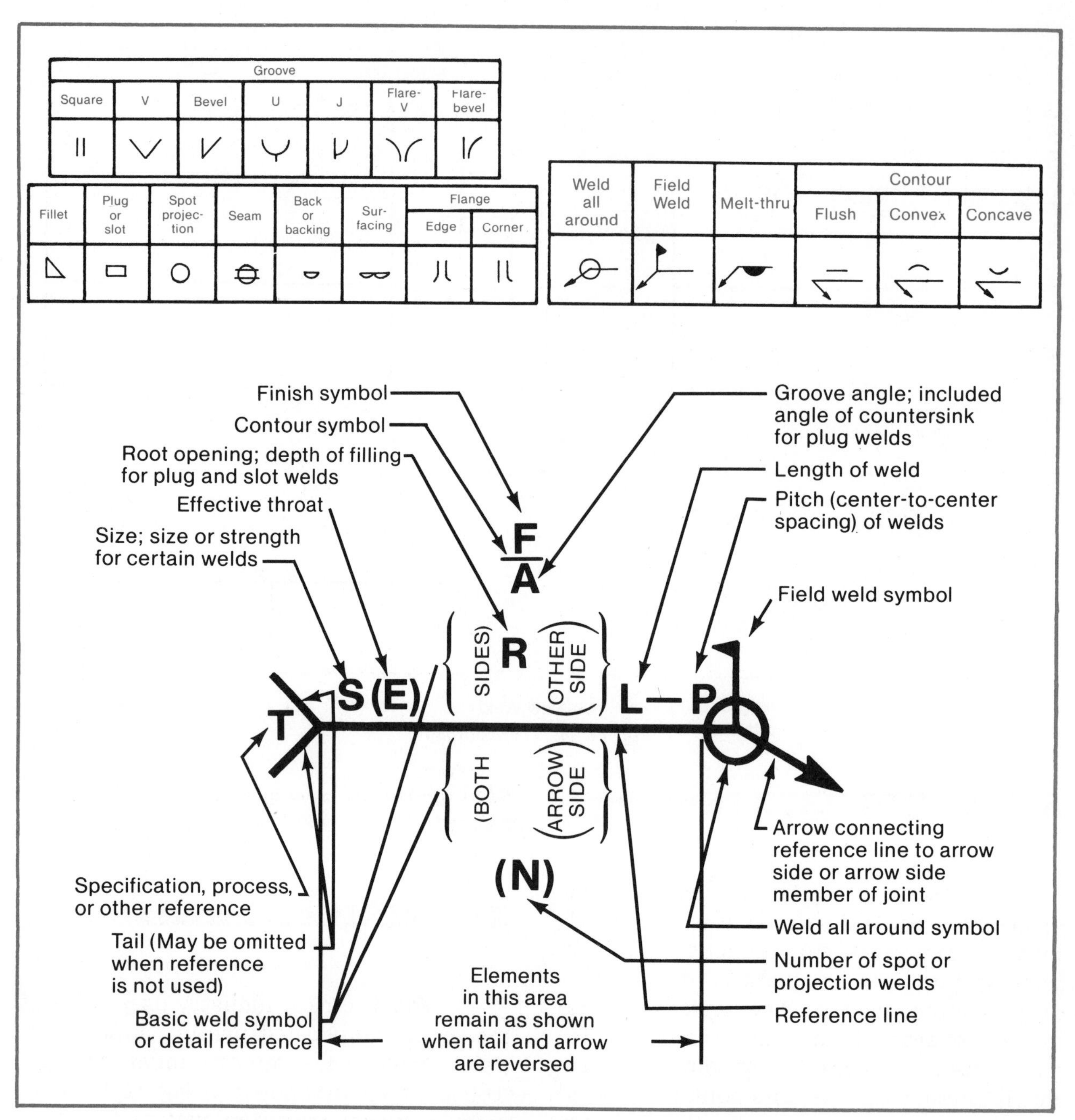

Fig. 14-25 Basic welding symbols *(Courtesy of American Welding Society)*

is indicated at the top. The complete schematic diagram for the entire unit is shown at the middle and upper left corner.

Electrical control prints may be hard to understand without sufficient training in the field. The print reader should recognize electrical motor control prints, and should be aware of the type of information that may be shown on these prints. The prints are used in the initial electrical wiring of equipment. They are also useful to maintenance electricians for troubleshooting and repair of equipment.

The symbols used in electrical schematic prints are illustrated in Table 14-2.

Letter designation	Welding and allied processes
AAC	air carbon arc cutting
AAW	air acetylene welding
ABD	adhesive bonding
AB	arc brazing
AC	arc cutting
AHW	atomic hydrogen welding
AOC	oxygen arc cutting
AW	arc welding
B	brazing
BB	block brazing
BMAW	bare metal arc welding
CAC	carbon arc cutting
CAW	carbon arc welding
CAW-G	gas carbon arc welding
CAW-S	shielded carbon arc welding
CAW-T	twin carbon arc welding
CW	cold welding
DB	dip brazing
DFB	diffusion brazing
DFW	diffusion welding
DS	dip soldering
EASP	electric arc spraying
EBC	electron beam cutting
EBW	electron beam welding
ESW	electroslag welding
EXW	explosion welding
FB	furnace brazing
FCAW	flux cored arc welding
FCAW-EG	flux cored arc welding—electrogas
FLB	flow brazing
FLOW	flow welding
FLSP	flame spraying
FOC	chemical flux cutting
FOW	forge welding
FRW	friction welding
FS	furnace soldering
FW	flash welding
GMAC	gas metal arc cutting
GMAW	gas metal arc welding
GMAW-EG	gas metal arc welding—electrogas
GMAW-P	gas metal arc welding—pulsed arc
GMAW-S	gas metal arc welding—short circuiting arc
GTAC	gas tungsten arc cutting
GTAW	gas tungsten arc welding
GTAW-P	gas tungsten arc welding—pulsed arc
HFRW	high frequency resistance welding
HPW	hot pressure welding
IB	induction brazing
INS	iron soldering
IRB	infrared brazing
IRS	infrared soldering
IS	induction soldering
IW	induction welding
LBC	laser beam cutting
LBW	laser beam welding
LOC	oxygen lance cutting
MAC	metal arc cutting
OAW	oxyacetylene welding
OC	oxygen cutting
OFC	oxyfuel gas cutting
OFC-A	oxyacetylene cutting
OFC-H	oxyhydrogen cutting
OFC-N	oxynatural gas cutting
OFC-P	oxypropane cutting
OFW	oxyfuel gas welding
OHW	oxyhydrogen welding
PAC	plasma arc cutting
PAW	plasma arc welding
PEW	percussion welding
PGW	pressure gas welding
POC	metal powder cutting
PSP	plasma spraying
RB	resistance brazing
RPW	projection welding
RS	resistance soldering
RSEW	resistance seam welding
RSW	resistance spot welding
ROW	roll welding
RW	resistance welding
S	soldering
SAW	submerged arc welding
SAW-S	series submerged arc welding
SMAC	shielded metal arc cutting
SMAW	shielded metal arc welding
SSW	solid state welding
SW	stud arc welding
TB	torch brazing
Automatic	AU
Machine	ME
Manual	MA
Semiautomatic	SA

Fig. 14-26 Standard welding abbreviations ***(Courtesy of American Welding Society)***

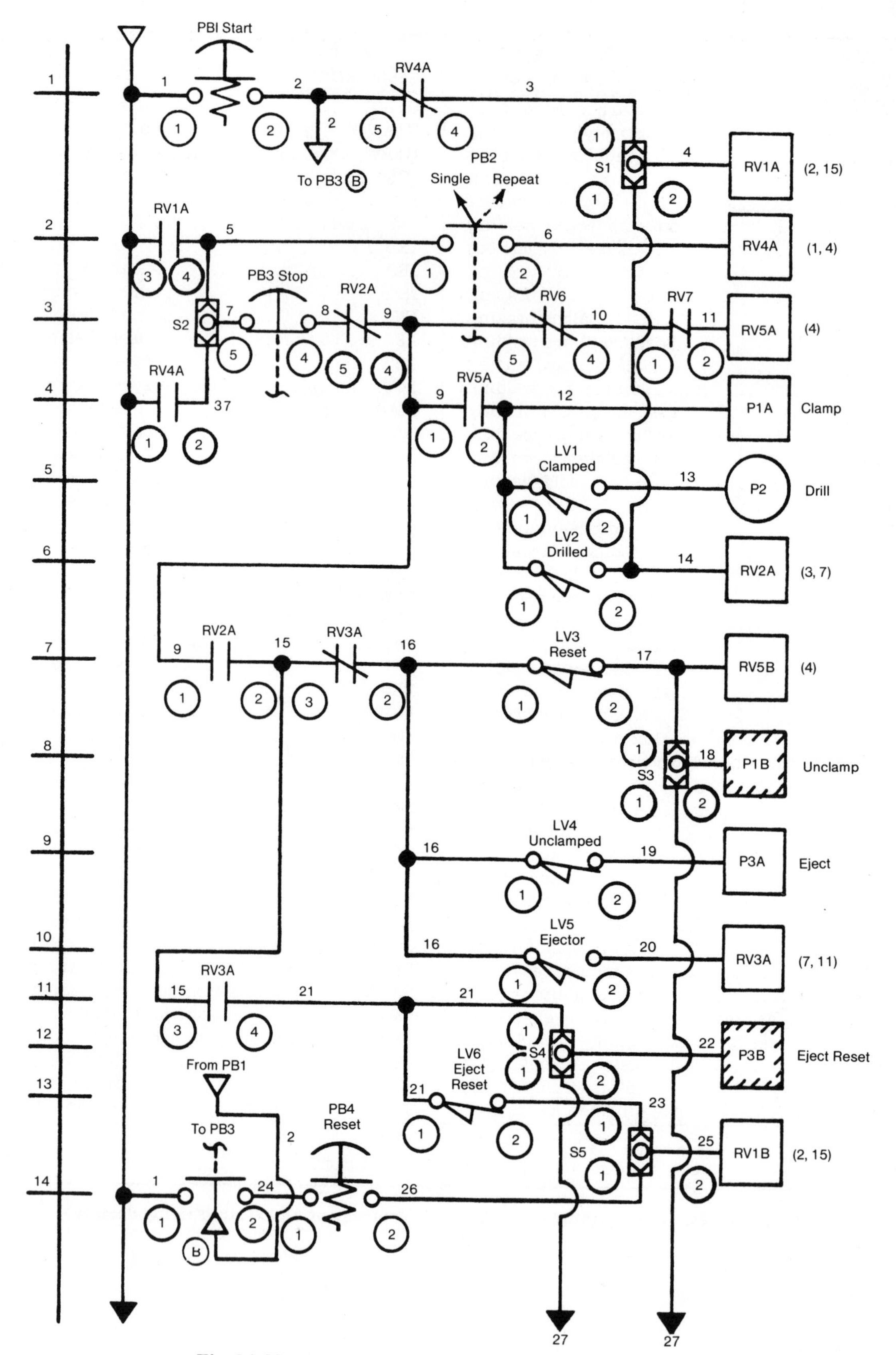

Fig. 14-27 *Air logic print (© 1974 Parker Hannifin Corp.)*

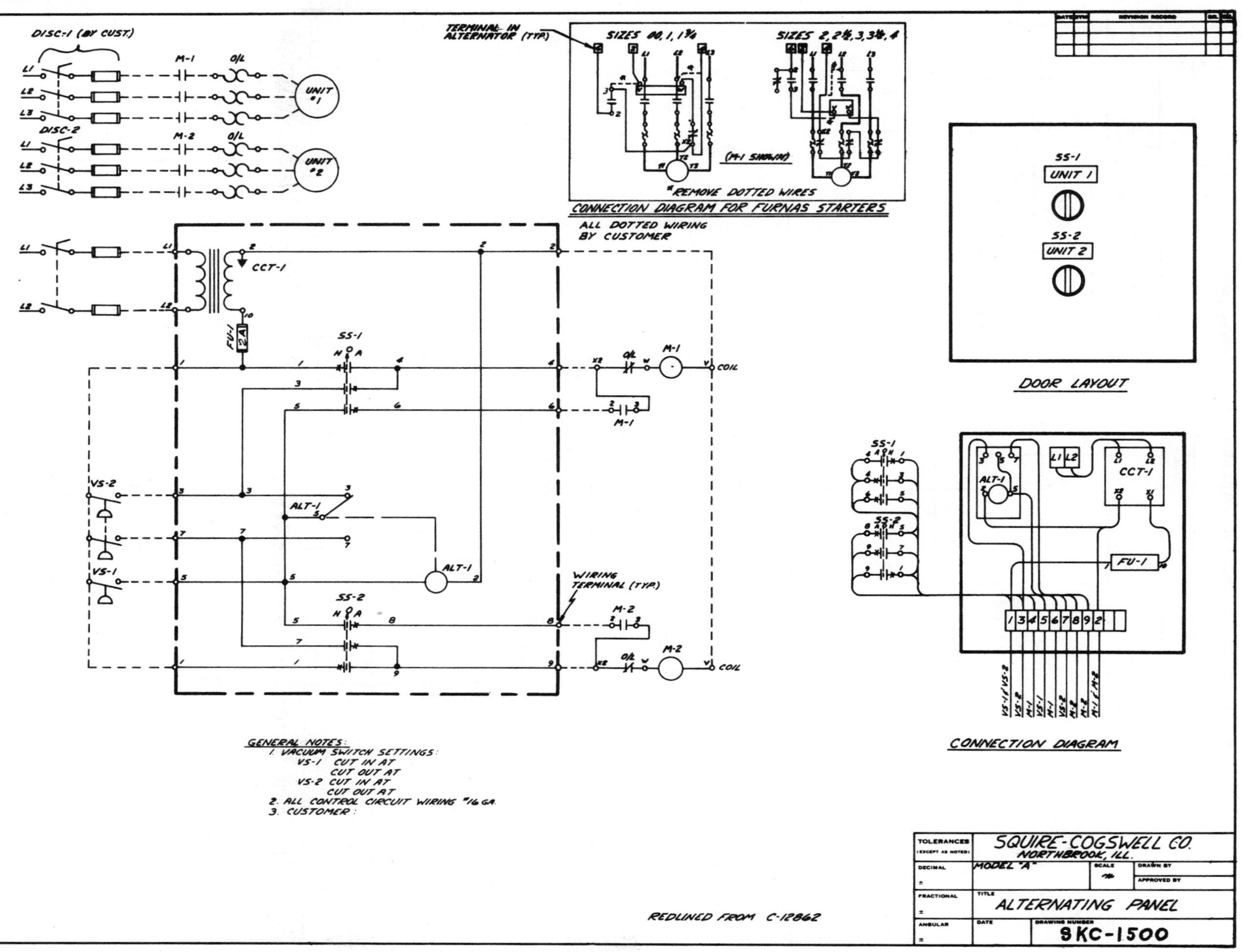

Fig. 14-28 Electrical print

Table 14–2 Standard Electrical Schematic Symbols

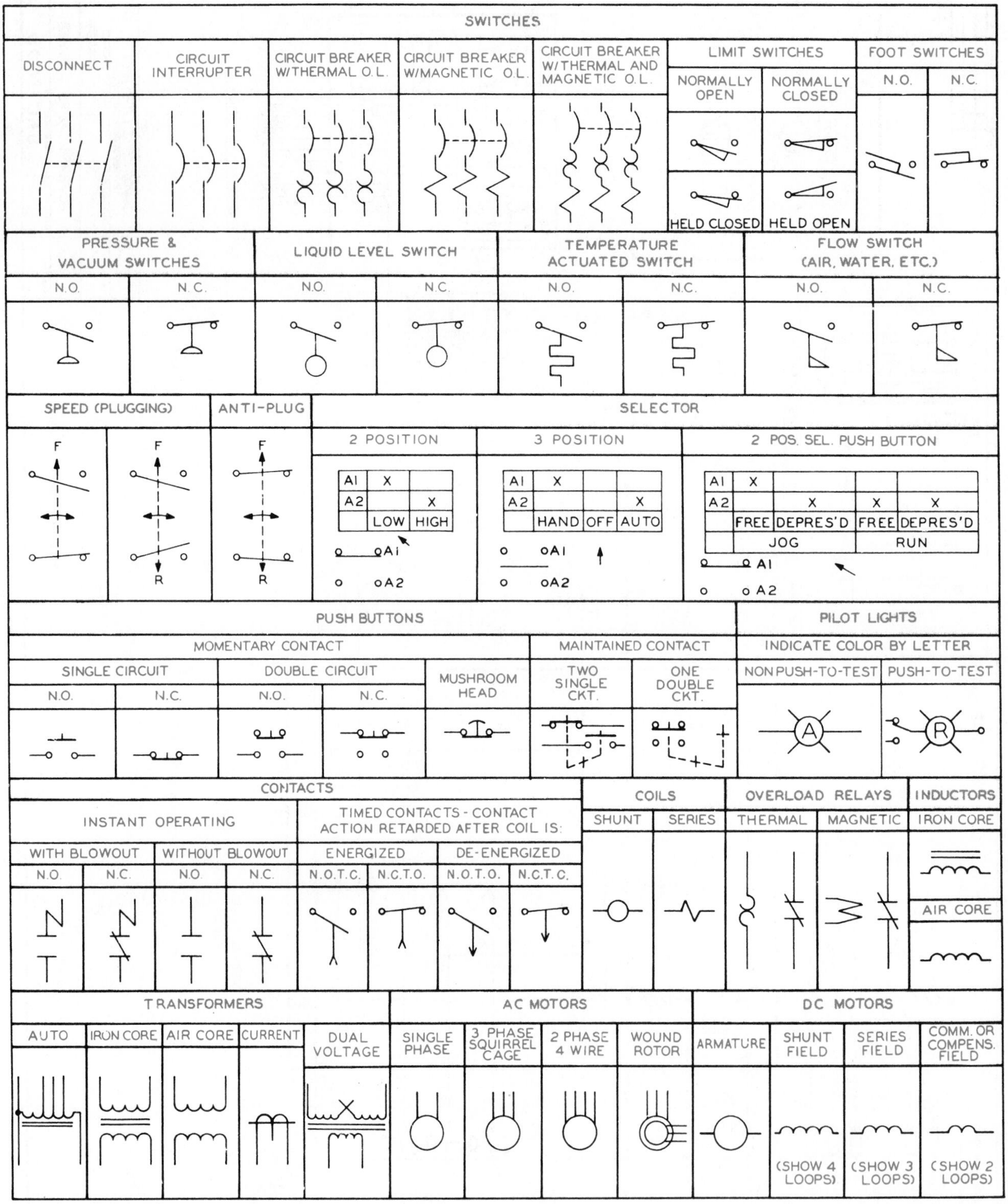

SWITCHES								
DISCONNECT	CIRCUIT INTERRUPTER	CIRCUIT BREAKER W/THERMAL O.L.	CIRCUIT BREAKER W/MAGNETIC O.L.	CIRCUIT BREAKER W/THERMAL AND MAGNETIC O.L.	LIMIT SWITCHES		FOOT SWITCHES	
					NORMALLY OPEN	NORMALLY CLOSED	N.O.	N.C.
					HELD CLOSED	HELD OPEN		

PRESSURE & VACUUM SWITCHES		LIQUID LEVEL SWITCH		TEMPERATURE ACTUATED SWITCH		FLOW SWITCH (AIR, WATER, ETC.)	
N.O.	N.C.	N.O.	N.C.	N.O.	N.C.	N.O.	N.C.

SPEED (PLUGGING)		ANTI-PLUG	SELECTOR		
F	F R	F R	2 POSITION	3 POSITION	2 POS. SEL. PUSH BUTTON

2 POSITION

A1	X	
A2		X
	LOW	HIGH

A1
A2

3 POSITION

A1	X		
A2			X
	HAND	OFF	AUTO

A1
A2

2 POS. SEL. PUSH BUTTON

A1	X			
A2		X	X	X
	FREE	DEPRES'D	FREE	DEPRES'D
	JOG		RUN	

A1
A2

PUSH BUTTONS							PILOT LIGHTS	
MOMENTARY CONTACT					MAINTAINED CONTACT		INDICATE COLOR BY LETTER	
SINGLE CIRCUIT		DOUBLE CIRCUIT		MUSHROOM HEAD	TWO SINGLE CKT.	ONE DOUBLE CKT.	NON PUSH-TO-TEST	PUSH-TO-TEST
N.O.	N.C.	N.O.	N.C.				A	R

CONTACTS								COILS		OVERLOAD RELAYS		INDUCTORS
INSTANT OPERATING				TIMED CONTACTS - CONTACT ACTION RETARDED AFTER COIL IS:				SHUNT	SERIES	THERMAL	MAGNETIC	IRON CORE
WITH BLOWOUT		WITHOUT BLOWOUT		ENERGIZED		DE-ENERGIZED						AIR CORE
N.O.	N.C.	N.O.	N.C.	N.O.T.C.	N.C.T.O.	N.O.T.O.	N.C.T.C.					

TRANSFORMERS					AC MOTORS				DC MOTORS			
AUTO	IRON CORE	AIR CORE	CURRENT	DUAL VOLTAGE	SINGLE PHASE	3 PHASE SQUIRREL CAGE	2 PHASE 4 WIRE	WOUND ROTOR	ARMATURE	SHUNT FIELD	SERIES FIELD	COMM. OR COMPENS. FIELD
										(SHOW 4 LOOPS)	(SHOW 3 LOOPS)	(SHOW 2 LOOPS)

Square D Company

REVIEW

This review is provided to serve as reinforcement study material. Fill in the appropriate word(s) to complete the sentences below.

1. A print that contains basic dimensions with no tolerances is usually (a) (an) ______________________ print.

2. A print that contains close dimensions with tolerances is usually (a) (an) ______________________ print.

3. The drawing size for individual part prints is usually indicated on the ______________________ drawing.

4. Dimensions are not usually shown on (a) (an) ______________________ drawing.

5. The bill of material usually contains the individual part ______________________, ______________________, ______________________, and ______________________ required.

6. Two types of hydraulic drawings are ______________________ and ______________________ diagrams.

7. Air logic control prints most resemble ______________________ prints.

8. Electrical prints may show not only part location but also ______________________ schematics.

9. The basic weld joint types are:

a. ______________________ **d.** ______________________

b. ______________________ **e.** ______________________

c. ______________________

(continued)

10. The actual types of welds are:

a. ______________________ **d.** ______________________

b. ______________________ **e.** ______________________

c. ______________________

NAME ______________________

DATE ______________________

SCORE ______________________

EXERCISE B14-1

Print #A16014-1

Answer the following questions after studying print #A160141 on page 257.

QUESTIONS	ANSWERS
1. What is this assembly?	1. ______
2. What are the size limits of the .656 holes?	2. ______
3. Where are the case hardness and depth checked?	3. ______
4. Identify the two types of welds used.	4. ______
5. What is the tolerance on the 1.306 dimension?	5. ______
6. How many changes were made to this print?	6. ______
7. What heat treatment is used?	7. ______
8. What is the thickness range of the case hardening?	8. ______
9. What is the surface hardness?	9. ______
10. What numbers identify the two parts that make up this assembly?	10. ______
11. How would the welder of this assembly be identified?	11. ______
12. When was this print released for production?	12. ______
13. What would be the maximum height location of the 1.218/1.210 hole from the base?	13. ______
14. What does the RH in the part name mean?	14. ______
15. What is the overall length of this assembly?	15. ______
16. What is the height of this assembly?	16. ______
17. What is the finish on the 1.218/1.210 hole?	17. ______

NAME ______

DATE ______

SCORE ______

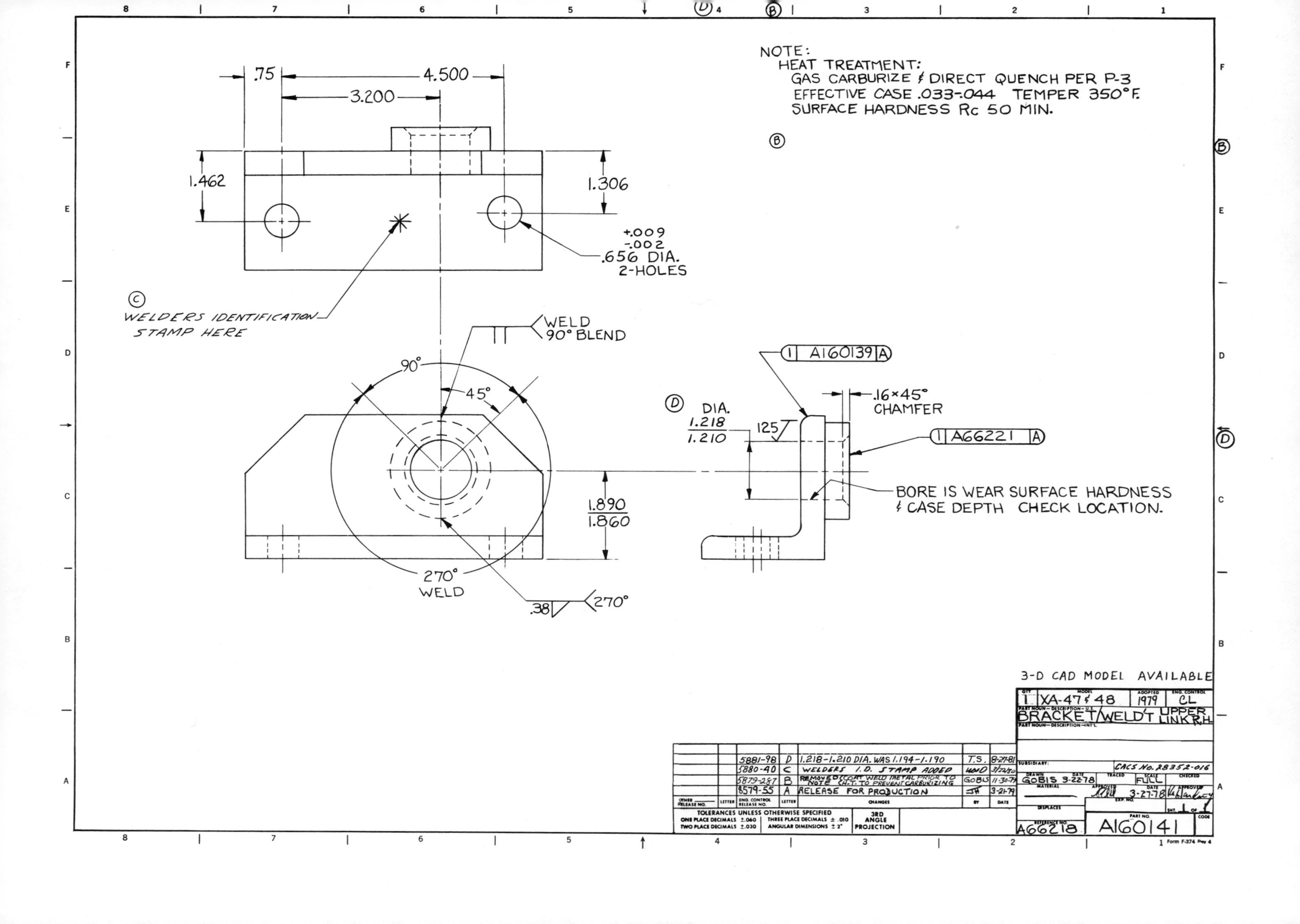
NOTE:
HEAT TREATMENT:
GAS CARBURIZE & DIRECT QUENCH PER P-3
EFFECTIVE CASE .033-.044 TEMPER 350°F.
SURFACE HARDNESS Rc 50 MIN.
.75
4.500
3.200
1.462
1.306
+.009
-.002
.656 DIA.
2-HOLES
WELDERS IDENTIFICATION STAMP HERE
WELD
90° BLEND
90°
45°
270°
WELD
.38
270°
1.890
1.860
DIA.
1.218
1.210
125
1 A160139 A
.16×45°
CHAMFER
1 A66221 A
BORE IS WEAR SURFACE HARDNESS
& CASE DEPTH CHECK LOCATION.
3-D CAD MODEL AVAILABLE
XA-47 & 48
1979
CL
BRACKET/WELD'T UPPER LINK R.H.
5881-98 D 1.218-1.210 DIA. WAS 1.194-1.190 T.S. 8-27-81
5880-40 C WELDERS I.D. STAMP ADDED 3/12/80
5879-287 B REMOVED CCAT WELD METAL PRIOR TO NOTE CH-T. TO PREVENT CARBURIZING GOBIS 11-30-79
8579-55 A RELEASE FOR PRODUCTION 3-21-79
TOLERANCES UNLESS OTHERWISE SPECIFIED
ONE PLACE DECIMALS ±.060 TWO PLACE DECIMALS ±.030
THREE PLACE DECIMALS ±.010 ANGULAR DIMENSIONS ±2°
3RD ANGLE PROJECTION
CACS No. 28352-016
GOBIS 3-22-78
FULL
3-27-78
A66218
A160141
Form F-374 Rev 4

CHAPTER 15

GEOMETRIC DIMENSIONING

OBJECTIVES

After studying this chapter, you will be able to:

- Explain why geometric dimensioning is needed.
- List the benefits of using symbols.
- Define the common geometric dimensioning terms.
- List the basic rules of geometric dimensioning.
- Describe how feature control frames (symbols) are displayed on industrial prints.

Geometric dimensioning and tolerancing is a means of dimensioning and tolerancing a drawing with respect to the actual function or relationship of part features. This drafting technique does not replace the conventional dimensioning system. Instead, it complements, or adds to, the conventional system.

The geometric dimensioning system is now used by approximately 75% of manufacturing companies.

THE NEED FOR GEOMETRIC DIMENSIONING

Due to the rapid technical changes in industry and the use of various manufacturing facilities, a more precise method of specifying information on prints was needed. The geometric dimensioning system fulfills this need. The system was established for the following purposes:

1. To provide a single drafting standard in the United States
2. To provide additional or extra tolerance, thereby reducing production costs
3. To ensure that design, dimensional, and tolerance requirements, as they relate to actual function, are specifically stated and thus carried out
4. To ensure interchangeability of mating parts at assembly

Additional advantages will become apparent during this discussion.

This chapter covers the basics of geometric dimensioning. More detailed information can be obtained from textbooks devoted especially to geometric dimensioning. The information in this chapter complies with the standard approved by the American National Standards Institute (ANSI) in *Dimensioning and Tolerancing (ANSI Y14.5M — 1982).*

Many changes were made in the 1982 standard that may not appear on industrial prints until 1984 or 1985. Therefore, it is essential that the print reader also be familiar with the procedures described in the older 1973 standard. The industrial prints using the geometric system that are shown in this book are in accordance with the 1973 standard. The print geometric displays have not been changed to conform to the 1982 standard because these prints are still being used in their present state and may not be revised for several years. However, comparisons of the former geometric dimensioning practices and the new standard practices are shown throughout this chapter.

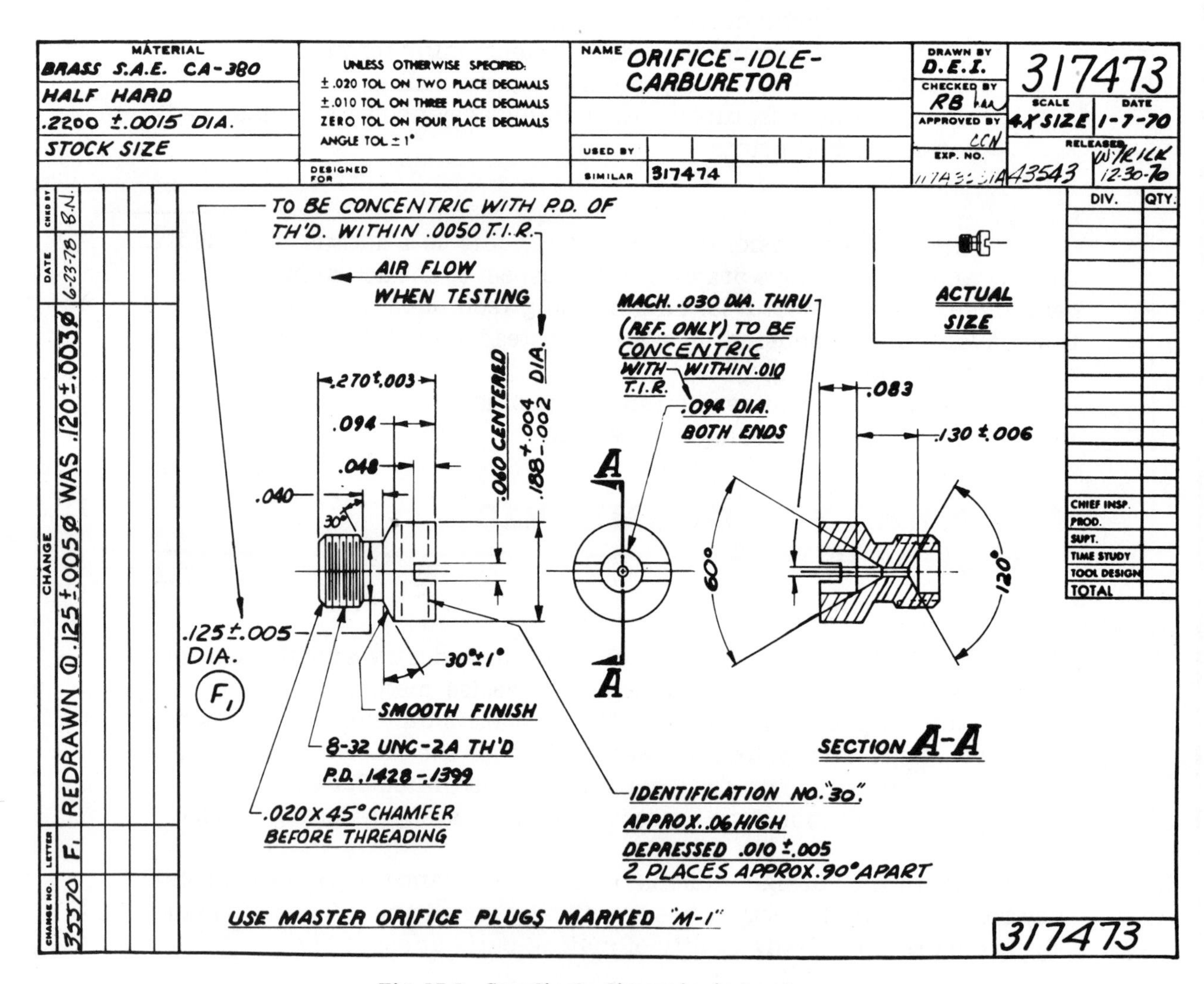

Fig. 15-1 Coordinate dimensioning system

BENEFITS OF USING SYMBOLS

A comparison of conventional coordinate dimensioning (Figure 15-1) to the newer geometric dimensioning (Figure 15-2) illustrates the basic difference between the two systems. (A print of a more complicated part than that shown in Figure 15-2 would display many more geometric symbols.)

The print reader is only required to properly interpret the meaning of the geometric notations on the print. The designer or engineer is responsible for determining which callout and tolerance to use in a particular part design.

The use of symbols provides many benefits, as follows:

1. Symbols are consistent with other established drafting systems, such as electrical and electronic, architectural, welding, and surface texture.

2. Symbols can be drawn easily with drafting templates, Figure 15-3, and are more legible than written notes.

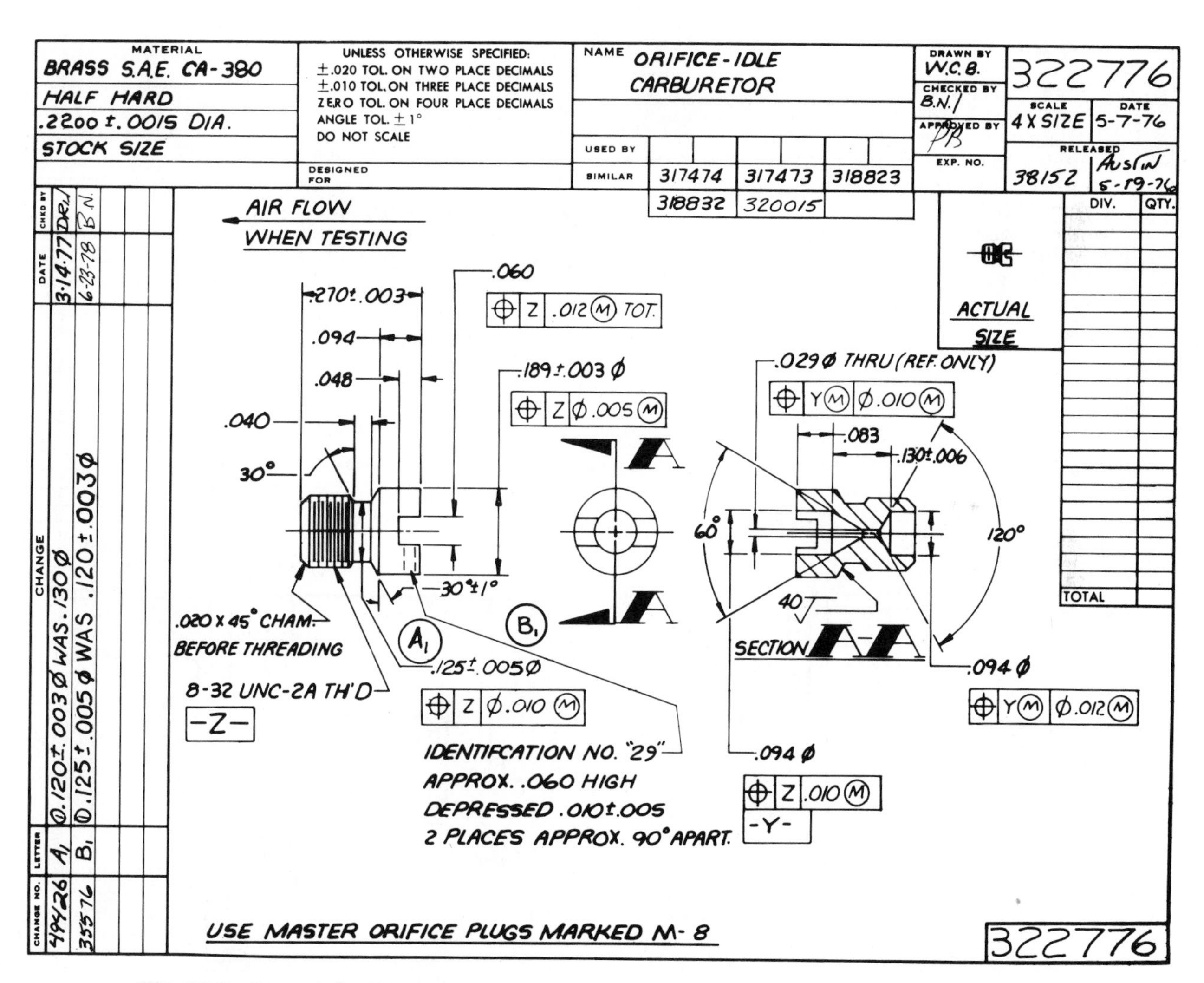

Fig. 15-2 Geometric dimensioning system (Conforms with ANSI Y14.5-1973 standard)

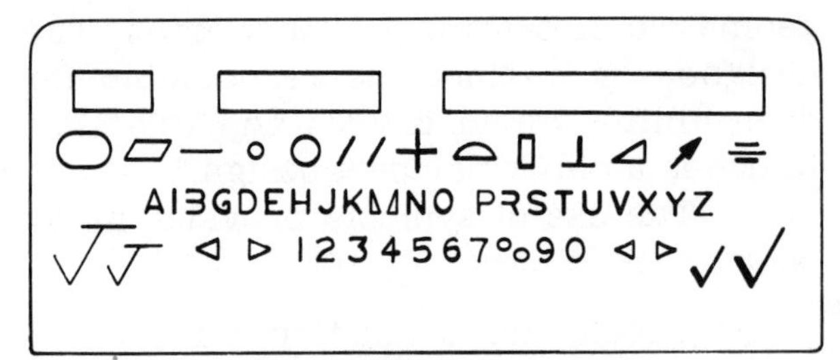

Fig. 15-3 Geometric template

3. The use of internationally standardized symbols can eliminate the many problems and errors possible in the translation of written notes.

4. Symbols use less print space. Therefore, they can be located closer to the part feature that needs to be controlled.

5. Symbols can give information which would be difficult to express by written notes.

Symbols may thus eliminate the "storybook" phase evident in prints not using the geometric symbols. See Figure 15-4.

GEOMETRIC NOMENCLATURE (TERMINOLOGY)

Knowledge of the following rules and concepts and their definitions is essential for studying the geometric dimensioning system.

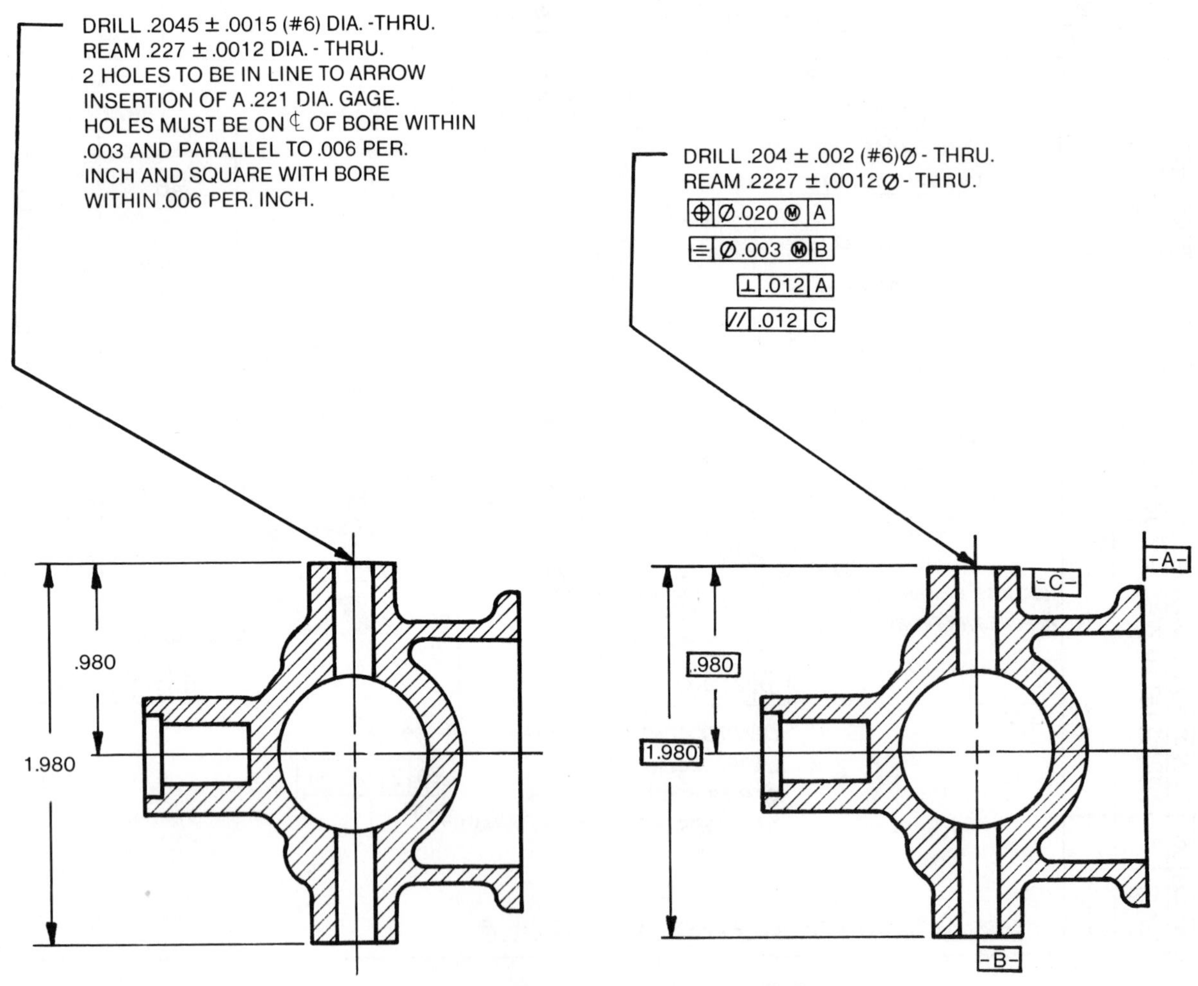

Fig. 15-4 Notes compared to symbols

Angularity—The condition of a surface or line which is at a specified angle (other than 90°) from the datum plane or axis. The symbol for angularity is ∠.

Basic Dimension—A dimension specified on a drawing as *BASIC* is a theoretical value used to describe the exact size, shape, or location of a feature. It is used as a basis from which permissible variations are established by tolerance on other dimensions or in notes. A basic dimension can be identified by the abbreviation *BSC*, or more readily by boxing in the dimension: [1.525].

Circularity (Roundness)—A tolerance zone bounded by two concentric circles within which each circular element of the surface must lie. The symbol for roundness is ○.

Concentricity—The condition in which the axes of all cross-sectional elements of a feature's surface of revolution are common. The symbol for concentricity is ◎.

Cylindricity—The condition of a surface of revolution in which all points of the surface are equidistant from a common axis, or for a perfect cylinder. The symbol for cylindricity is ⌭.

Datum—A point, line, plane, cylinder, etc., assumed to be exact for purposes of computation from which the location or geometric relationship of other features of a part may be established. A datum identification symbol contains a letter (except *I*, *O*, and *Q*) placed inside a rectangular box: [-A-].

Datum Target—A specified point, line, or area on a part used to establish a special datum for manufacturing or inspection purposes. An example of the symbol for a datum target is (Ø.250 / A 1).

Feature—Any component portion of a part that can be used as a datum, such as a hole, thread, surface, or chamfer.

Feature Control Frame—A rectangular box containing the geometric symbol, datum reference, tolerance, and modifiers relating to the specific controlled feature. An example of such a symbol is [// | .002Ⓜ | A].

Fit—The general term used to specify the range of tightness or looseness which results from application of a specific combination of allowance and tolerance in mating parts. The four general types of fits are *clearance, interference, transition,* and *line.*

Flatness—The condition of a surface having all elements in one plane. A flatness tolerance specifies a tolerance zone confined by two parallel planes within which the surface must lie. The symbol for flatness is ▱.

Form Tolerance—This tolerance controls the condition of straightness, flatness, roundness, and cylindricity. The amount of tolerance indicates how much the controlled surface or feature may vary from the desired form, as stated on the print.

Least Material Condition (LMC)—The condition of a part feature when it contains the least amount of material. The term is opposite from maximum material condition. The symbol of this modifier is Ⓛ.

Location Tolerance—This tolerance controls the condition of true position and concentricity. The indicated tolerance limits the amount the finished part may vary from the desired dimension.

Maximum Material Condition (MMC)—The condition of a part feature when it contains the maximum amount of material. The symbol of this modifier is Ⓜ.

Modifier—The term used to indicate the use of the least material condition, maximum material condition, or regardless of feature size principles.

Parallelism—The condition of a surface or axis which is equidistant at all points from a datum plane or axis. The symbol for parallelism is // or ‖.

Perpendicularity—The condition of a surface, line, or axis which is at a right angle (90°) from a datum plane or datum axis. The symbol for perpendicularity is ⊥.

Profile Of Any Line — The condition limiting the amount of profile variation along a line element of a feature. Its symbol is ⌒.

Profile Of Any Surface — Similar to profile of any line, but this condition relates to the entire surface. Its symbol is ⌓.

Projected Tolerance Zone — A zone applied to a hole in which a pin, stud, screw, etc., is to be inserted. It controls the perpendicularity of any hole which controls the fastener's position; this will allow the adjoining parts to be assembled. The symbol is Ⓟ.

Regardless of Feature Size (RFS) — A condition in which the tolerance of form or position must be met, regardless of where the feature is within its size tolerance. The symbol of this modifier is Ⓢ.

Runout — The maximum permissible surface variation during one complete revolution of the part about the datum axis. This is usually detected with a dial indicator. The symbol for circular runout is ↗ and for total runout is ⌰.

Straightness — The condition in which a feature of a part must be a straight line. Its symbol is —.

Symmetry — A condition wherein a part or feature has the same contour and size on opposite sides of a central plane. The symbol for symmetry is ⌯.

Three-plane Concept — Most geometric dimensioning makes reference to datum reference features. Therefore, it is important to understand how parts are related to a reference system composed of three mutually perpendicular planes. The designer selects, as starting places for dimensions, those surfaces or other features most important in the functioning of the part. When magnified, flat surfaces of manufactured parts have many geometric irregularities, Figure 15-5. This means that in order to apply the three-plane concept, it is necessary to determine the minimum number of high points that would have to touch the three planes if placed in contact with them. An example of how datum planes are established from the part surfaces is shown in Figure 15-6.

Tolerance — The total amount a specific dimension may vary; the tolerance is the difference between the limits.

Total Indicator Reading (TIR) — The total movement of a dial indicator as applied to a surface when the part is rotated about its datum axis. The term *total indicator reading* (TIR) means the same as the terms *full indicator reading* (FIR) and *full indicator movement* (FIM).

True Position — This term denotes the theoretically exact position of a feature. Its symbol is ⊕.

ILLUSTRATION OF TERMS

The following illustrations and explanations are helpful in more precisely defining the terms used in geometric tolerancing.

Basic Dimensions

In Figure 15-6, the abbreviation *BSC* is encased in rectangular boxes. Such notations identify basic dimensions, which are theoretically exact. They are used to describe the location, size, or shape of a feature, from which permissible variations

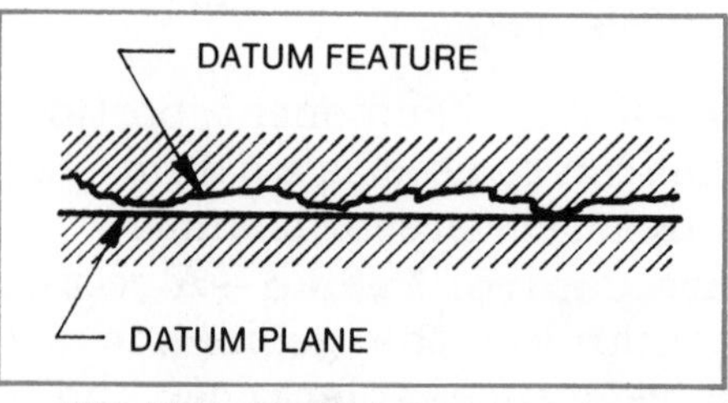

Fig. 15-5 Surface irregularities

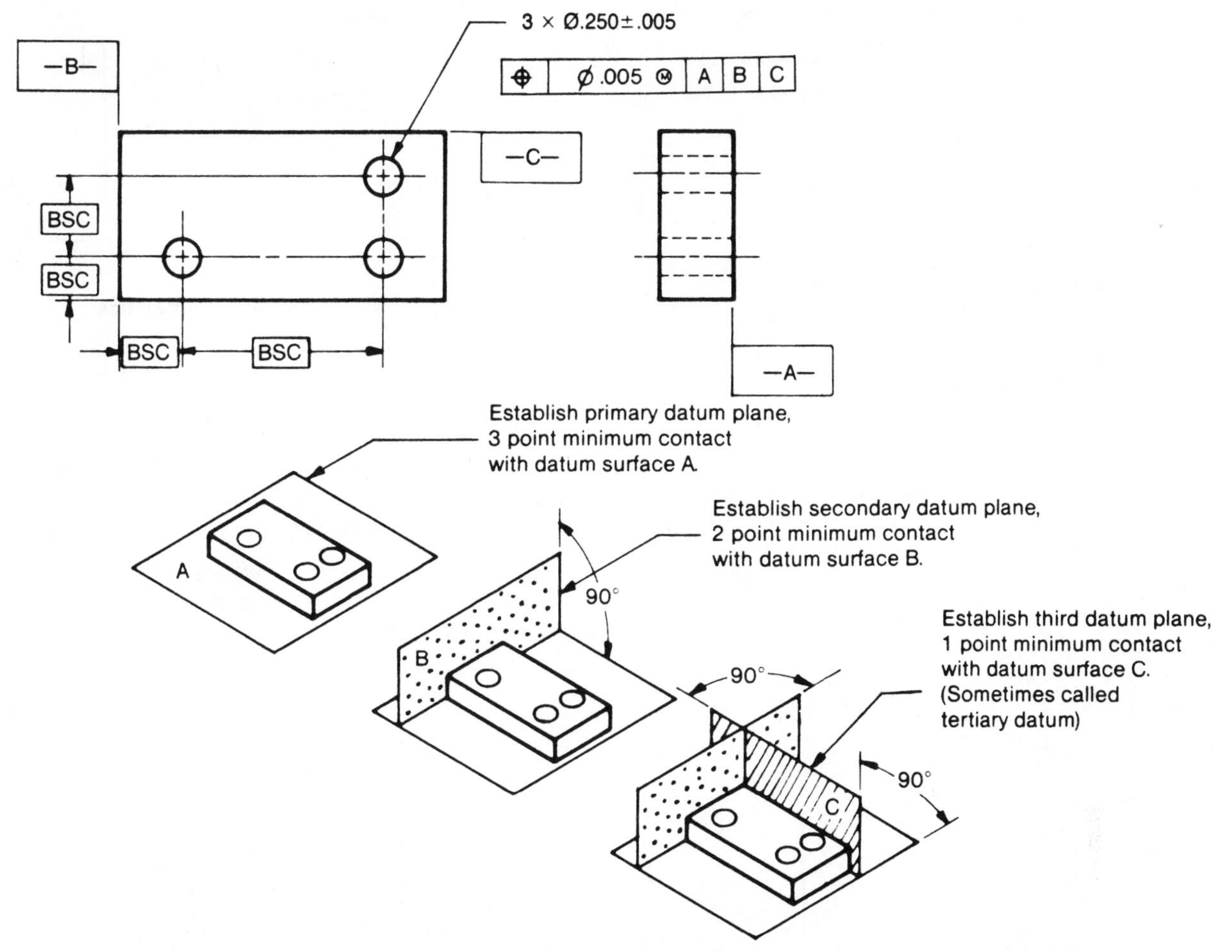

Fig. 15-6 Datum—three-plane concept. It is from these three mutually perpendicular planes that the designer dimensions surfaces or other features of the part. Datum planes are assumed to exist, not on the part itself, but on the more precisely made manufacturing and inspection equipment, such as machine tables, surface plates, angle plates, and sine bars. These represent the theoretically perfect counterparts of the specified features on a part.

are established by tolerances on other dimensions or notes. An example is [1.750] or [BSC].

For parts that are cylindrical or that have cylindrical surfaces, center lines are used as reference datums. This method is used particularly for parts which are machined between centers. See Figure 15-7.

A short cylindrical part may use one flat surface as the primary datum. The surface selected will most likely be that surface which will contact its mating part. See Figure 15-8, datum [-K-].

Datum Targets

Sometimes a surface condition such as rough casting, warpage, or weld projection is so rough that the entire surface cannot be used as a datum feature. In such cases, points, lines, or small areas

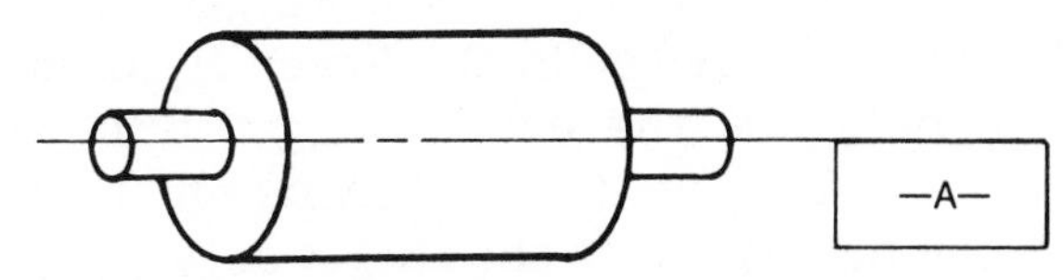

Fig. 15-7 Center line as a reference datum

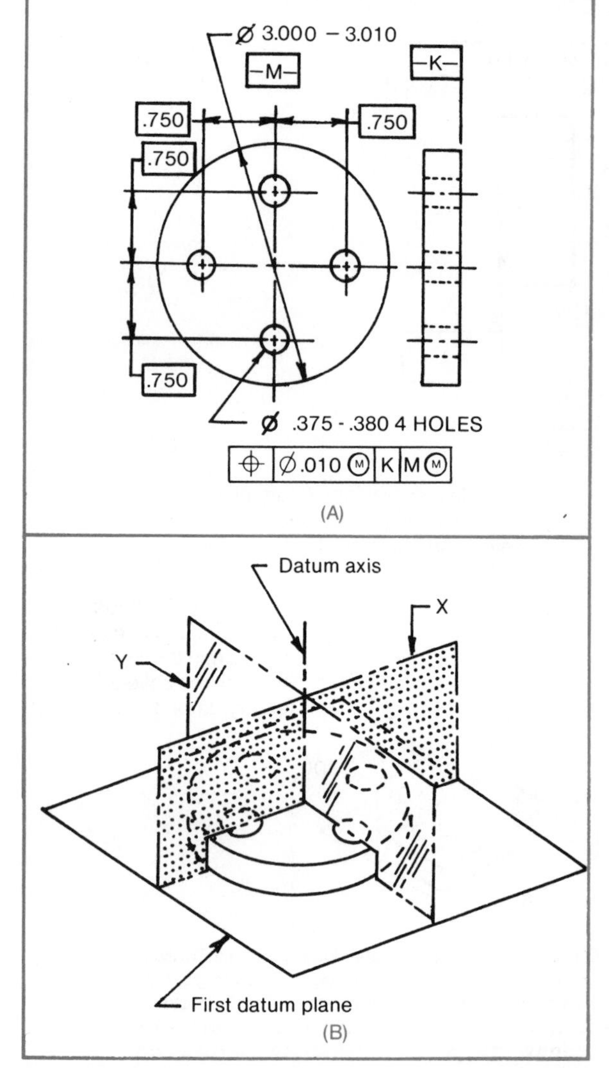

Fig. 15-8 Example of a part with a cylindrical datum feature (*Courtesy of ASME, extracted from ANSI Y14.5m-1982*)

on a part will be used for datums. These are called *datum targets*.

An example of a datum target symbol is (Ø.250 / A 1). The letter in the lower left corner of the symbol identifies the datum surface on which the datum target will be found. The number in the lower right corner indicates the datum target number, such as 1, 2, or 3. The dimension in the upper half specifies the target area size, if applicable.

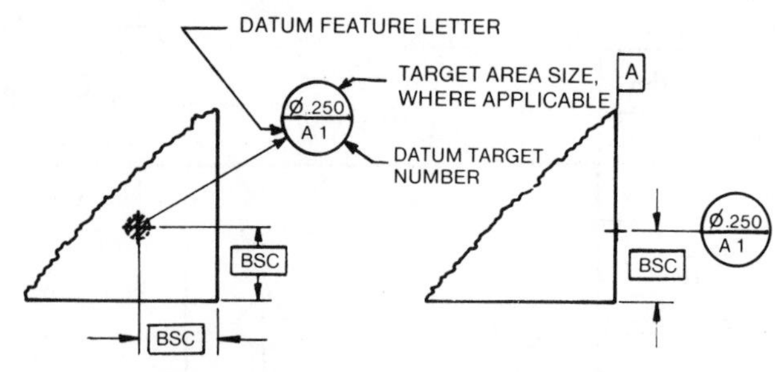

Fig. 15-9 Datum target notations

The locations of datum targets are controlled by basic or untoleranced dimensions, and imply exactness within standard tooling or gaging practices. See Figure 15-9. Datum target contacts are shown in Figures 15-10, 15-11, and 15-12.

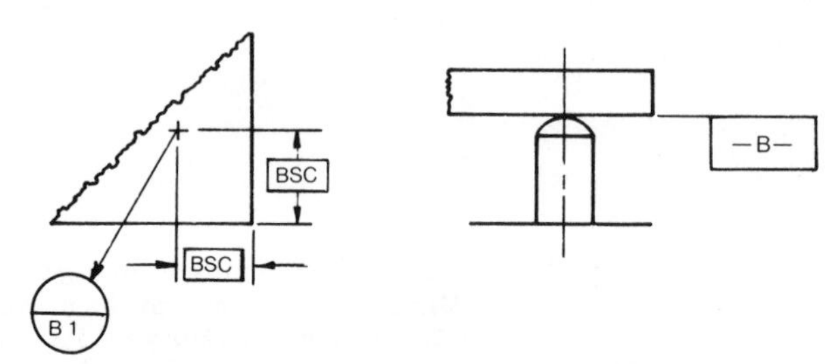

Fig. 15-10 Datum target point contact

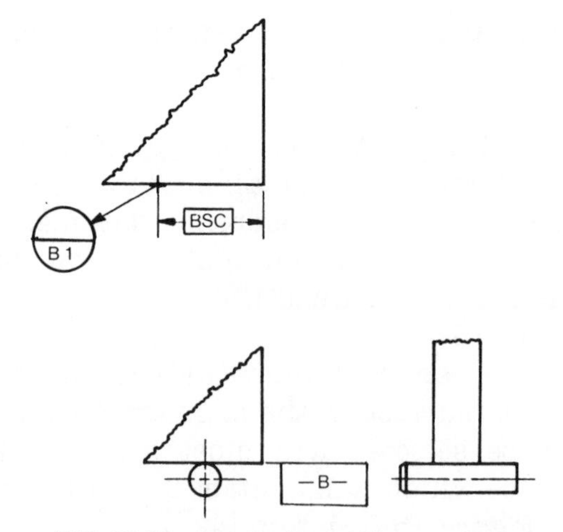

Fig. 15-11 Datum target line contact

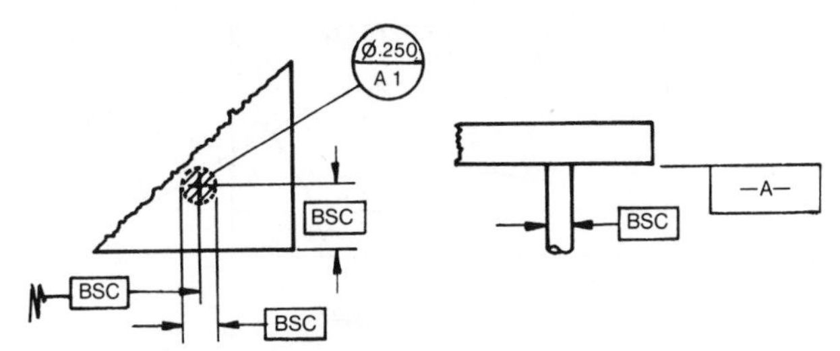

Fig. 15-12 Datum target area contact

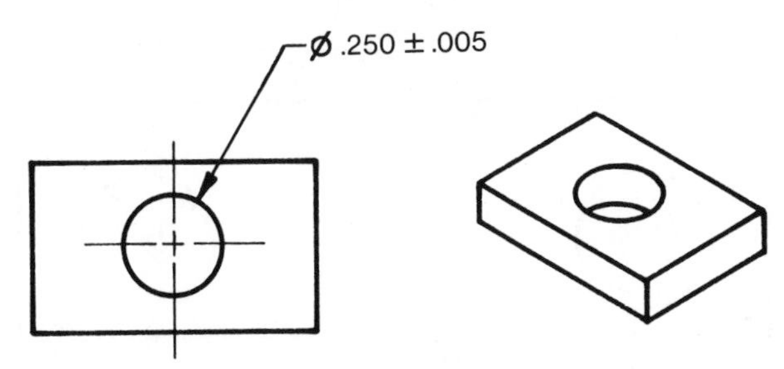

Fig. 15-14 Maximum material condition of a hole

Modifiers

Maximum material condition (MMC), a modifier, is one of the most important principles involved in geometric dimensioning and tolerancing. A good understanding of this principle is important to completely comprehend geometric dimensioning. Maximum material condition is the condition of a part feature when it contains the maximum amount of material. An example is the maximum shaft diameter.

The symbol for maximum material condition is Ⓜ or *MMC.* A feature identified as MMC is permitted greater positional or form tolerance. As the feature's size changes from MMC size, the additional tolerance is equal to the amount of size change. (It is the "green light" for more tolerance on that feature.)

The MMC size of the shaft in Figure 15-13 is .255. The MMC size of the hole in Figure 15-14 is .245.

Use of the MMC concept provides machining and assembly advantages. Part features are generally toleranced so that they will assemble when mating features are at MMC of both size and location. If bolt holes are machined larger than maximum material condition, the bolts will have more clearance. Therefore, the holes can be farther from the ideal location and still permit assembly of mating parts. In the coordinate tolerancing system, however, the print does not provide for this additional tolerance. Therefore, usable parts may be rejected. Refer to Figures 15-15, 15-16, and 15-17.

Regardless of feature size (RFS) is the condition in which tolerance of form or position must be met, regardless of where the feature lies within its size tolerance. Unlike MMC, the RFS principle permits no additional positional or form tolerance regardless of the size at which the related features are produced. (It indicates a "red light" because no additional tolerance is allowed.) The symbol for regardless of feature size is Ⓢ or *RFS.*

Least material condition (LMC) is the condition of a part feature when it contains the minimum amount of material. The 1982 standard has given approval to the LMC concept; the symbol is Ⓛ or *LMC.*

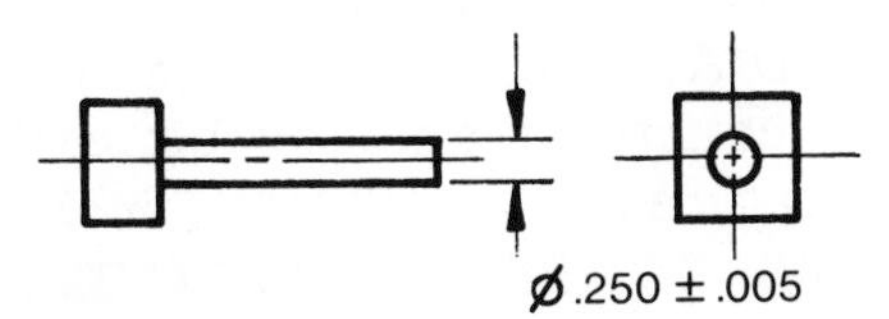

Fig. 15-13 Maximum material condition of a shaft

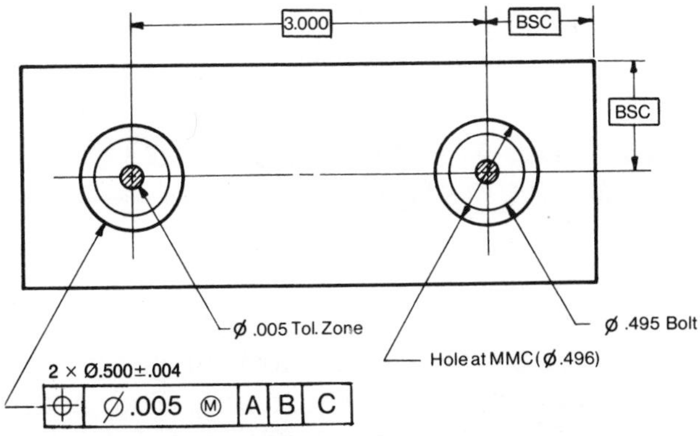

Fig. 15-15 Perfect hole location at MMC (smallest hole)

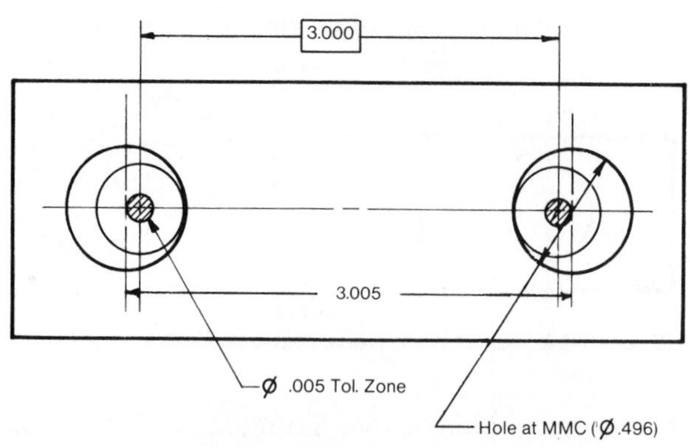

Fig. 15-16 Holes offset at MMC (smallest hole)

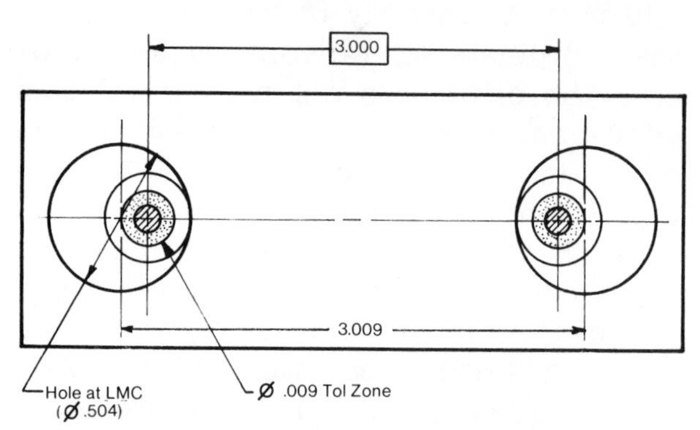

Fig. 15-17 Holes offset at LMC (largest hole)

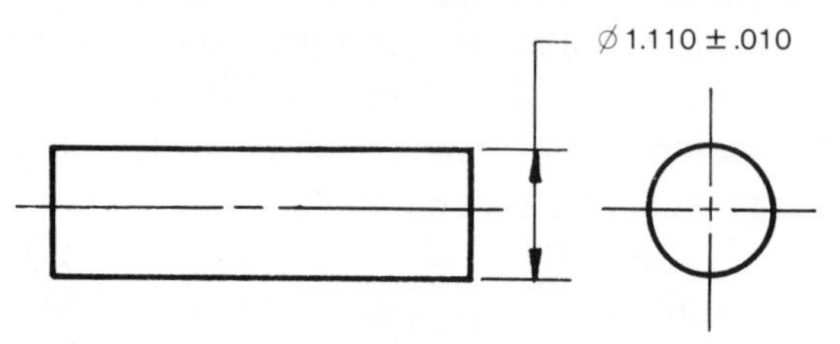

Fig. 15-18 No form specified–MMC. If the entire shaft diameter were at 1.120 size, then it should be perfectly straight and would pass through a 1.120 diameter straightness gauge, regardless of the part length.

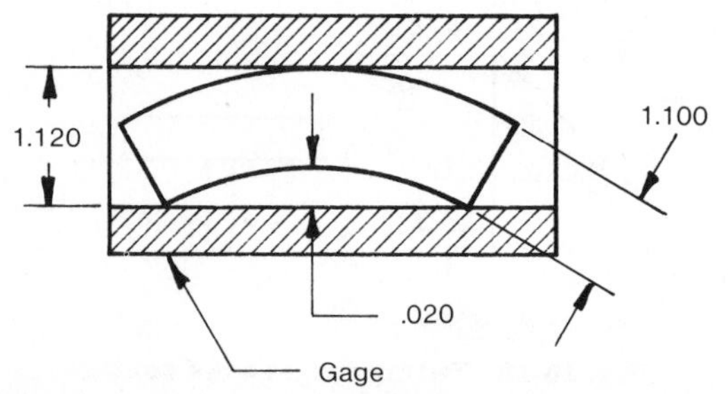

Fig. 15-19 No form specified–LMC. If the shaft diameter was at 1.100 size, then it could be out of straightness by as much as 0.020 and still be within tolerance.

RULES OF GEOMETRIC DIMENSIONING

Like most new conventions, some basic rules were established for geometric dimensioning and tolerancing. The *ANSI Y14.5M—1982* standard contains five general rules that should be understood.

RULE #1: NO FORM SPECIFIED. Where no tolerance of form is specified, the toleranced dimension for the size of the feature controls the form as well as the size, Figures 15-18 and 15-19. No element of the actual feature shall extend beyond the specified high or low limits of size, Figure 15-20.

RULE #2: POSITION TOLERANCE. When tolerance of position ⊕ is being controlled, the RFS, MMC, or LMC modifier must be specified on the drawing in respect to the individual tolerance or datum reference.

RULE #3: NO MODIFIER. When no modifier is specified, the RFS Ⓢ condition applies, Figure 15-21. The MMC Ⓜ modifier must be specified where required, Figure 15-22. Specifying the RFS is optional, Figure 15-23.

RULE #4: SCREW THREADS. Each tolerance of orientation, position, and datum reference specified for a screw thread applies to the axis of the thread derived from the pitch cylinder. When an exception to this practice is necessary, the specific feature of the screw thread (*MINOR DIA* or *MAJOR DIA*) shall be stated beneath the feature control frame or beneath the datum feature symbol, Figure 15-24.

RULE #5: GEARS AND SPLINES. Each tolerance of orientation, location, and datum reference for gears and splines

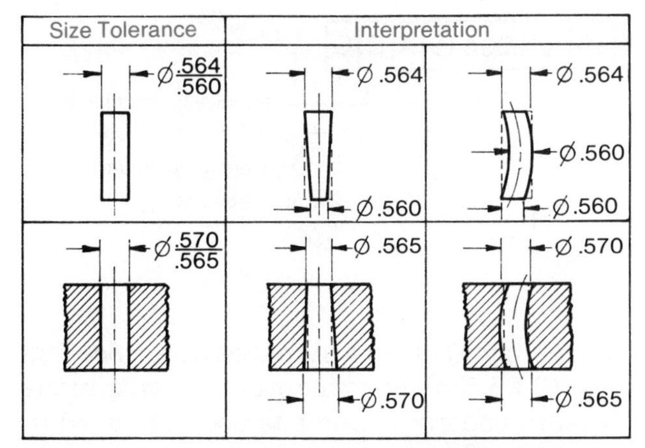

Fig. 15-20 No form specified—examples of high and low form limits. MMC indicated by phantom lines.

must indicate the specific feature of the gear or spline to which each applies (*MAJOR DIA*, *PITCH DIA*, or *MINOR DIA*). This notation shall be stated beneath the feature control frame or below the datum feature symbol, Figure 15-25.

FEATURE CONTROL FRAME

When a tolerance of position or form is related to a datum, this relationship is stated in the *feature control*

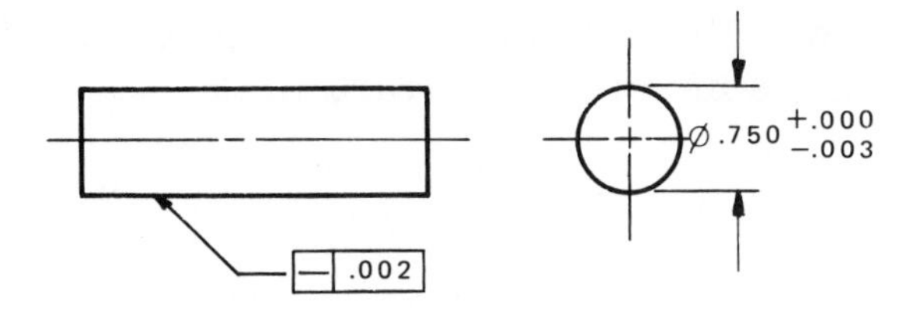

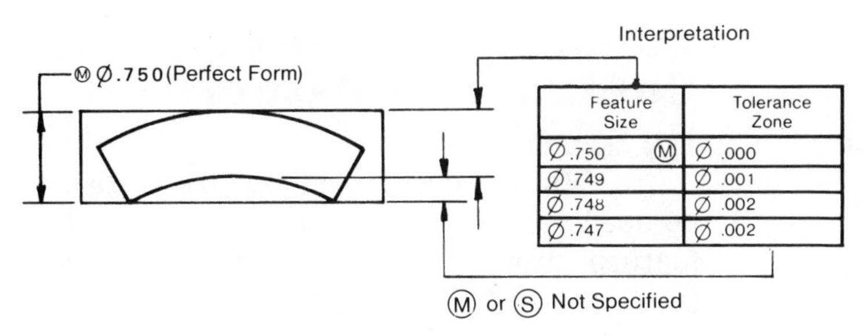

Feature Size	Tolerance Zone
Ø.750 Ⓜ	Ø.000
Ø.749	Ø.001
Ø.748	Ø.002
Ø.747	Ø.002

Fig. 15-21 Tolerance zone interpretation—no modifier specified

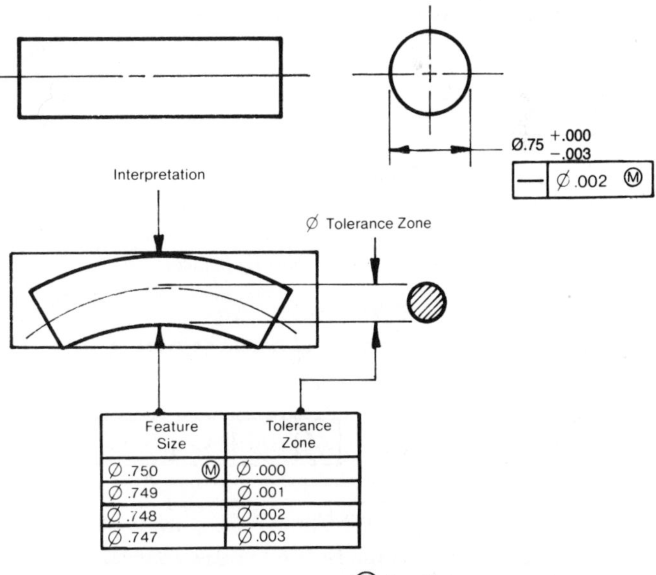

Feature Size	Tolerance Zone
Ø.750 Ⓜ	Ø.000
Ø.749	Ø.001
Ø.748	Ø.002
Ø.747	Ø.003

Fig. 15-22 Tolerance zone interpretation—MMC specified

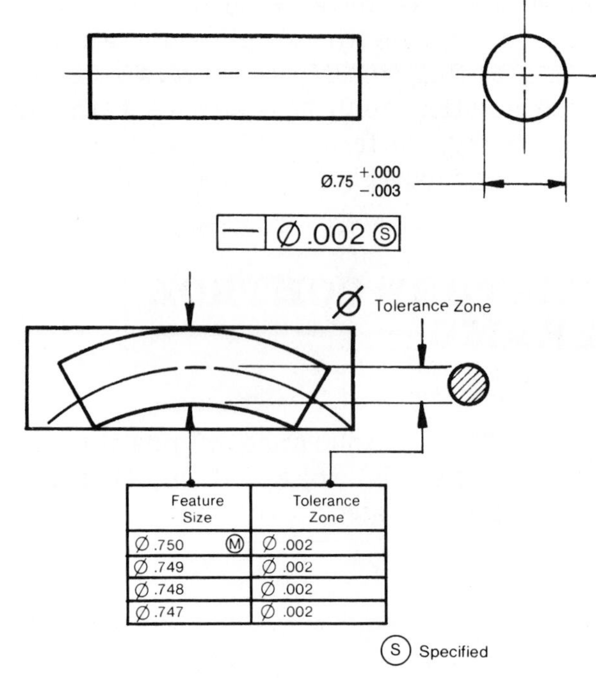

Fig. 15-23 Tolerance zone interpretation—RFS specified

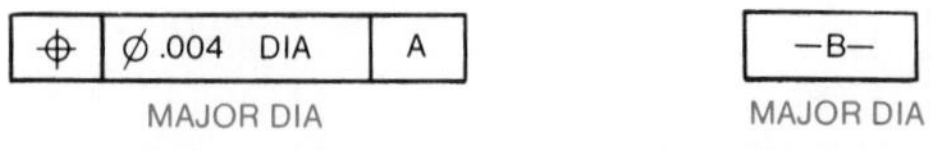

Fig. 15-24 Screw thread datum example

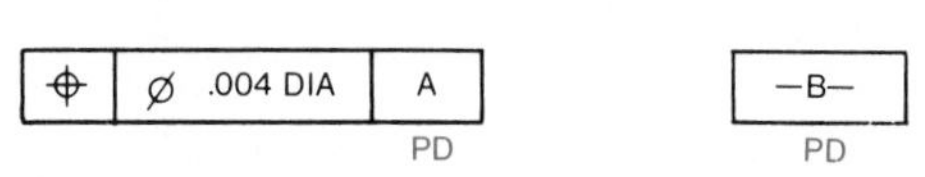

Fig. 15-25 Gear or spline datum example

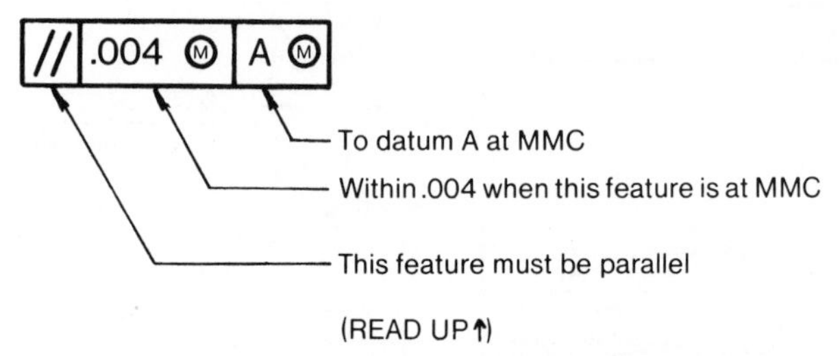

Fig. 15-26 Feature control frame (symbol)

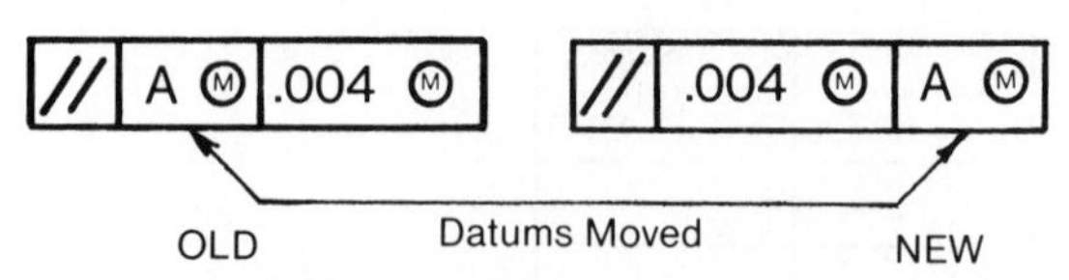

Fig. 15-27 1982 change in the feature control entries

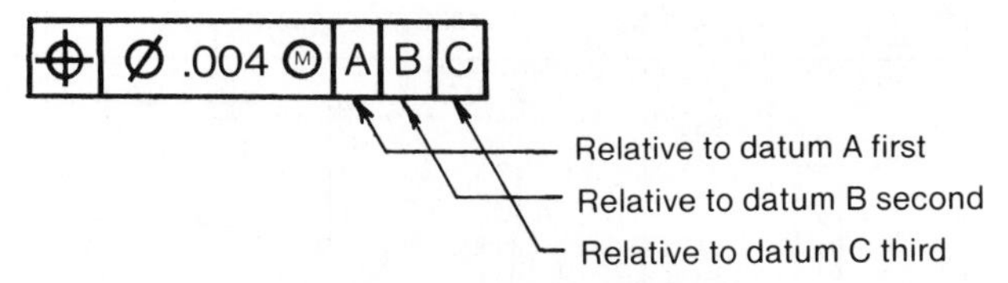

Fig. 15-28 Datum sequence

frame (symbol) in a standard order. See Figure 15-26. The sequence of entries within the feature control frame has been changed in the 1982 standard to agree with the International Organization for Standardization (ISO) standard, Figure 15-27.

When more than one datum is shown, the datums are listed in order of relative importance, Figure 15-28.

Combined Feature Control and Datum Symbol

When a controlled feature is also to be used for a datum for another controlled feature, this datum is listed below the feature control frame. See Figure 15-29.

Composite Feature Control Symbol

When more than one tolerance of a given geometric characteristic apply to the same feature, a composite feature control symbol is used. See Figure 15-30.

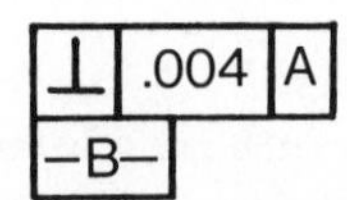

Fig 15-29 Combined feature control and datum feature frame. Interpretation: This feature must be ⊥ to datum "A" within .004 total and will be used as datum "B" for another feature.

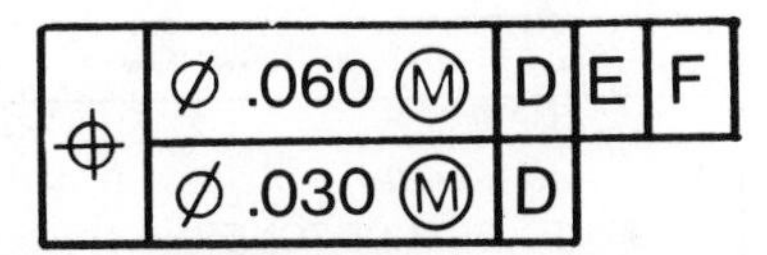

Fig. 15-30 Composite feature control frame. Interpretation: Two or more holes to be held within Ø.060 of true position relative to datum "D," "E," and "F" at Ⓜ and to be held to each other within Ø.030 at true position with respect to datum "D" at Ⓜ.

Special attention should be given to prints that may be used outside of the United States. In Europe and in other areas, datum symbols are usually placed after the tolerance notation, rather than preceding the notation. See Figure 15-31. The datum reference locations as shown in the top of Figure 15-31 are still what appear on most United States prints.

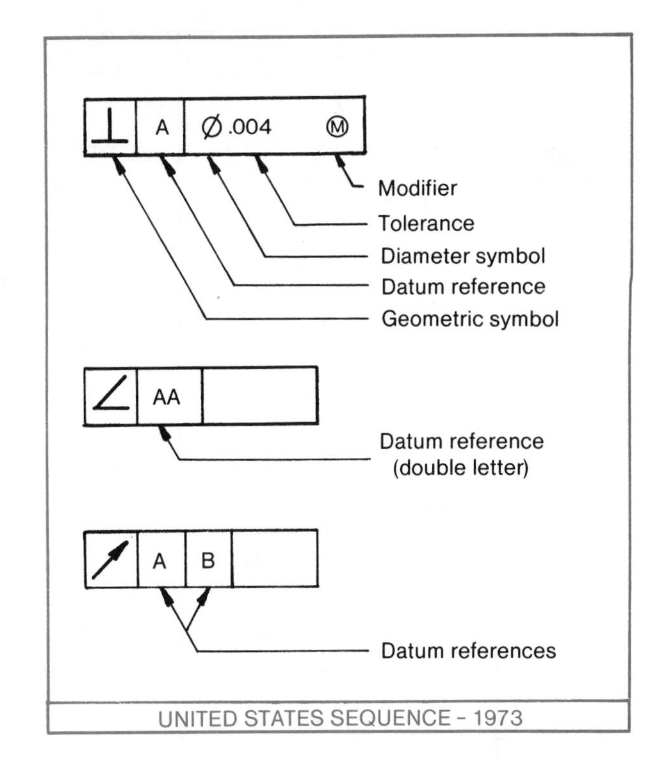

GEOMETRIC FORM CHARACTERISTICS

When tolerances of form are noted in the feature frame, the indicated amount of tolerance controls the amount the surface or features can vary from the desired form, as noted on the print. The form characteristics and tolerance zone interpretations are illustrated in Figure 15-32.

GEOMETRIC PROFILE CHARACTERISTICS

The tolerances of profile are not used as often as other geometric tolerance forms, but they do have special applications for curved surfaces. These symbols appear frequently on casting or forging part prints. The profile characteristics and their tolerance zone interpretations are illustrated in Figure 15-33.

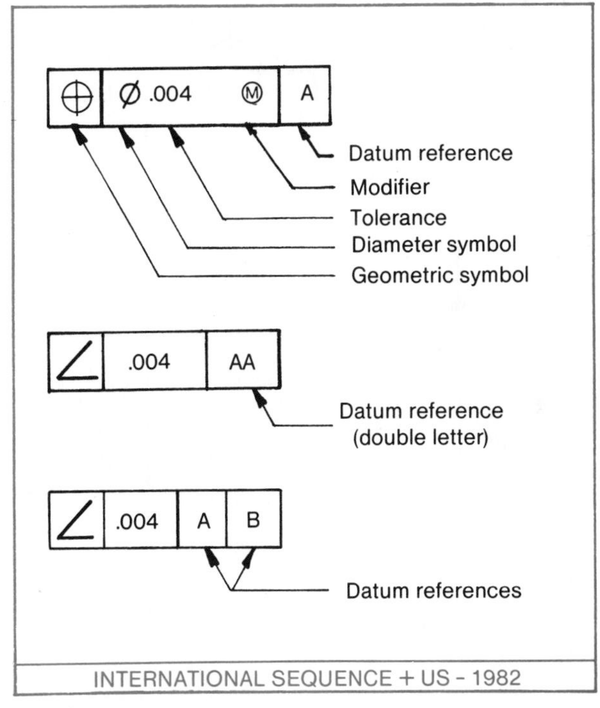

Fig. 15-31 United States and international datum locations

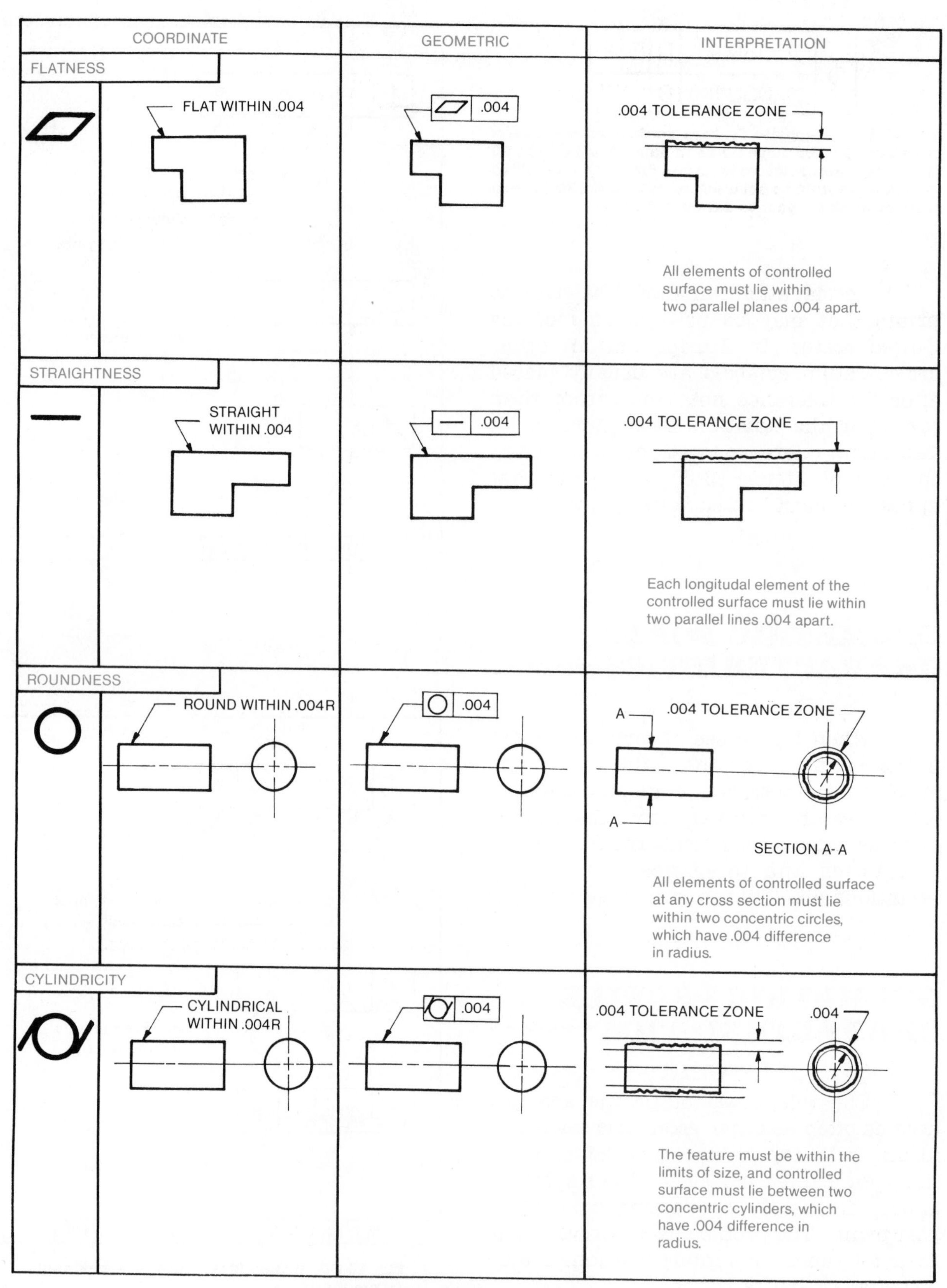

Fig. 15-32 Tolerance of form

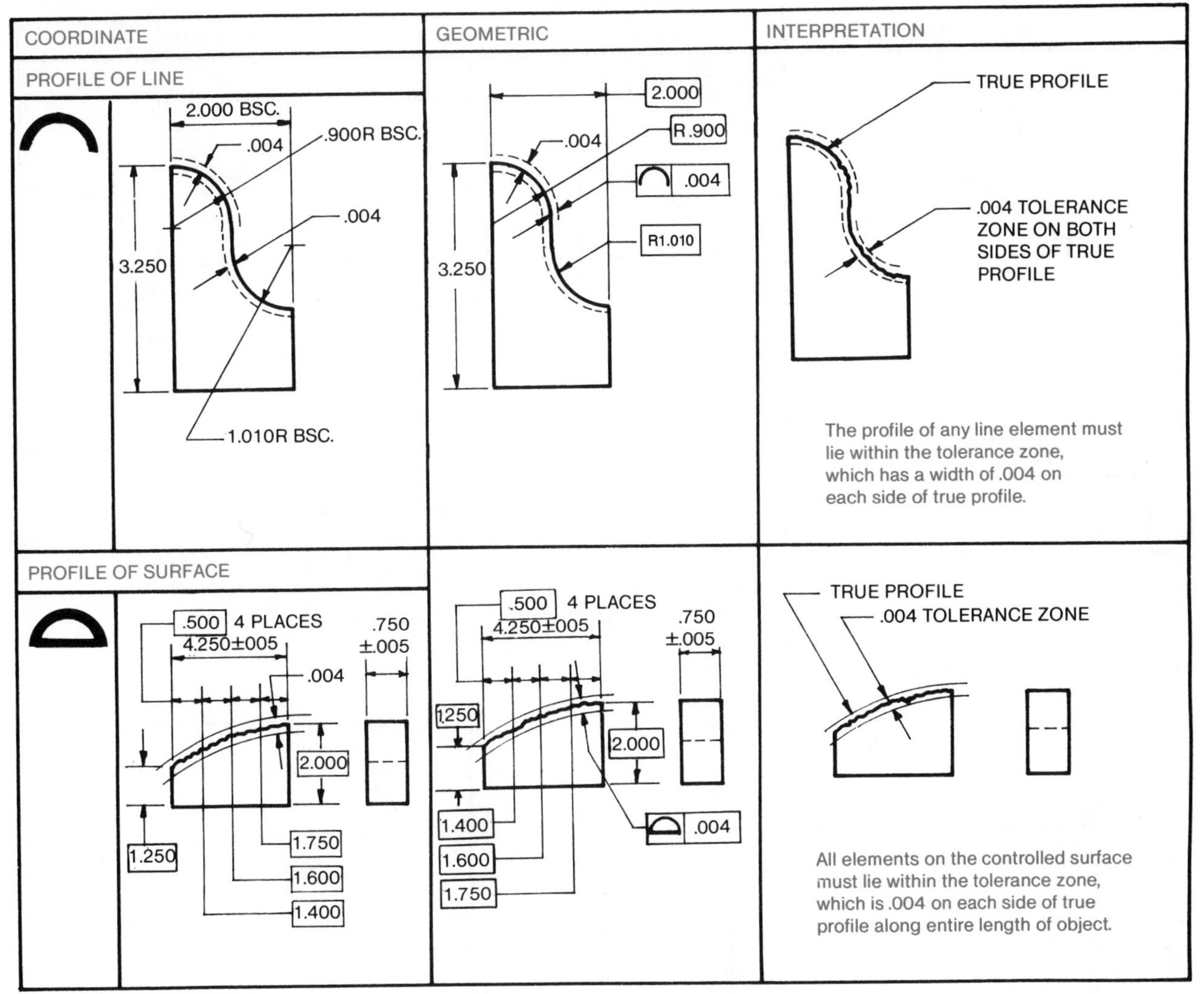

Fig. 15-33 Tolerance of profile

GEOMETRIC ORIENTATION CHARACTERISTICS

∠ ⊥ //

The tolerances of orientation are widely used on part prints which require extensive machining operations. The orientation characteristics and their tolerance zone interpretations are illustrated in Figure 15-34.

When the symbol Ⓜ is indicated in the feature control frame, a possible additional tolerance is allowed due to the actual part feature size. An example of this "bonus" tolerance is shown in Figure 15-35. Other geometric controls would have similar extra tolerance due to the same or different variables.

GEOMETRIC LOCATION CHARACTERISTICS

⊕ ◎ ⌯

When tolerances of location are noted in the feature control frame, the amount of tolerance indicated controls

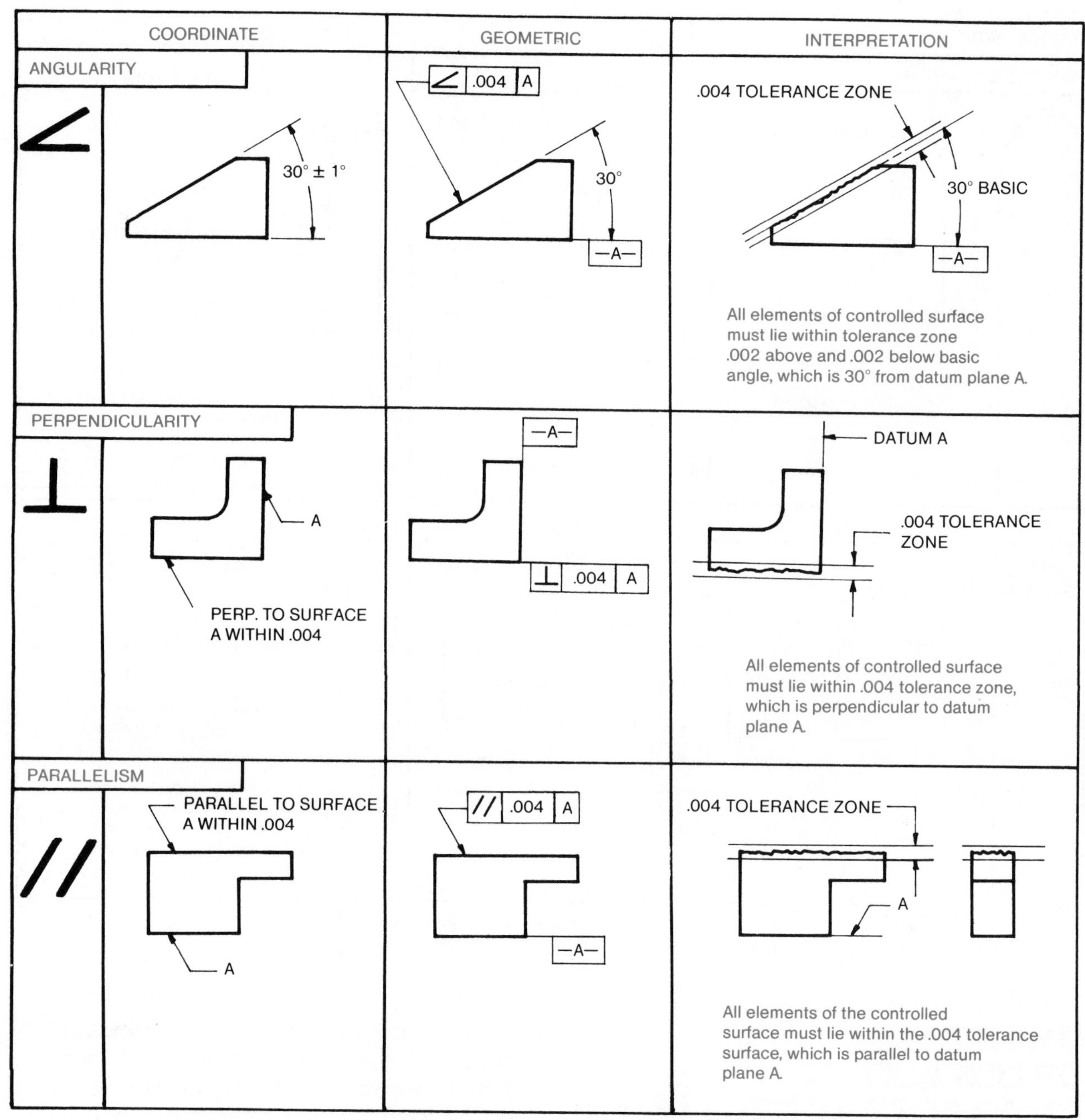

Fig. 15-34 Tolerance of orientation

how far the position of a feature may vary from the ideal or exact position.

True Position

The term *true position* denotes the theoretically exact position of the feature. True position, symbolized by ⊕, is the most frequently used locational characteristic.

A part made with all of its features produced within the print's limits of size is useless if its features are not located in their proper position. Often it is more difficult to control the position of features than to control their size. As previously

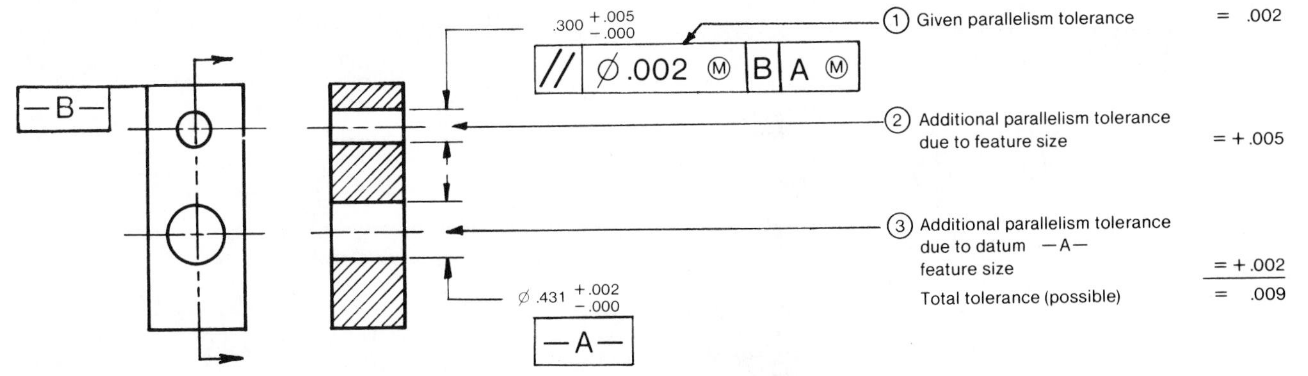

Fig. 15-35 Possible total tolerance

mentioned, a hole machined to its large limit of size will allow a greater positional variation of the fastener used in the hole. The geometric dimensioning system allows for both the size and position variables in determining the acceptability of finished parts.

True Position Tolerancing

True position tolerancing is a system used to specify the amount that a feature may vary from the true position established by basic location dimensions.

The illustrations in Figure 15-36 compare the tolerance zones in the coordinate and true position dimensioning systems. The holes and their respective size and location tolerances are the same for both parts. In the coordinate drawing (A), the ±.005 permissible in the horizontal and vertical directions results in the square positional tolerance zones. In the true position drawing (B), the positional tolerance is placed in the feature control frame. It is round, as indicated by the diameter symbol Ø preceding the .003.

The two tolerance zones from Figure 15-36 are superimposed in Figure 15-37. The square tolerance zone results from the ± coordinate system. Note that the .007 diameter zone results from constructing a circle of the same diameter

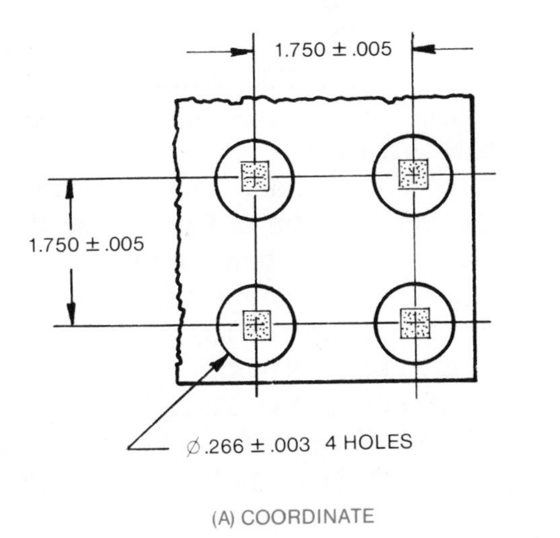

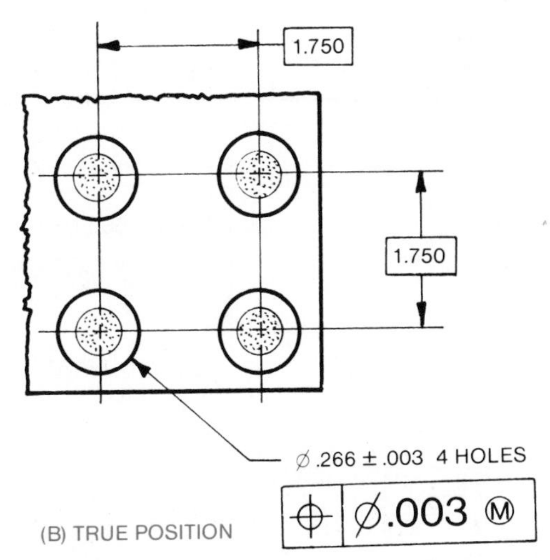

Fig. 15-36 Tolerance zones

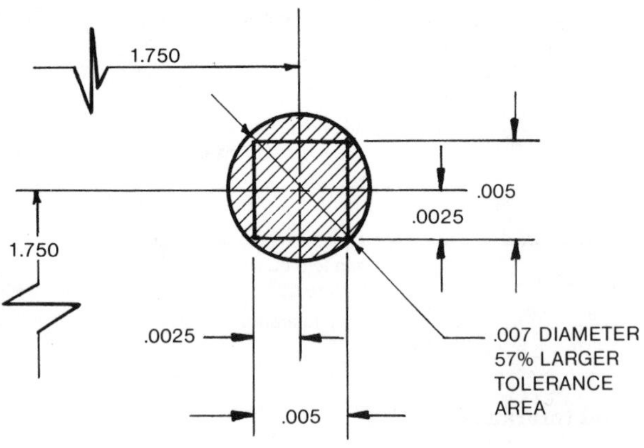

Fig. 15-37 Tolerance zones compared

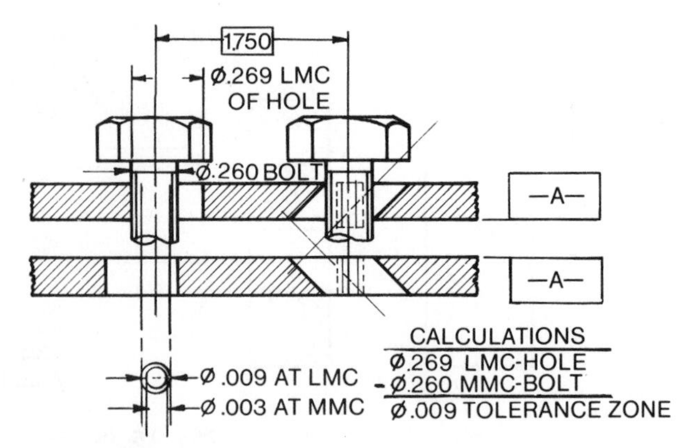

Fig. 15-39 Tolerance zone variations

passing through all four corners of the square tolerance zone.

It can easily be seen that the true position tolerance zone permits a larger area of tolerance. This tolerance calculates out to be 57% larger than that provided by the coordinate dimensioning system. Remember, the illustrations in Figures 15-36 and 15-37 are true when the features are at maximum material condition.

As the features depart from MMC, an additional tolerance will become available. This will increase the tolerance zone in the geometric system to .009, triple the original .003.

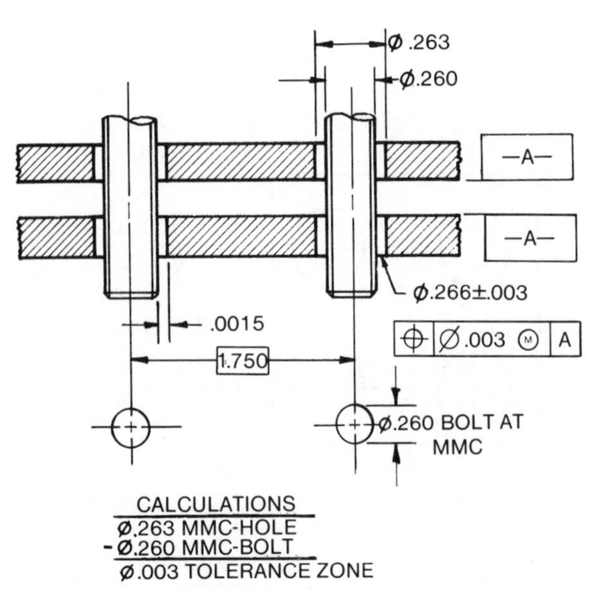

Fig. 15-38 Tolerance zone calculations

Figures 15-38 and 15-39 are visual analyses of mating features and provide justification for the tolerance zones. The two mating parts, which are to be bolted together, must satisfy the requirements of having the holes in both parts located at true position relative to their respective -A- datums within a diameter of .003 at Ⓜ. See Figure 15-38. Note that the mating surfaces of both parts are specified as datums.

How were the .003 diametral tolerance zones at Ⓜ established? With the holes and bolts all at their MMC and perfectly located, the positional tolerance permitted at those conditions can be calculated. The .263 Ⓜ hole size minus the .260 Ⓜ bolt size equals .003, which is the total diametral clearance as indicated in the feature control frame.

Therefore, under the conditions of MMC, perfect location of the indicated feature, and placing the bolts at true position, a .0015 clearance would exist between the surface of the holes and the surfaces of the bolts in all radial directions from the respective true positions. Thus, the center of each hole would be permitted to vary from its respective true position within a .003 diameter tolerance zone, and would still be functional.

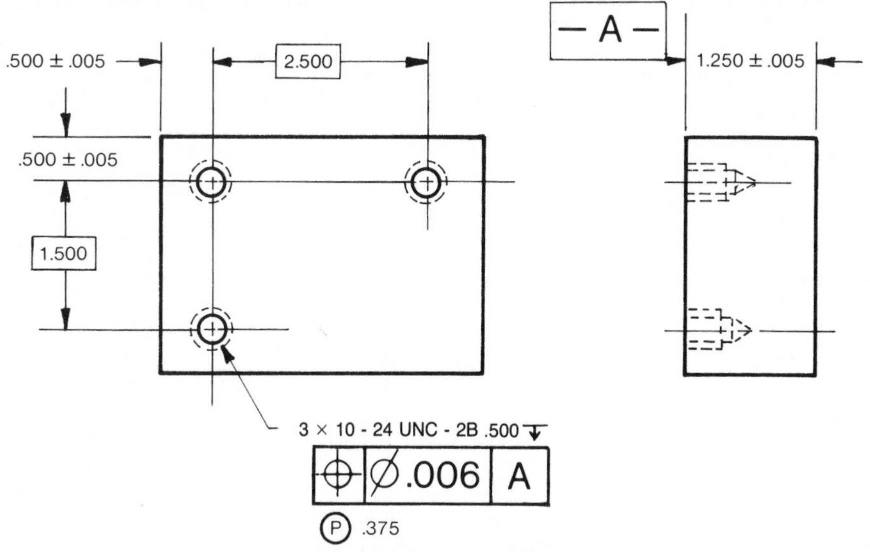

Fig. 15-40 True position

The .003 diameter positional tolerance would be required only when the features are at Ⓜ size. Additional positional tolerance would become available as the holes approach the Ⓛ of size. A total positional tolerance of .009 diameter would be available for each hole when at its high limit of size. This total tolerance zone equals the specified positional tolerance of .003 plus the total size tolerance of .006 for the feature. See Figure 15-39.

The attitude (in this case, perpendicularity) of the feature's holes, relative to primary datums -A-, may vary within their respective positional tolerance zones. Therefore, the holes in part one and part two could be out of perpendicularity, relative to their respective datum -A- features. This is possible if their theoretical center lines do not exceed the boundaries of the tolerance zones, as illustrated.

Now that the theory of true positional tolerancing has been discussed, it will be useful to illustrate how the principle is used on prints.

A typical illustration is shown in Figure 15-40. Note that the location dimensions between the holes are basic. They are coordinate dimensioned from the edges of the part. The threaded holes also carry a projection tolerance zone, the length of which is equal to the thickness of its mating part. (More information on projected tolerance zones is provided later in this chapter.)

The hole pattern with three datums is illustrated in Figure 15-41. Datum *A* is indicated first to show from which surface the datums *B* and *C* and holes are located.

Datum *B* is located from coordinate ± dimensions, but the four-hole pattern is located by basic dimensions related back to the center of the large hole. Each hole in the four-hole pattern may deviate from its true position location by the amount that the datum *B* departs from MMC. Note (a) at the top of Figure 15-41 indicates how the note would appear in a coordinate dimensioning system drawing.

When design requirements allow features to vary within a larger hole-pattern positional tolerance, a composite positional feature control symbol is used. See Figure 15-42. The top entry of the feature control symbol indicates the overall positional tolerance of the hole pattern,

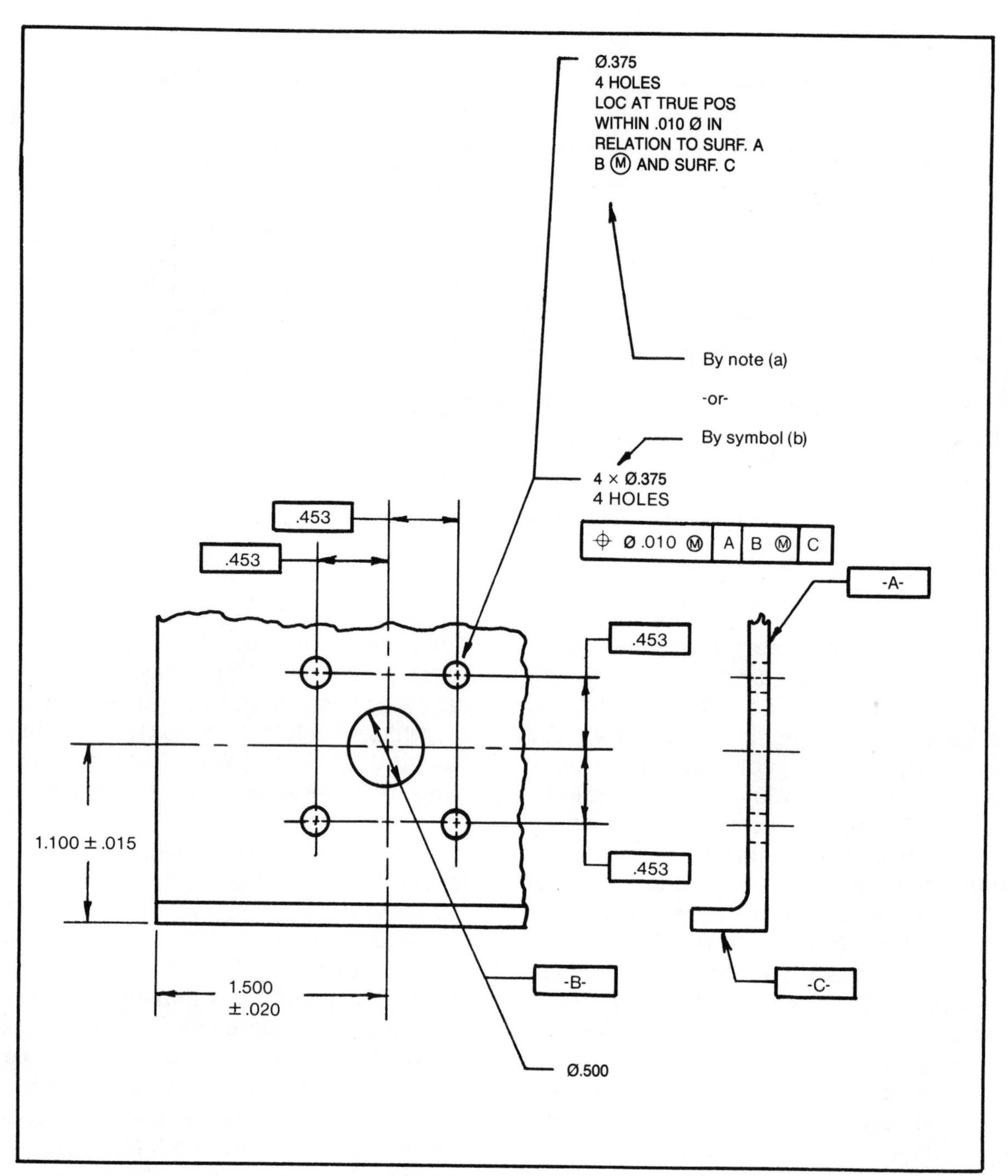

Fig. 15-41 Hole pattern with three datums

whereas the lower entry indicates the permissible tolerance variations from hole to hole.

Figure 15-43 illustrates a method used to show true position with more tolerance in one direction than the other. This is a very common usage for elongated holes.

The location characteristics and their tolerance zone interpretations are illustrated in Figure 15-44. Symmetry has been deleted in the 1982 standard. The rare need for its use is being satisfied by true position.

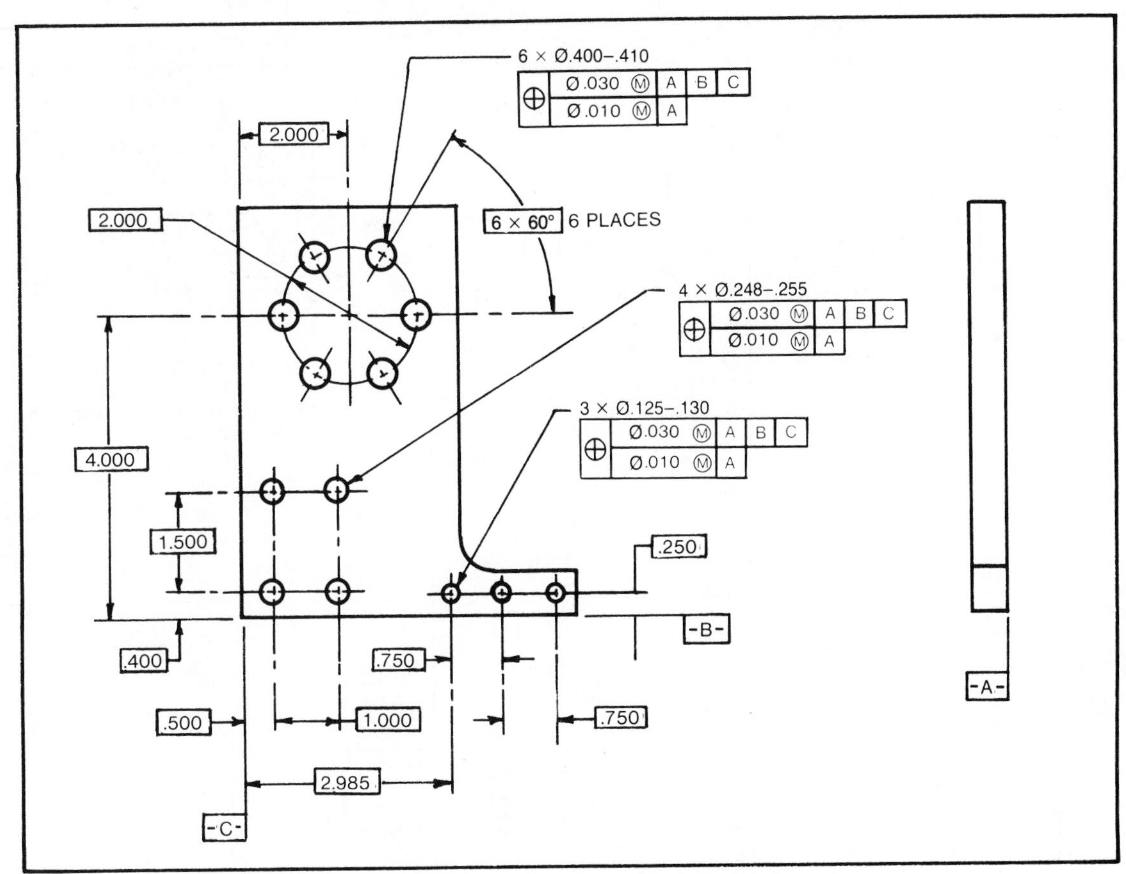

Fig. 15-42 Hole patterns located by composite positional tolerancing (*Courtesy ASME, extracted from ANSI Y14.5M–1982*)

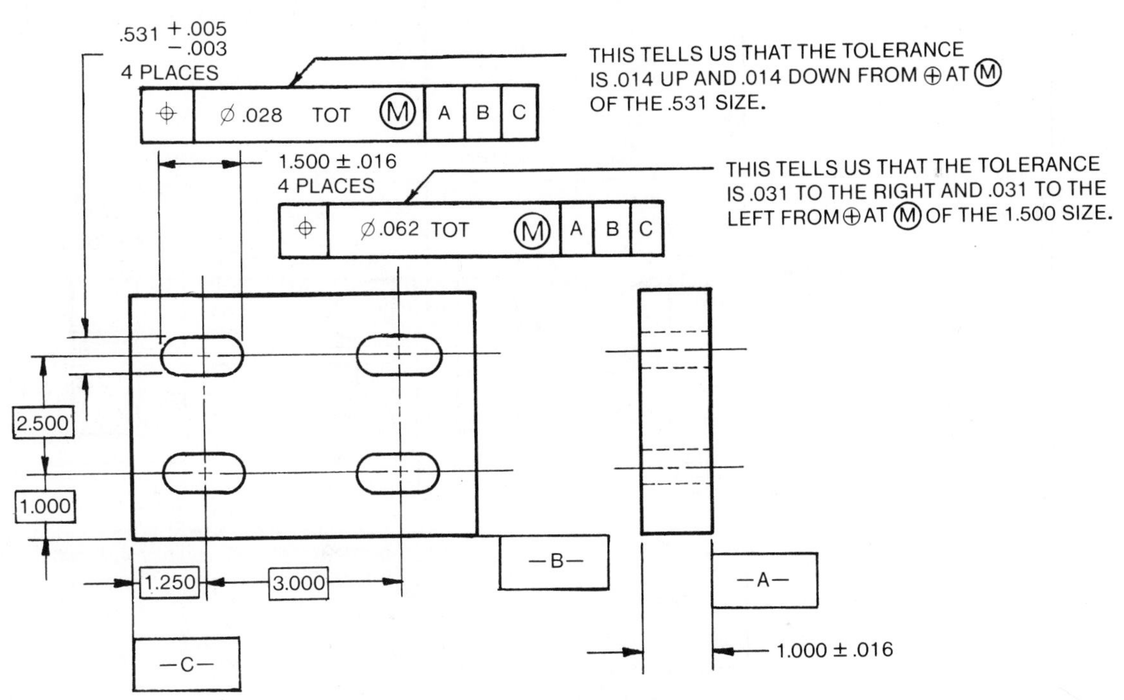

Fig. 15-43 Variable tolerances

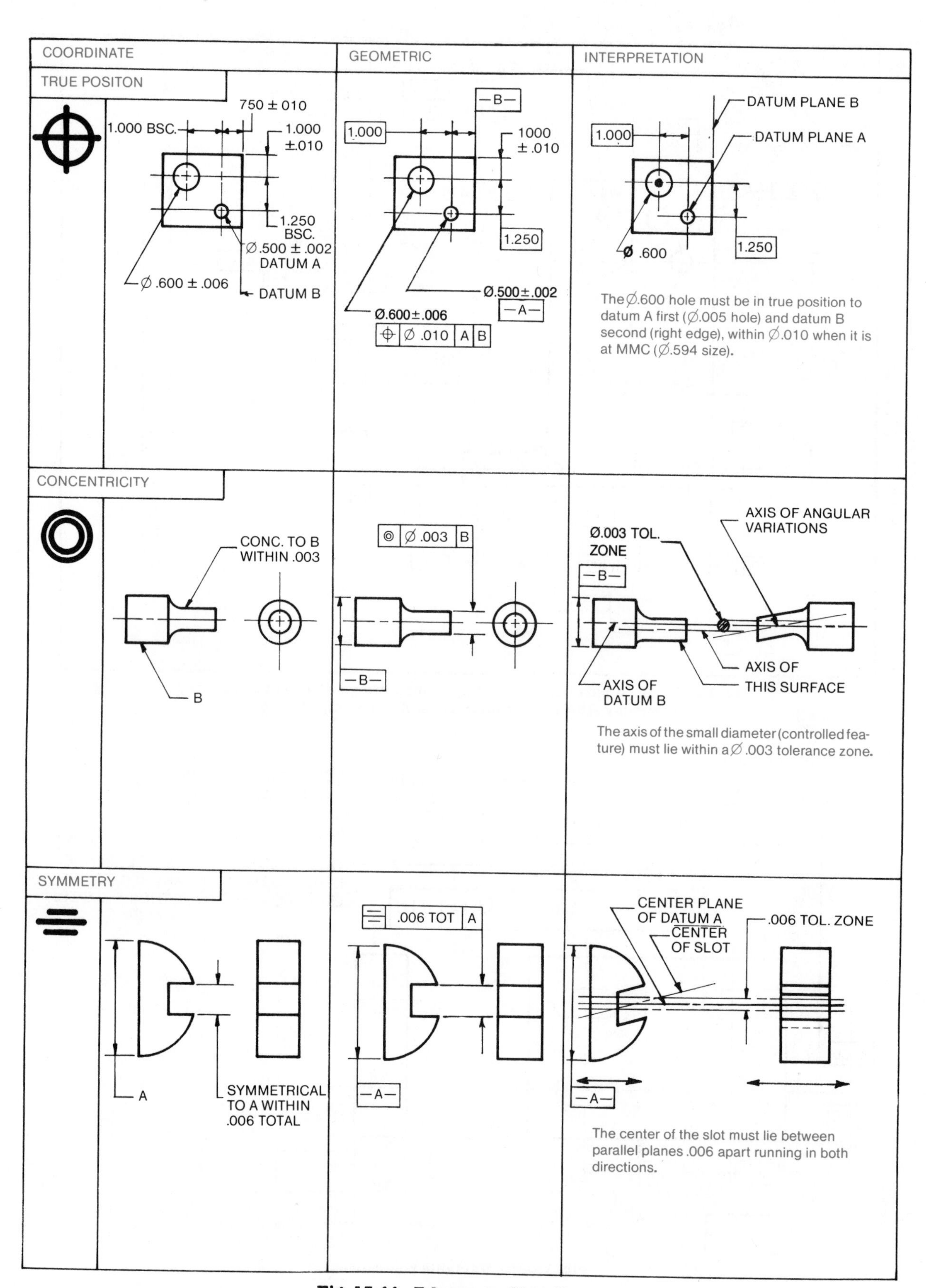

Fig. 15-44 Tolerance of location

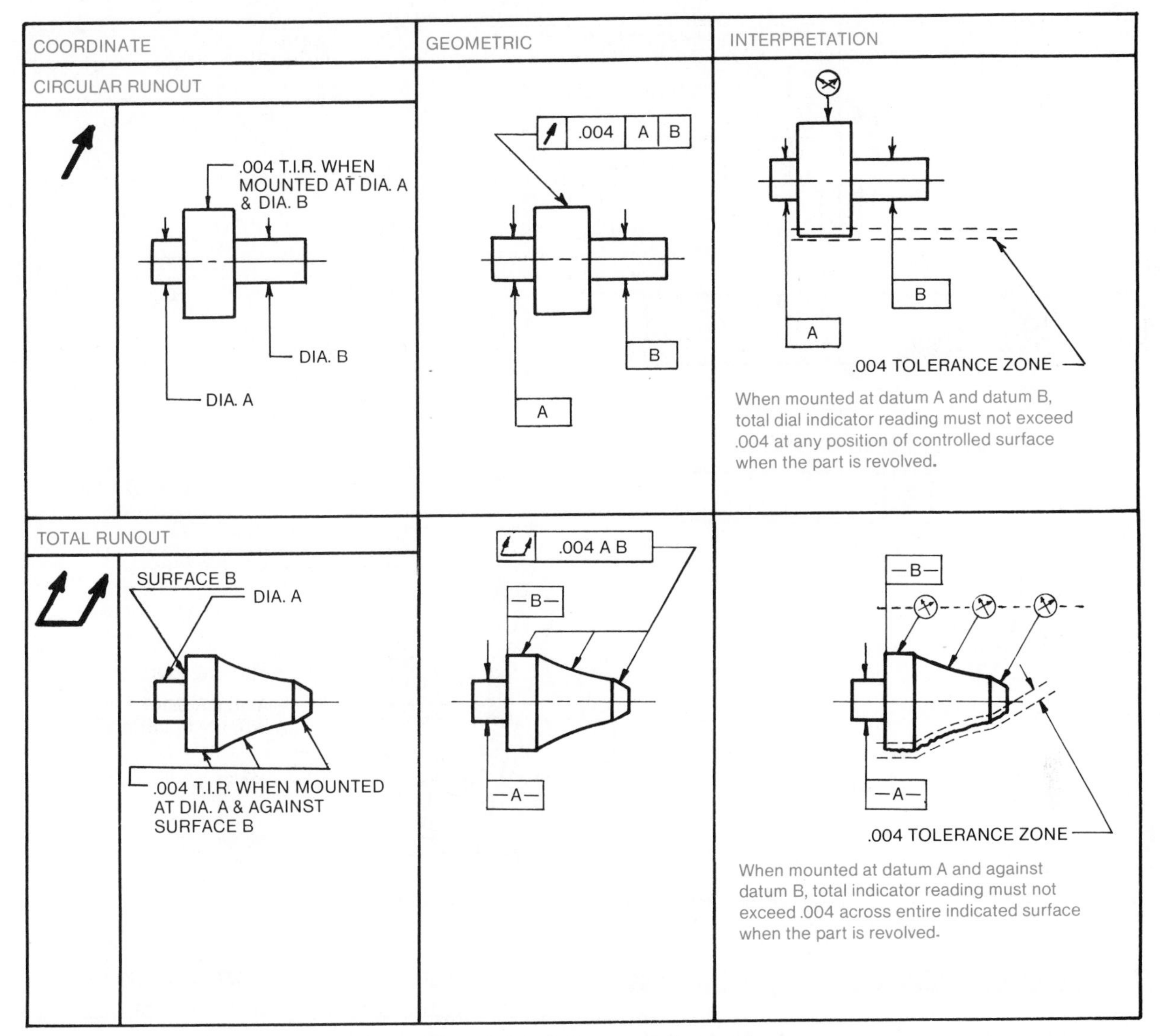

Fig. 15-45 Tolerance of runout

RUNOUT—CIRCULAR AND TOTAL

↗ ↗↗

The two runout tolerances are used extensively when machined shafts and bearing holes are features of a part. The fits on these types of features are usually very close. When components have close fits, the outer surface of a shaft or the inner surface of a hole must run true to their centers. A dial indicator is usually used to measure runout. When this dial indicator is read at a fixed position, it is called *circular runout.* When the dial indicator is moved across the surface to be measured, while the part is rotated, it is called *total runout.*

The symbol for circular runout is ↗. The 1973 standard notation for total runout was a single arrow with the word TOTAL placed below the arrow. The 1982 standard use a notation of a double arrow ↗↗ for total runout.

The runout characteristics and their tolerance zone interpretations are illustrated in Figure 15-45.

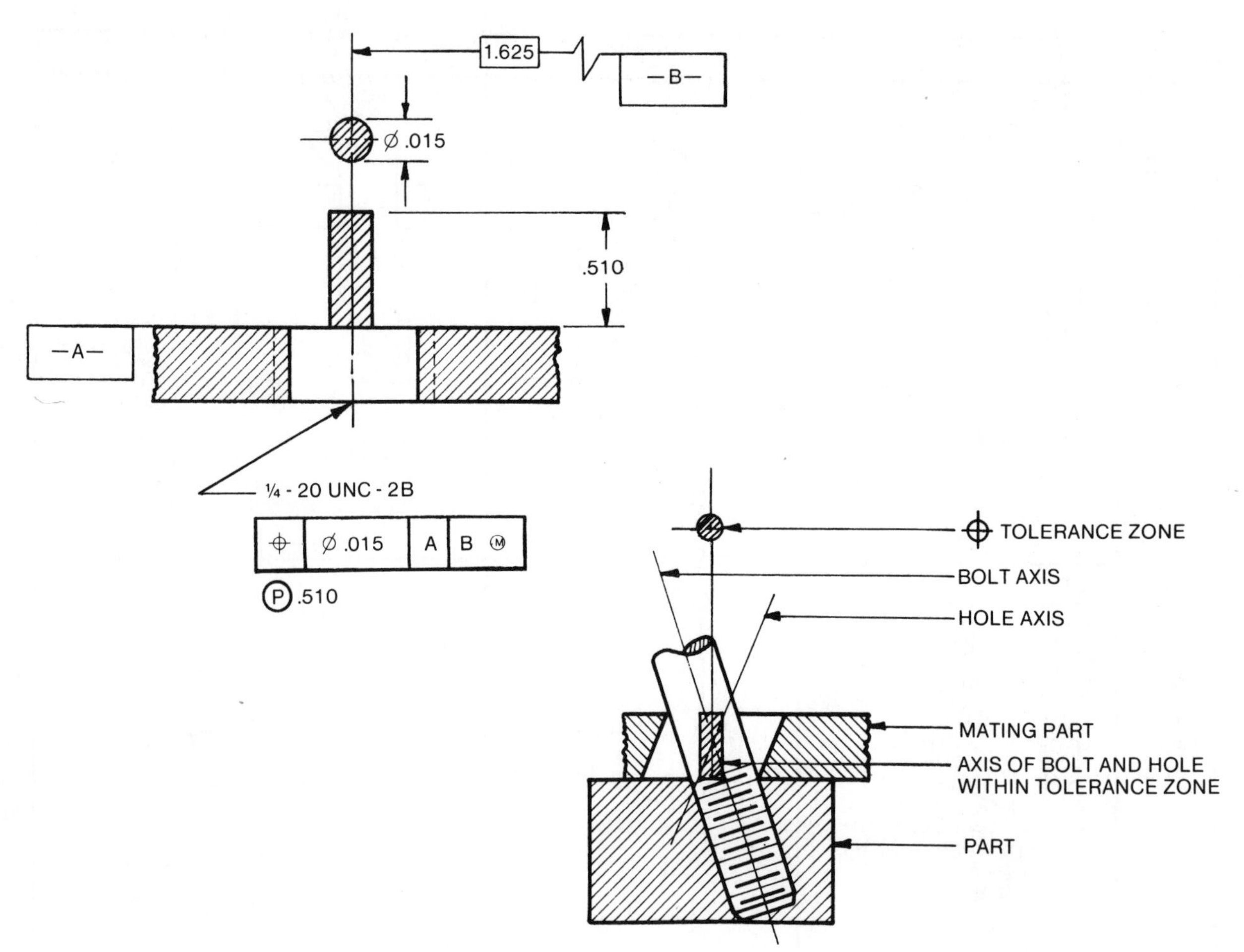

Fig. 15-46 Projected tolerance zone

Projected Tolerance Zone

The *projected tolerance zone* principle is used when a fastener device, such as a bolt or screw, must pass through a hole in a part that is attached to another part, usually threaded to the second part. The symbol for projected tolerance zone is Ⓟ. The symbol is followed by a dimension that is equal to the thickness of the mating part. This tolerance zone controls the perpendicularity of the hole to the extent of the projection from the hole and as it relates to the mating part thickness.

The projected tolerance zone extends *above* the surface of the part to the functional length of the fastener device, relative to its assembly with the mating part. See Figure 15-46.

Many companies which use the geometric dimensioning and tolerancing system have developed techniques to help those reading their prints to better understand this relatively new drafting system. Some companies display the geometric symbols immediately below the title block. Basic dimension information is placed in tolerance blocks on all of the companies' prints. See Figure 15-47. (The draft symbols D, Ð, and Ɖ are common only to this company. They are not ANSI-approved symbols.) Some companies also provide employee training to familiarize their employees with this new drafting system. A sample of the training aids used is shown in Figure 15-48.

A listing of the geometric characteristic symbols of the 1982 standard is

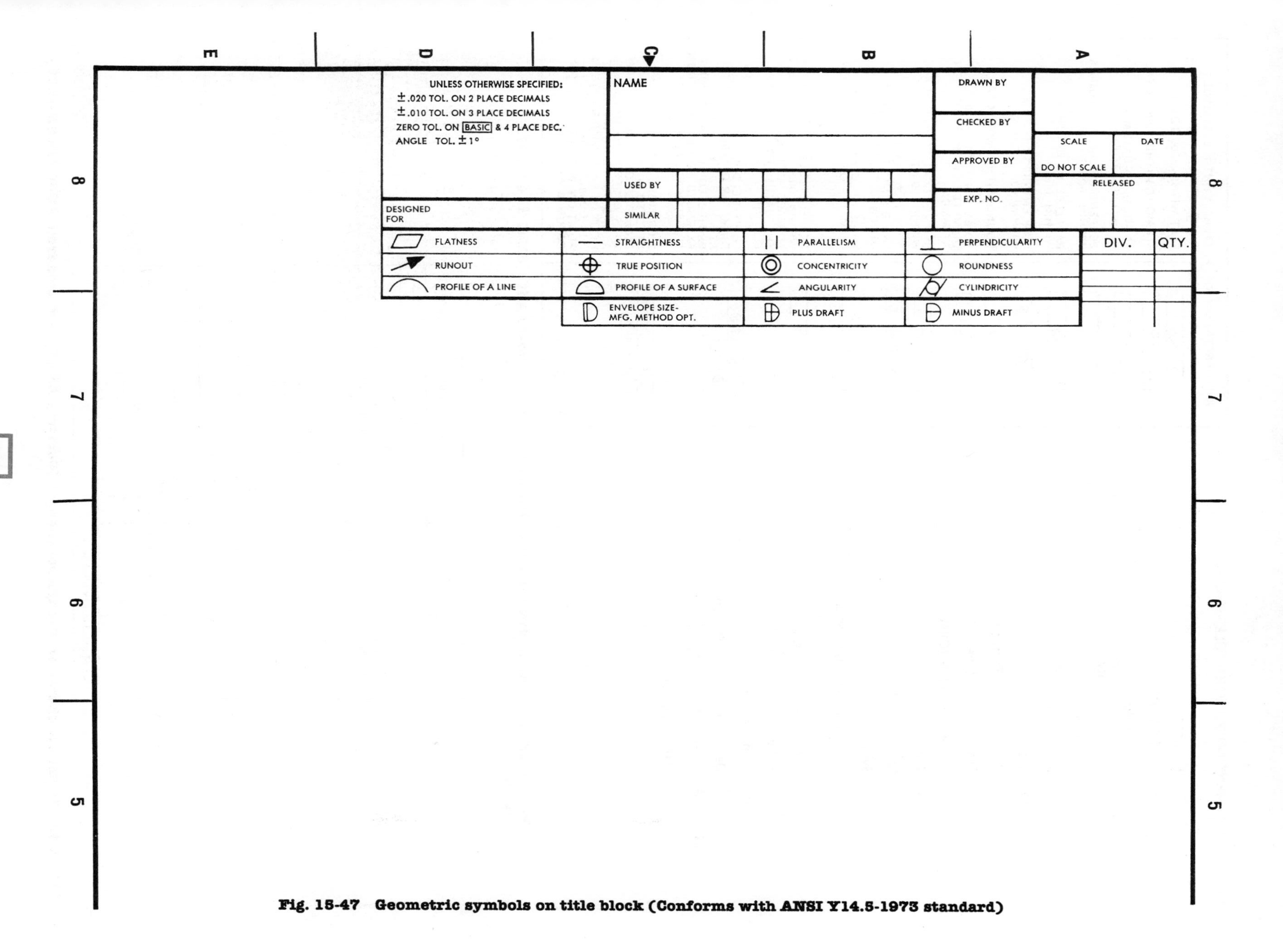

Fig. 15-47 Geometric symbols on title block (Conforms with ANSI Y14.5-1973 standard)

- ▱ Flatness
- — Straightness
- ∠ Angularity
- ⊥ Perpendicularity
- ‖ Parallelism
- ○ Roundness
- /○/ Cylindricity
- ⌓ Profile of Any Surface
- ⌒ Profile of Any Line
- ↗ Runout (TOT) & (CIR)
- ⊕ True Position
- ◎ Concentricity
- ⌯ Symmetry
- Ⓜ Maximum Material Condition (MMC)
- Ⓢ Regardless of Feature Size (RFS)
- Ⓟ Projected Tolerance Zone
- Ø Diameter
- -A- Datum Identifying Symbol
- ⊕ A Ø.001Ⓜ Feature Control Symbol
- Datum Target

INTERPRETATION OF FEATURE CONTROL SYMBOLS	
▱ .005	FLAT WITHIN .005 TOTAL
— .005	STRAIGHT WITHIN .005 TOTAL
∠ A .002	ANGULAR TO DATUM A WITHIN .002 TOTAL
⊥ A .005	PERPENDICULAR TO DATUM A WITHIN .005 TOTAL
‖ A .002	PARALLEL TO DATUM A WITHIN .002 TOTAL
◎ AⓈ .002 Ⓢ	CONCENTRIC TO DATUM A WITHIN .002 TOTAL RFS
○ .002	ROUND WITHIN .002 TOTAL
⊕ Ø.010 Ⓜ	TO BE HELD WITHIN .010 DIAMETER OF TRUE POSITION AT MMC
⌯ A .005 Ⓢ	SYMMETRICAL TO DATUM A WITHIN .005 TOTAL RFS
↗ AⓈ-BⓈ .002 Ⓢ CIR	RUNOUT TO DATUMS A AND B SIMULTANEOUSLY WITHIN .002 T.I.R. RFS
⊕ Ø.010 Ⓢ	TO BE HELD WITHIN .010 DIAMETER OF TRUE POSITION RFS
⊥ A .002 -Y-	PERPENDICULAR TO DATUM A WITHIN .002 TOTAL AND THIS SURFACE ALSO ESTABLISHES THE DATUM FOR ANOTHER FEATURE
⊕ Ø.005 Ⓜ Ⓟ.500	TO BE HELD WITHIN .005 DIAMETER OF TRUE POSITION AT MMC BUT WITHIN A PROJECTED TOLERANCE ZONE EXTENDED .50 OUTWARD FROM THE SURFACE OF THE PART
⊕ A B C Ø.060Ⓜ / A Ø.010Ⓜ	2 OR MORE HOLES TO BE HELD TO WITHIN .060 DIAMETER OF TRUE POSITION WITH RESPECT TO DATUMS A, B, & C AT MMC AND TO BE HELD TO EACH OTHER WITHIN .010 DIAMETER OF TRUE POSITION WITH RESPECT TO DATUM A AT MMC
11/73	

Fig. 15-48 Geometric dimensioning training aids (Conforms with ANSI Y14.5-1973 standard)

	TYPE OF TOLERANCE	CHARACTERISTIC	SYMBOL
FOR INDIVIDUAL FEATURES	FORM	STRAIGHTNESS	—
		FLATNESS	▱
		CIRCULARITY (ROUNDNESS)	○
		CYLINDRICITY	/○/
FOR INDIVIDUAL OR RELATED FEATURES	PROFILE	PROFILE OF A LINE	⌒
		PROFILE OF A SURFACE	⌓
FOR RELATED FEATURES	ORIENTATION	ANGULARITY	∠
		PERPENDICULARITY	⊥
		PARALLELISM	//
	LOCATION	POSITION	⊕
		CONCENTRICITY	◎
	RUNOUT	CIRCULAR RUNOUT	↗*
		TOTAL RUNOUT	⇗*

*Arrowhead(s) may be filled in.

Fig. 15-49 Geometric characteristic symbol changes (Courtesy ASME, extracted from ANSI Y14.5M-1982)

SYMBOL FOR:	ANSI Y14.5	ISO
STRAIGHTNESS	—	—
FLATNESS	▱	▱
CIRCULARITY	○	○
CYLINDRICITY	⌭	⌭
PROFILE OF A LINE	⌒	⌒
PROFILE OF A SURFACE	⌓	⌓
ALL AROUND–PROFILE	↙⌀	NONE
ANGULARITY	∠	∠
PERPENDICULARITY	⊥	⊥
PARALLELISM	//	//
POSITION	⌖	⌖
CONCENTRICITY/COAXIALITY	◎	◎
SYMMETRY	NONE	⌯
CIRCULAR RUNOUT	*↗	↗
TOTAL RUNOUT	*⌰	⌰
AT MAXIMUM MATERIAL CONDITION	Ⓜ	Ⓜ
AT LEAST MATERIAL CONDITION	Ⓛ	NONE
REGARDLESS OF FEATURE SIZE	Ⓢ	NONE
PROJECTED TOLERANCE ZONE	Ⓟ	Ⓟ
DIAMETER	Ø	Ø
BASIC DIMENSION	[50]	[50]
REFERENCE DIMENSION	(50)	(50)
DATUM FEATURE	[-A-]	OR [A]
DATUM TARGET	Ø6 / A1	Ø6 / A1
TARGET POINT	X	X

Fig. 15-50 Comparison of ANSI Y14.5M-1982 to ISO standard (Courtesy of ASME extracted from ANSI Y14.5M-1982)

shown in Figure 15-49. The two main changes from the 1973 standard are the deletion of symmetry ⌯ and the addition of total runout ⌰.

A comparison of the *ANSI Y14.5M —1982* symbols with the ISO symbols is shown in Figure 15-50. Also, a symbol to indicate "all around" has been added for use in profile tolerance applications.

REVIEW

This review is provided to serve as reinforcement study material. Fill in the appropriate word(s) to complete the sentences below.

1. Geometric dimensioning provides ______________________ tolerance than the coordinate dimensioning system does.

2. Three advantages of using symbols are:

a. ______________________

b. ______________________

c. ______________________

3. A basic dimension has a tolerance of ______________________.

4. The primary plane of projection requires ______________________ contact points.

5. A part machined at its maximum material condition would have the

______________________ weight allowed.

6. The ______________________ block in the feature control frame contains the geometric form or position character.

7. The three orientation geometric symbols are:

a. ______________________

b. ______________________

c. ______________________

8. The symbol for the diameter is ______________________.

9. Identify the following symbols:

a. ⏥ ______________________

b. — ______________________

c. ∠ ______________________

d. ⊥ ______________________

e. // ______________________

f. ○ ______________________

g. ⌭ ______________________

h. ⌓ ______________________

i. ⌒ ______________________

j. ↗ ______________________

NAME ______________________

DATE ______________________

SCORE ______________________

EXERCISE B15-1

Print #321013

(Conforms with ANSI Y14.5-1973 standard)

Answer the following questions after studying Print #321013 found after this exercise.

QUESTIONS	ANSWERS
1. What part does Print #321013 describe?	**1.** ______________________
2. What drawing scale is used?	**2.** ______________________
3. What are the reference datums?	**3.** ______________________
4. How many basic dimensions are indicated?	**4.** ______________________
5. Draw one datum as shown on the print.	**5.** ______________________
6. What is the tolerance allowed on basic dimensions?	**6.** ______________________
7. What do the arrows A-A and the lines connecting these arrows represent?	**7.** ______________________
8. Prior to 10-4-74, what was the dimension .645 shown on section A-A? What is the tolerance for this dimension?	**8.** ______________________ ______________________
9. What is the typical wall thickness for this part? What is the tolerance for this wall thickness?	**9.** ______________________ ______________________
10. The surface identified as -Y- must have a maximum surface finish of what? This surface must be flat within what?	**10.** ______________________ ______________________

NAME ______________________

DATE ______________________

SCORE ______________________

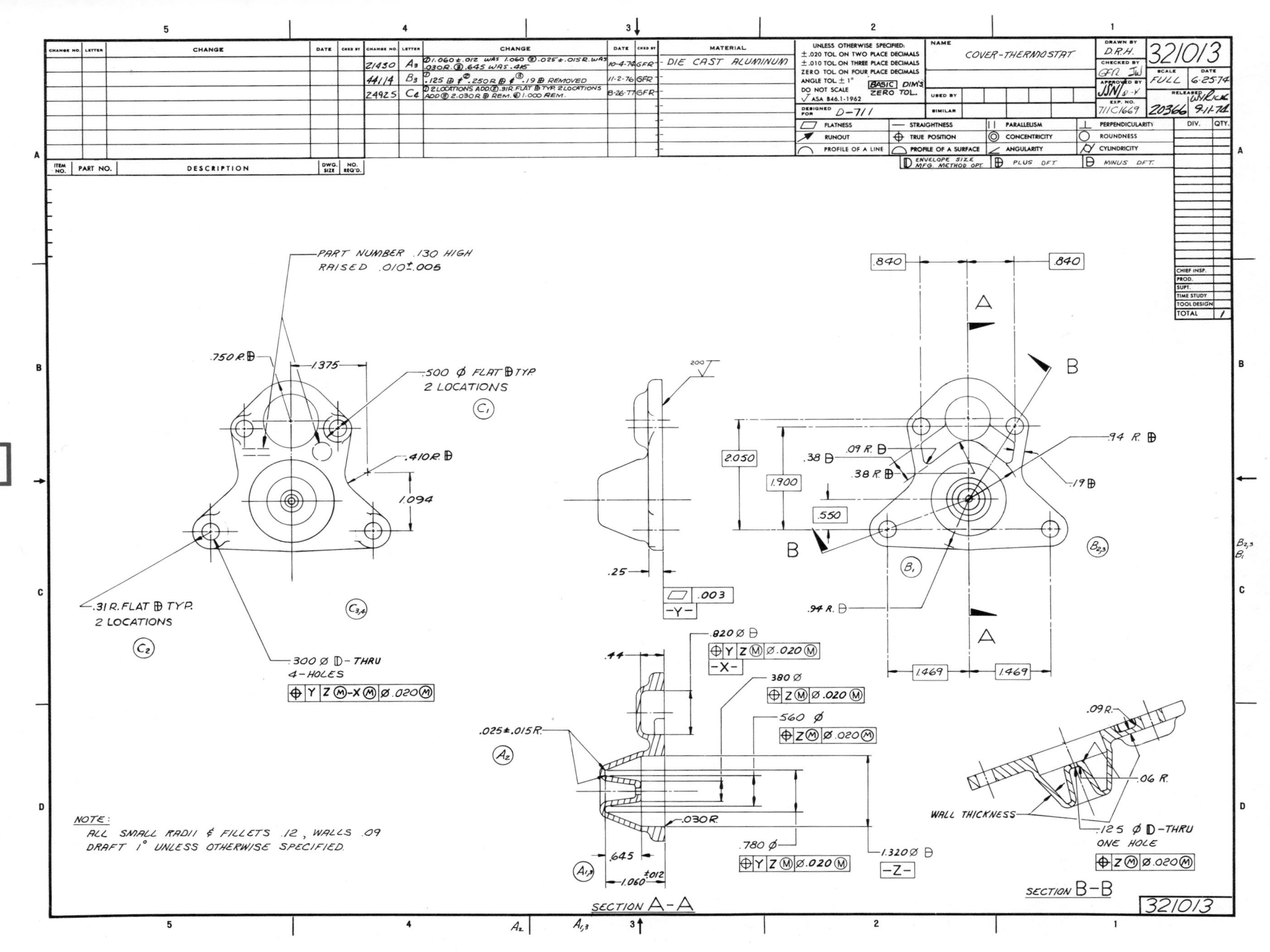

COVER-THERMOSTAT
321013
MATERIAL
DIE CAST ALUMINUM
UNLESS OTHERWISE SPECIFIED:
±.020 TOL. ON TWO PLACE DECIMALS
±.010 TOL. ON THREE PLACE DECIMALS
ZERO TOL. ON FOUR PLACE DECIMALS
ANGLE TOL. ± 1°
DO NOT SCALE
BASIC DIM'S ZERO TOL.
ASA B46.1-1962
DESIGNED FOR D-711
DRAWN BY D.R.H.
SCALE FULL
DATE 6-25-74
EXP. NO. 711C1669
20366
21430 A3 10-4-74
44114 B3 11-2-76
24925 C4 8-26-77
FLATNESS
RUNOUT
PROFILE OF A LINE
STRAIGHTNESS
TRUE POSITION
PROFILE OF A SURFACE
PARALLELISM
CONCENTRICITY
ANGULARITY
PERPENDICULARITY
ROUNDNESS
CYLINDRICITY
ENVELOPE SIZE MFG. METHOD OPT.
PLUS DFT
MINUS DFT.
ITEM NO.
PART NO.
DESCRIPTION
DWG. SIZE
NO. REQ'D.
CHANGE
DATE
LETTER
CHANGE NO.
CHIEF INSP.
PROD.
SUPT.
TIME STUDY
TOOL DESIGN
TOTAL 1
DIV.
QTY.
PART NUMBER .130 HIGH RAISED .010±.005
.750 R.
1.375
.500 Ø FLAT TYP 2 LOCATIONS
.410 R.
1.094
.31 R. FLAT TYP. 2 LOCATIONS
.300 Ø - THRU 4-HOLES
.25
.003
-Y-
.840
.840
2.050
1.900
.550
.09 R.
.38
.38 R.
.94 R.
.19
1.469
1.469
.820 Ø
-X-
380 Ø
560 Ø
.025±.015 R.
.44
.030 R.
.645
1.060 ±.012
.780 Ø
1.320 Ø
-Z-
SECTION A-A
.09 R.
.06 R.
WALL THICKNESS
.125 Ø -THRU ONE HOLE
SECTION B-B
NOTE:
ALL SMALL RADII & FILLETS .12, WALLS .09
DRAFT 1° UNLESS OTHERWISE SPECIFIED.

CHAPTER 16

NEW DRAFTING SYSTEMS

OBJECTIVES

After studying this chapter, you will be able to:

- Describe the ordinate and tabular dimensioning systems.
- Explain the basics of numerical control and how the data is displayed on industrial prints.
- Discuss the difference between fixed zero and floating zero numerical control systems.
- Describe the numerical control program sheet and how it relates to the numerical control punched tape.
- Describe the dual dimensioning systems.
- Explain the basics of first angle and third angle projections.
- Describe the product liability notations that may appear on prints.
- Discuss the impact of computer aided drafting on the drafting professions.

In this chapter we discuss some of the newer drafting systems introduced in Chapter 6. These systems include the different methods of indicating dimensions on prints.

The newer drafting systems are being introduced to achieve cost reduction in drafting departments. They are also used to better accommodate newer machining operations, particularly numerical control machining.

ORDINATE DIMENSIONING

The *ordinate dimensioning system* is a type of rectangular datum dimensioning. In ordinate dimensioning, all location dimensions start from two or three perpendicular datum planes, Figure 16-1. The datum lines are indicated as *zero coordinates.* Dimensions are shown at the end of the extension or center lines, without the use of dimension lines

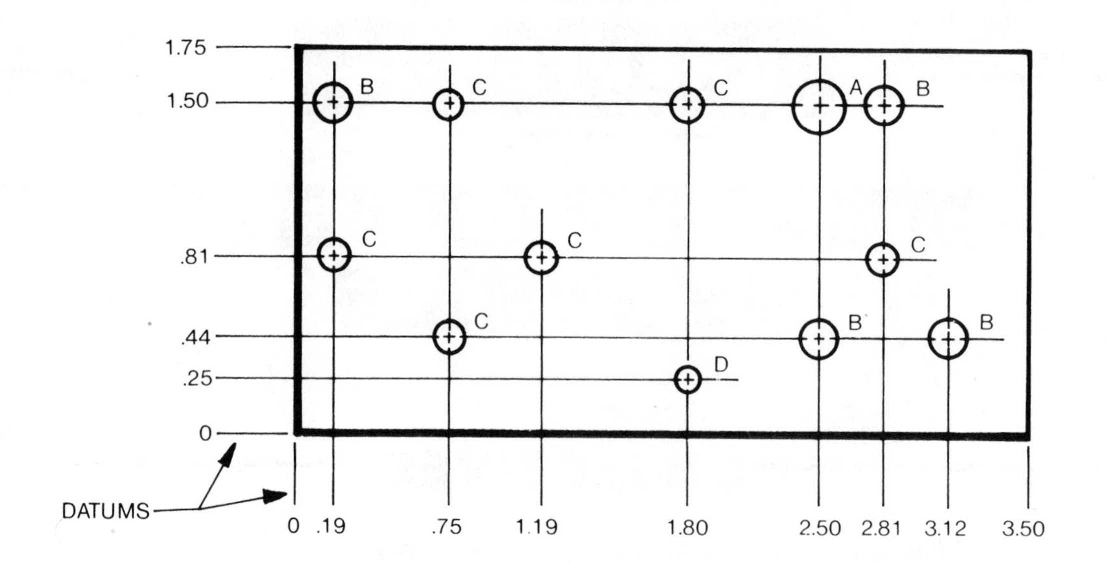

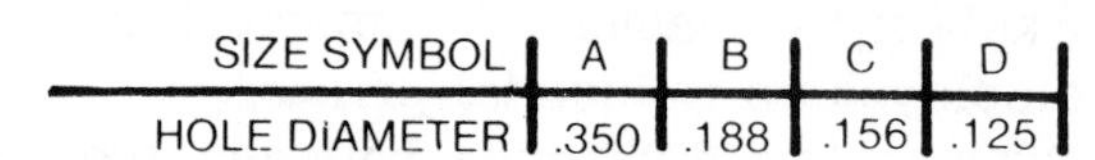

SIZE SYMBOL	A	B	C	D
HOLE DIAMETER	.350	.188	.156	.125

Fig. 16–1 Ordinate dimensioning

or arrowheads. All dimensions originate from the intersection of the lower horizontal and left vertical base lines. These dimensions locate the part features—the holes, in the case of Figure 16-1. The hole diameters are also indicated in a table located on the print next to the part drawing.

TABULAR DIMENSIONING

Tabular dimensioning is similar to the ordinate system, except that the location dimensions are not shown next to the features on the part drawing. The feature locations and sizes are indicated on a separate table, Figure 16-2, which is usually placed in a corner of the print.

Zero coordinate, the starting point, is usually located at the lower left corner

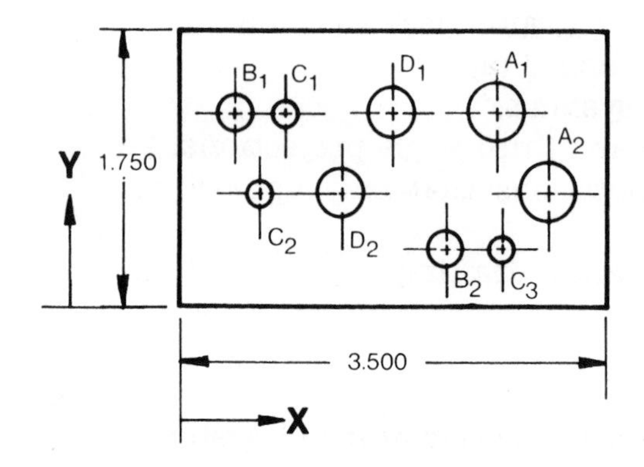

	REQD.	2	2	3	2
	HOLE DIA.	.250	.190	.150	.200
POSITION		HOLE SYMBOL			
X →	Y ↑	A	B	C	D
3.000	1.500	A_1			
3.250	.780	A_2			
.500	1.500		B_1		
2.750	.500		B_2		
.780	1.500			C_1	
.680	.780			C_2	
3.100	.500			C_3	
2.250	1.500				D_1
1.500	.780				D_2

Fig. 16–2 Tabular dimensioning

of the part. The *X* coordinate is located to the right and the *Y* coordinate is located above the starting point. If a feature has a depth dimension, such as a hole passing only partway through the object, this will be indicated by a *Z* coordinate on the table.

Both the tabular and ordinate dimensioning systems are used on drawings which require a large number of similarly shaped features. These systems are also used on drawings of parts which are produced by numerical control machining operations.

NUMERICAL CONTROL WITH CONVENTIONAL PRINTS

Often, production departments will start with a standard part print when producing parts on numerical control (N/C) machines. (They are sometimes called *numerical controlled* or *numerically controlled machines.*) Personnel who understand these programmable machines will prepare a special document to be sent with the part print. The document will contain machine control information required by the machine operator. The information is then put into a coded, punched tape; this tape controls the movements of the metal cutting machine. At the end of the work cycle, the tape will signal the machine to shut down.

Holding and locating fixtures for the part are usually required. The fixtures allow the N/C machine to always start from a fixed reference point.

NUMERICAL CONTROL PART PRINTS

Industrial prints may not be labeled as numerical control prints. However, the dimensioning system used will sometimes indicate whether the part should be machined on an N/C metal cutting machine. The ordinate and tabular dimensioning systems are often used for numerical control machining.

The basic dimensional information on the part print must be programmed for the N/C machine operator. The machines used most often utilize the *floating zero system* or *fixed zero system.* These systems define the zero point and machine starting point. The part to be machined will be located relative to these zero points.

CARTESIAN COORDINATE NUMBERING SYSTEM

The *cartesian coordinate numbering system* is a basic system of measurement which relates to the *X*, *Y*, and *Z* axes. Understanding the cartesian coordinate system is helpful for learning about N/C machining, especially the floating zero system.

X Coordinates

An equally divided line with each division numbered is called a *numbered line.* If this line is horizontal as viewed from the front view of an orthographic projection, it is concerned with the *X* axis. The starting point, labeled 0, can be located anywhere on this line.

Numbers to the right of the starting point are labeled + (plus). Numbers to the left of the starting point are labeled − (minus). In Figure 16-3, the point *A* has an *X* value of +3. Point *B* has an *X* value of −2.

Y Coordinates

A vertical line as viewed from the top view of an orthographic projection

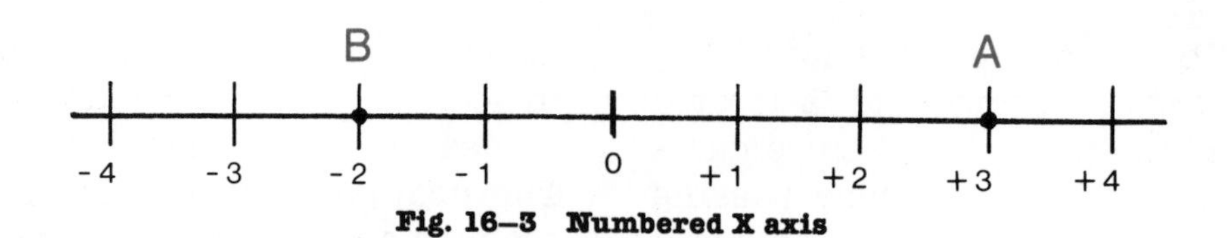

Fig. 16–3 Numbered X axis

would be concerned with the *Y* axis. Numbers above the 0 starting point would be labeled +. Numbers below the starting point would be labeled −. In Figure 16-4, the point *A* has a *Y* value of +2. Point *B* has a *Y* value of −4.

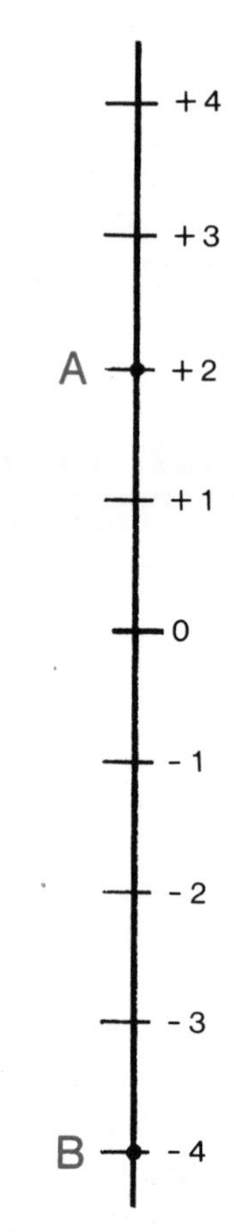

Fig. 16–4 Numbered Y axis

Z Coordinates

The third axis is positioned perpendicular to the *X* and *Y* axes, Figure 16-5. The third axis is called the *Z* axis. This line would be in the vertical position, as shown in the front and side views of an orthographic projection. An example is the depth of a drilled hole.

Quadrants

When the *X* and *Y* axis lines intersect perpendicularly at the 0 point, they form a graph called a *cartesian grid*. These two lines form four segments. The segments are called *quadrants*, and are numbered I, II, III, and IV. See Figure 16-6.

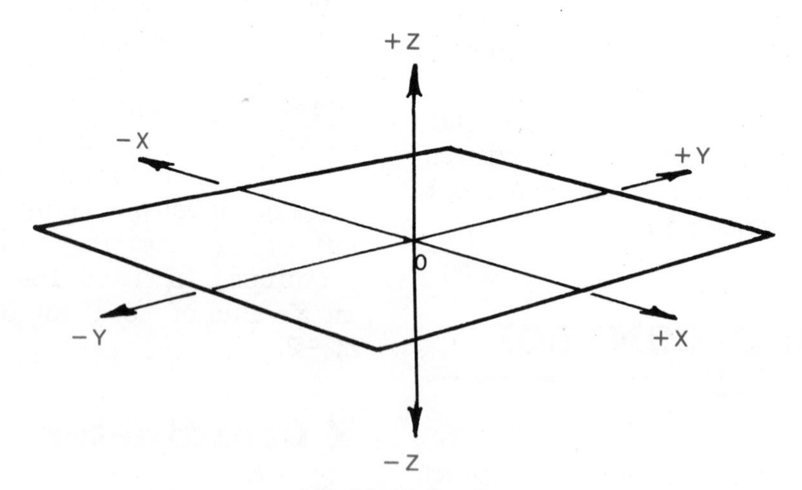

Fig. 16–5 Z axis

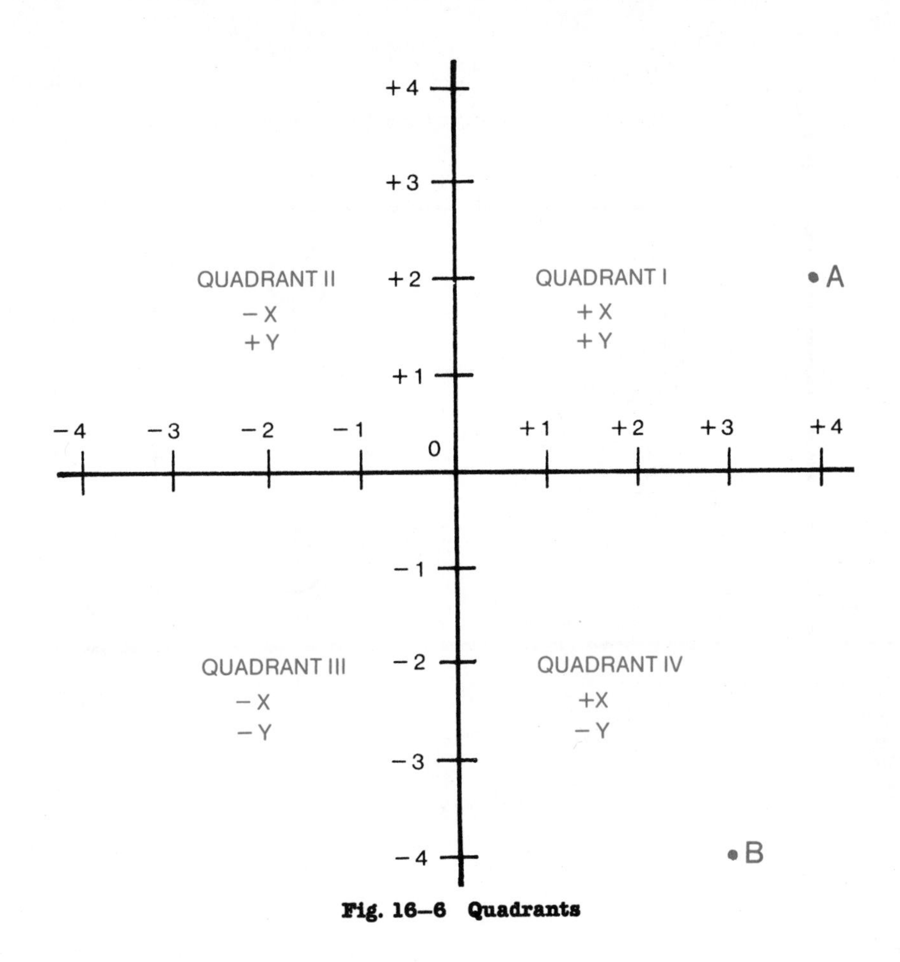

Fig. 16–6 Quadrants

The *X* and *Y* values of any point are determined by the point's position from the starting point 0. The coordinates of point *A* would have values of $X = +4$ and $Y = +2$. The coordinates of point *B* would have values of $X = +3$ and $Y = -4$. Therefore, coordinate values of a point in quadrant I would have a value of + for *X* and + for *Y*. A point in quadrant II would have a value of − for *X* and + for *Y*. A point in quadrant III would have a value of − for both *X* and *Y* coordinates. A point in quadrant IV would have a value of + for *X* and − for *Y*.

The *Z* coordinate values would be + when above and − when below starting point 0. (See Figure 16-5.) From this explanation, it can be seen that the starting point will determine the plus or minus numerical control values.

FLOATING ZERO SYSTEM

In the floating zero system, the cutting tool can be located anywhere that is convenient to the machine table or part. Quite often the cutting tool will be located at a corner of the part. See Figure 16-7A. In this case the cutting tool is a drill, and the floating zero point is adjusted to coincide with the set-up point. All movements of the cutting tool will be relative to the $+X$ and $+Y$ base lines.

In Figure 16-7A, since the starting point O is located in the lower left corner of the part, all dimensions are in quadrant I. Therefore, the *X* and *Y* coordinates are both plus. The shape of the part or feature location will usually determine

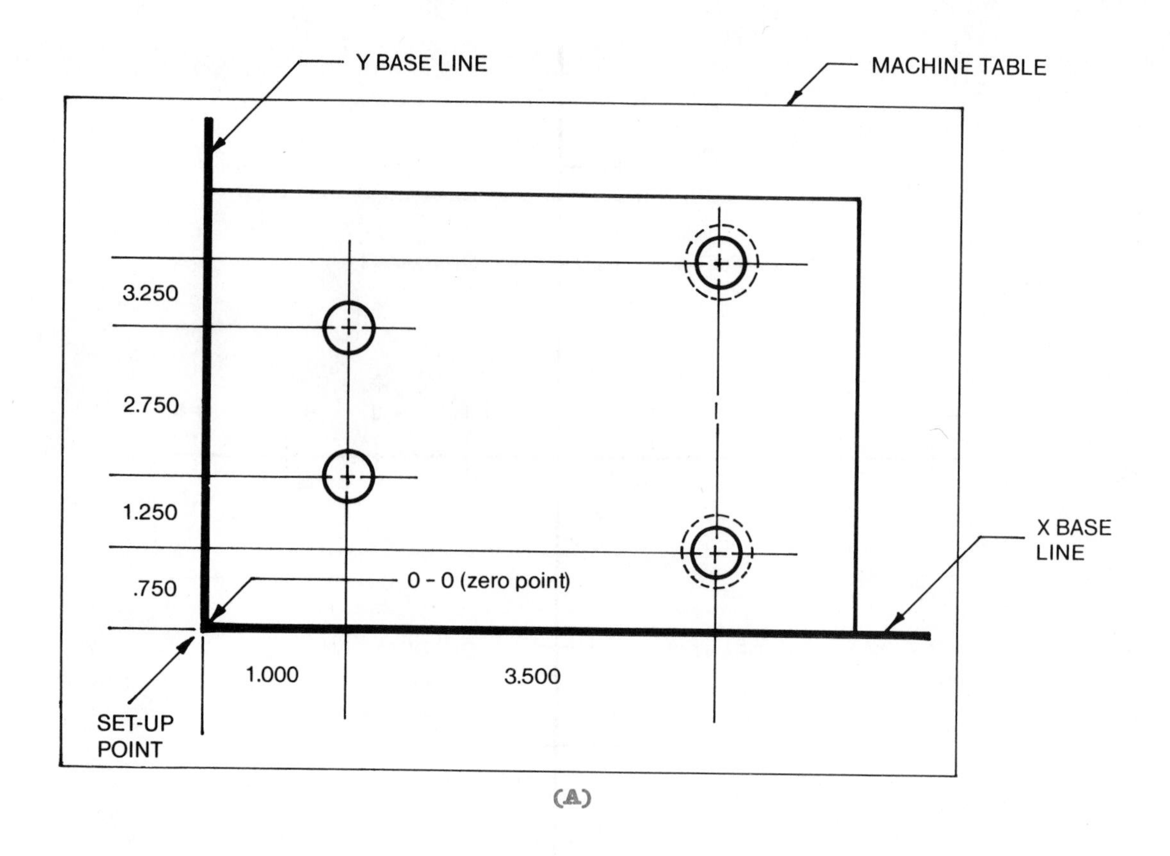

(A)

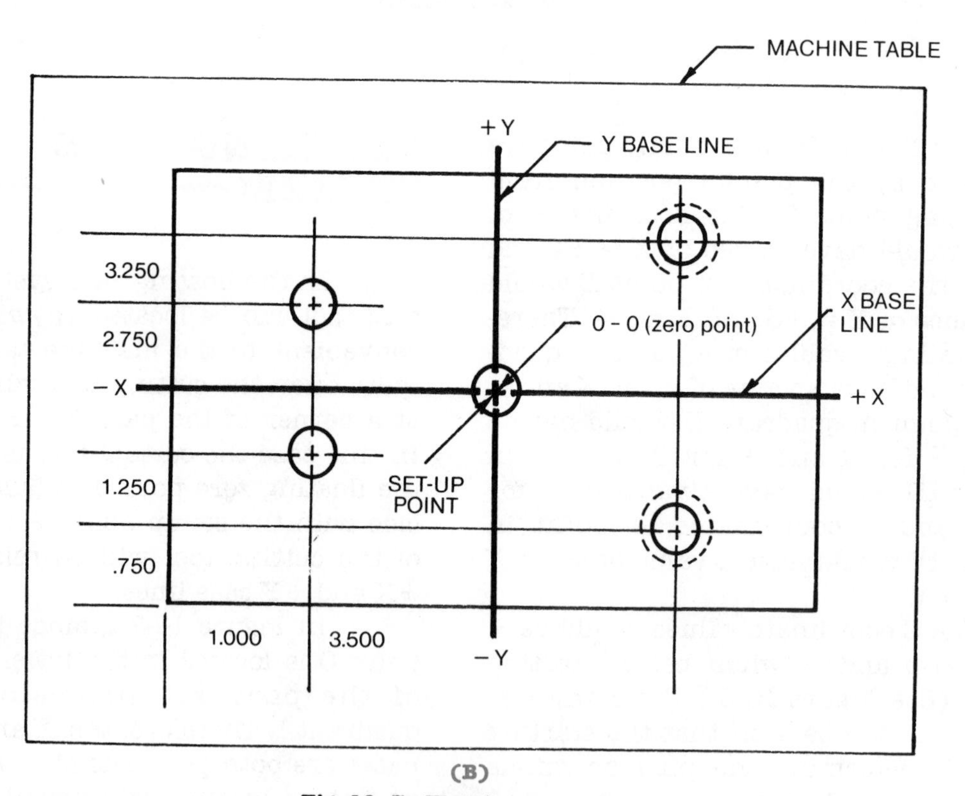

(B)

Fig. 16–7 Floating zero system

the most suitable numerical control starting point.

The machining of a similar part, but one with an additional hole in the center of the part, may suggest a more suitable N/C starting point. See Figure 16-7B. In this example, the location of the two drilled and tapped holes may be critical, relative to the center hole. The center of this hole was therefore selected as the numerical control starting point. The *X* and *Y* coordinates will lie in all quadrants and will have both plus and minus values.

The dimensions on the print may be similar to those in Figure 16-7A, but the numerical control program will be different. The starting point will have changed, and the programmer will have to recalculate values from the part print dimensions.

FIXED ZERO SYSTEM

Some of the N/C machines now in use have a fixed zero point. This point is usually located at one of the four corners of the machine table. When the machine's cutting tool spindle is centered over the fixed point, the machine controls will be adjusted for zero (0,0) on the *X* and *Y* coordinates. If the fixed zero point is located at the lower left corner of the machine table, then the spindle movement away from that point will be in a plus direction for both the *X* and *Y* axes. See Figure 16-8.

The codes used for this program are the same as those used for the floating zero system, but the start position will enter as the first item on the program

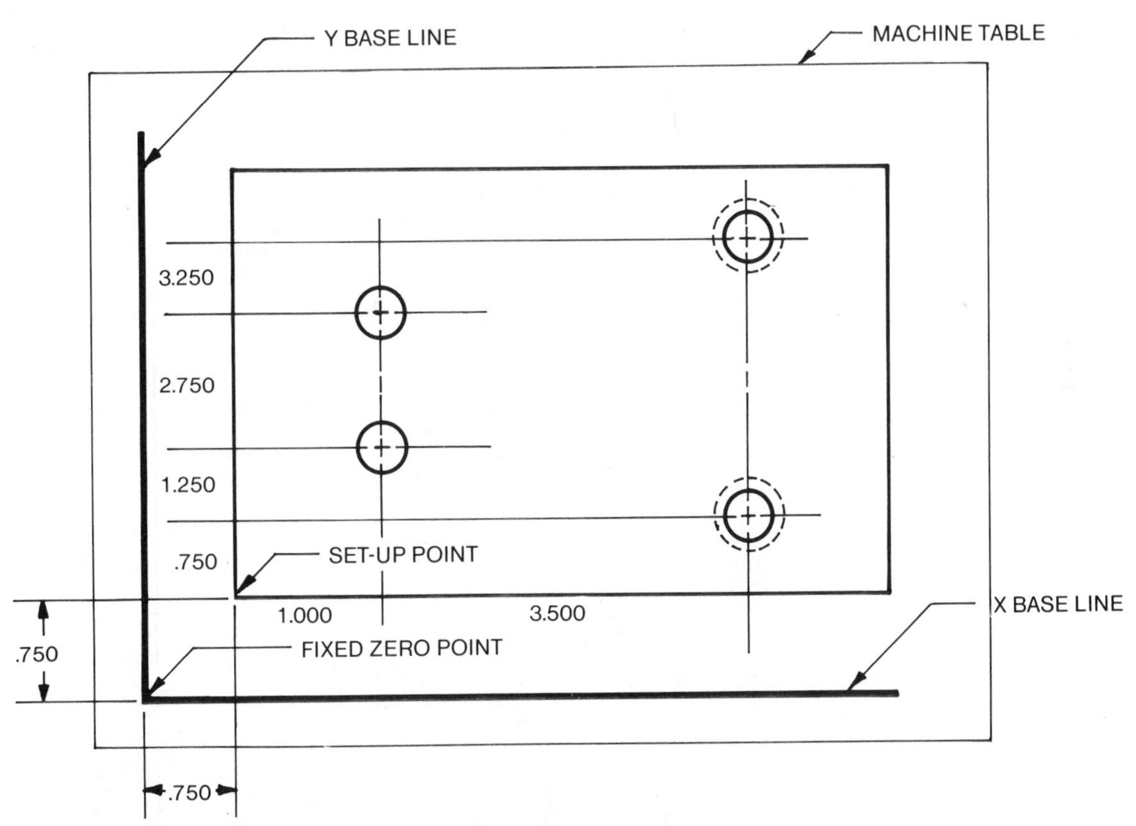

Fig. 16–8 Fixed zero system

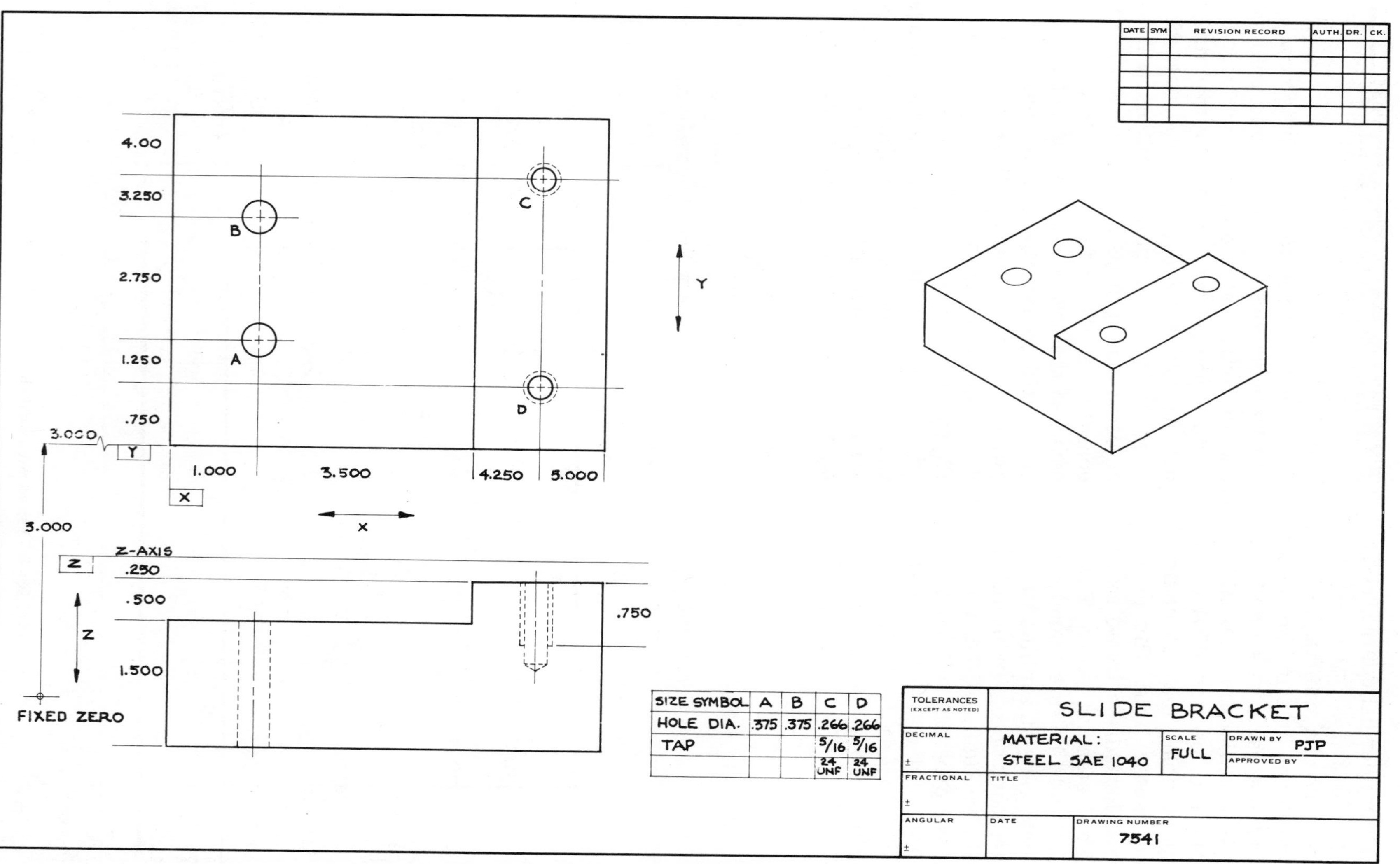

Fig. 16–9 N/C print

sheet. The machine operator must position the part so that the set-up point will be at the specified distance from the fixed zero point.

In Figure 16-8, the lower left corner of the part (the set-up point) is +.750 inch on the *X* coordinate and +.750 inch on the *Y* coordinate. If a large number of these parts are to be machined, then a part-holding fixture may be clamped to the machine table. This will allow the part to be quickly and consistently positioned at that location. (The number of parts to be machined will determine whether the cost of making a part-holding fixture is feasible.)

The production engineering department is usually responsible for programming a part that will be processed through numerical control machines. The following discussions illustrate the techniques used in N/C machining.

SLIDE BRACKET PRINT #7541

The part print #7541, shown in Figure 16-9, uses ordinate dimensioning. The set-up point is the lower left corner of the part. The machining operations will consist of drilling two .375 through holes (*A* and *B*), drilling two .266 (*H*) tap drill holes (*C* and *D*) 1 inch deep, and tapping these holes with a 5/16-24 UNF tap, 3/4 inch deep.

The logical first operation is to drill the hole closest to the set-up point, the lower .375 hole (*A*). The second operation is to drill the second .375 hole (*B*). The third operation is to drill the .266 upper tap hole (*C*). The fourth operation is to drill hole *D*. The fifth operation is to tap hole *D*. The sixth operation is to go back and tap hole *C*.

At the end of the program, the spindle will return to the fixed zero point, as indicated on the print. The location of the fixed zero point is not normally indicated on the part print. This is because the fixed zero point may vary depending on the type of machine on which this part will be produced. The *Z* axis is also indicated because the depth of the tapped holes must be programmed, for both drill and tap operations.

PROGRAM SHEET FOR PRINT #7541

Many companies which use N/C machines develop their own program documents (sheets) to satisfy their particular machining operations. However, their program documents are similar to the one used for print #7541 (Figure 16-9). This program document is shown in Figure 16-10. Fixed zero: $X = 3.000$, $Y = 3.000$, $Z = 2.250$.

The *program document* is prepared primarily to assist in producing the *perforated*, or *punched, tape*. This tape will activate the controls of the N/C machine. The punched tape for the part on print #7541 is 36 inches long. It contains all of the numerical control language needed by the electronic reader of the N/C machine. The first several inches of this tape are shown in Figure 16-11. (The callouts included in this figure refer to the coded printout in Figure 16-15.)

The number and position of the punched holes determine the coded numbers and letters. The *program codes,* Figure 16-12, are derived from standardized punched tape codes, Figure 16-13. The original EIA code is not used now as much as the newer ASCII/ISO code. However, most tape punch machines can produce tapes using either standardized code. The EIA tapes have an odd number of punched holes in each row across the tape. The ASCII/ISO tapes have an even number of punched holes in each row across the tape.

Part No. 7541	Part Name. SLIDE BRACKET			Material. SAE1040		Page 1 of Page 1			By.
Operation. HOLES	Seq. No.	Prep Function	X Position	Y Position	Z Position	Misc. Func.	Speed RPM	Feed	Remarks
Rapid to Hole A	N001	G01	X 40000	Y 42500				FO	3/8 Drill
Drill Hole A	N002	G81			Z 25000	M03	S84	F185	
Retract Hole A	N003				Z 0000			FO	
Rapid to Hole B	N004	G01		Y 57500				FO	
Drill Hole B	N005	G81			Z 25000			F185	
Retract Hole B	N006				Z 0000	M06		FO	Tool Change
Rapid to Hole C	N007	G01	X 72500	Y 62500				FO	Let. "I" Drill
Drill Hole C	N008	G81			Z 12500	M03	S11	F200	
Retract Hole C	N009				Z 0000			FO	
Rapid to Hole D	N010	G01		Y 37500				FO	
Drill Hole D	N011	G81			Z 12500			F200	
Retract Hole D	N012				Z 0000	M06		FO	Tool Change
Tap Hole D	N013	G84			Z 10000	M03	S32	F10	5/16 Tap
Retract Hole D	N014				Z 0000	M04		F10	
Rapid to Hole C	N015	G01		Y 62500				FO	
Tap Hole C	N016	G84			Z 10000	M03	S32	F10	
Retract Hole C	N017				Z 0000	M04			
Rapid to Fixed Zero	N018	G01	X 0000	Y 0000		M02		FO	End of Program Tool Change

Fig. 16–10 N/C program document

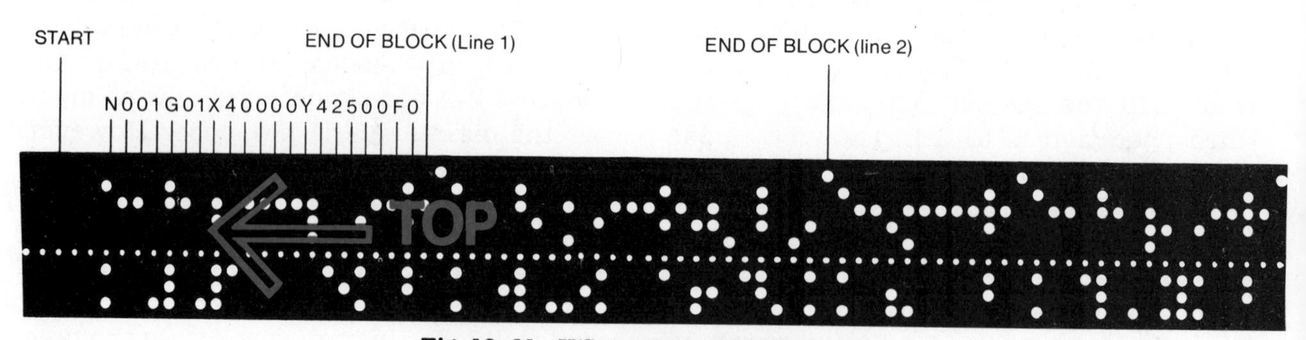

Fig. 16–11 N/C punched tape (EIA code)

Coding For Print No. 7541

Spindle Movement
X Axis - Left To Right
Y Axis - Front To Back
Z Axis - Up And Down

F Word - Feed Rate
S Word - Spindle Speed

G Word - **Mode Cycle**
G01 - Linear Move
G81 - Drill
G84 - Tap

M Word - **Function**
M02 - End Of Program
M03 - Spindle On - Clockwise
M04 - Spindle On - Counter Clockwise
M06 - Tool Change

Fig. 16–12 Program code (EIA)

The *tape punch machine*, Figure 16-14, not only produces the punched tape but also provides a coded printout, Figure 16-15.

The part shown on the print in Figure 16-9 was programmed to be machined on an N/C machine.

METRIC DIMENSIONS—

The *meter* (m) is the standard metric unit of measurement. One meter equals approximately 3¼ feet. The meter is subdivided into smaller length units in multiples of 10. The *millimeter* (mm), which is

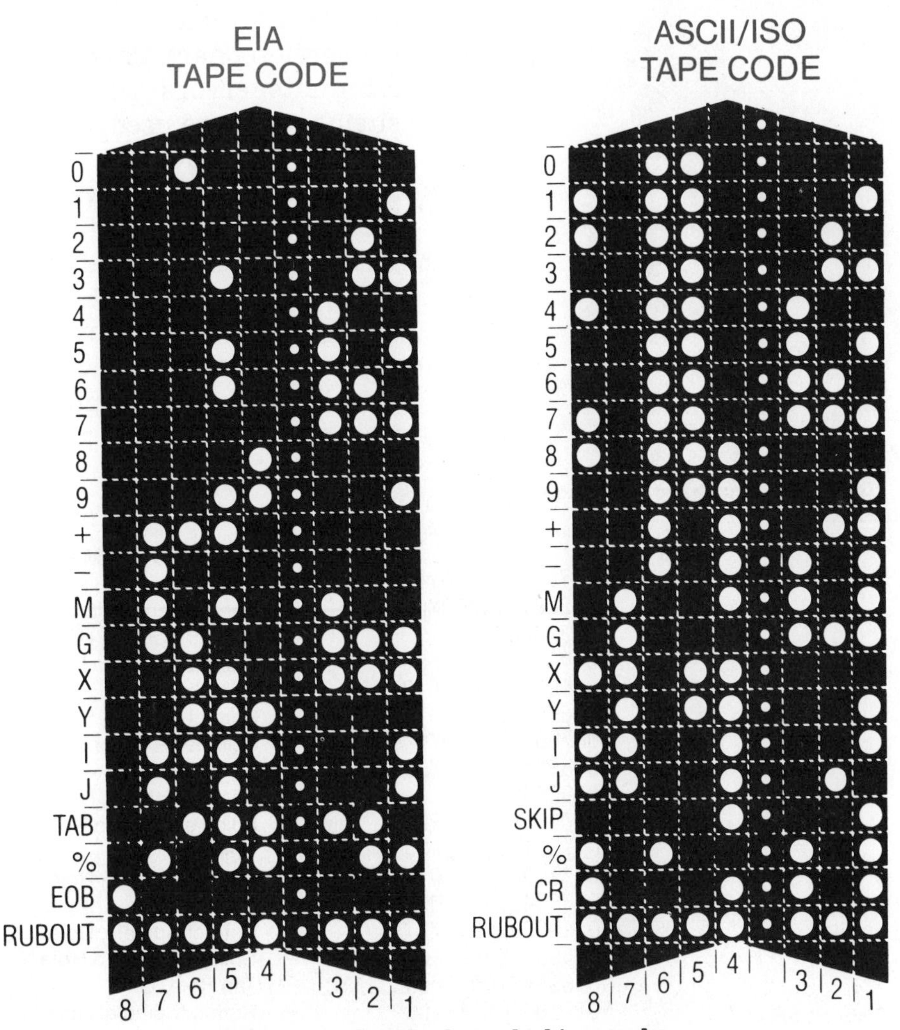

Fig. 16–13 Standard punched tape codes

Fig. 16–14 Tape punch machine (*Dependable Educational Program Co.*)

1/1000 of a meter, is used as a comparison to the inch. One inch is equivalent to 25.4 mm. Usually a millimeter measurement will not be indicated with more than two decimal places.

Some United States companies are converting to the metric system. This is particularly true of those which conduct international trade. Such companies are also designing their newer products using metric units; often the prints for these products are completely metric, Figure 16-16. Metric prints can be recognized easily because the word METRIC will be located close to the title block in bold letters. Also, a general note will be included: "Unless otherwise specified, dimensions are in millimeters." The print in Figure 16-16 includes a millimeter-to-inch conversion chart. This chart is intended to eliminate errors when nonmetric measuring tools are used during machining or inspection of the part.

Table 16-1 shows conversions from inch fractions and decimals to millimeters. The widespread availability of pocket calculators has lessened the need for such conversion tables. Many calculator producers have marketed calculators which convert English measurements to metric as well as handle many other common and scientific units.

		PROGRAM EXPLANATION
N001G01X40000Y42500F0	—	Spindle Move To Hole "A"
N002G81Z25000M03S84F185	—	Speed, Feed, and Drill Hole "A"
N003Z0000F0	—	Retract Spindle, Hole "A"
N004G01Y57500F0	—	Spindle Move To Hole "B"
N005G81Z25000F185	—	Drill Hole "B"
N006Z0000M06F0	—	Retract Spindle, Hole "B"
N007G01X72500Y62500F0	—	Spindle Move To Hole "C" - Tool Change
N008G81Z12500M03S11F200	—	Drill Hole "C"
N009Z0000F0	—	Retract Spindle, Hole "C"
N010G01Y37500F0	—	Spindle Move To Hole "D"
N011G81Z12500F200	—	Drill Hole "D"
N012Z0000M06F0	—	Retract Spindle, Hole "D" - Tool Change
N013G84Z10000M03S32F10	—	Tap Hole "D"
N014Z0000M04F10	—	Retract Spindle, Hole "D"
N015G01Y62500F0	—	Spindle Move To Hole "C"
N016G84Z10000M03S32F10	—	Tap Hole "C"
N017Z0000M04F10	—	Retract Spindle, Hole "C"
N018G01X0000Y0000M02F0	—	Spindle Return To Fixed Zero - Tool Change

Fig. 16–15 N/C coded printout for punched tape

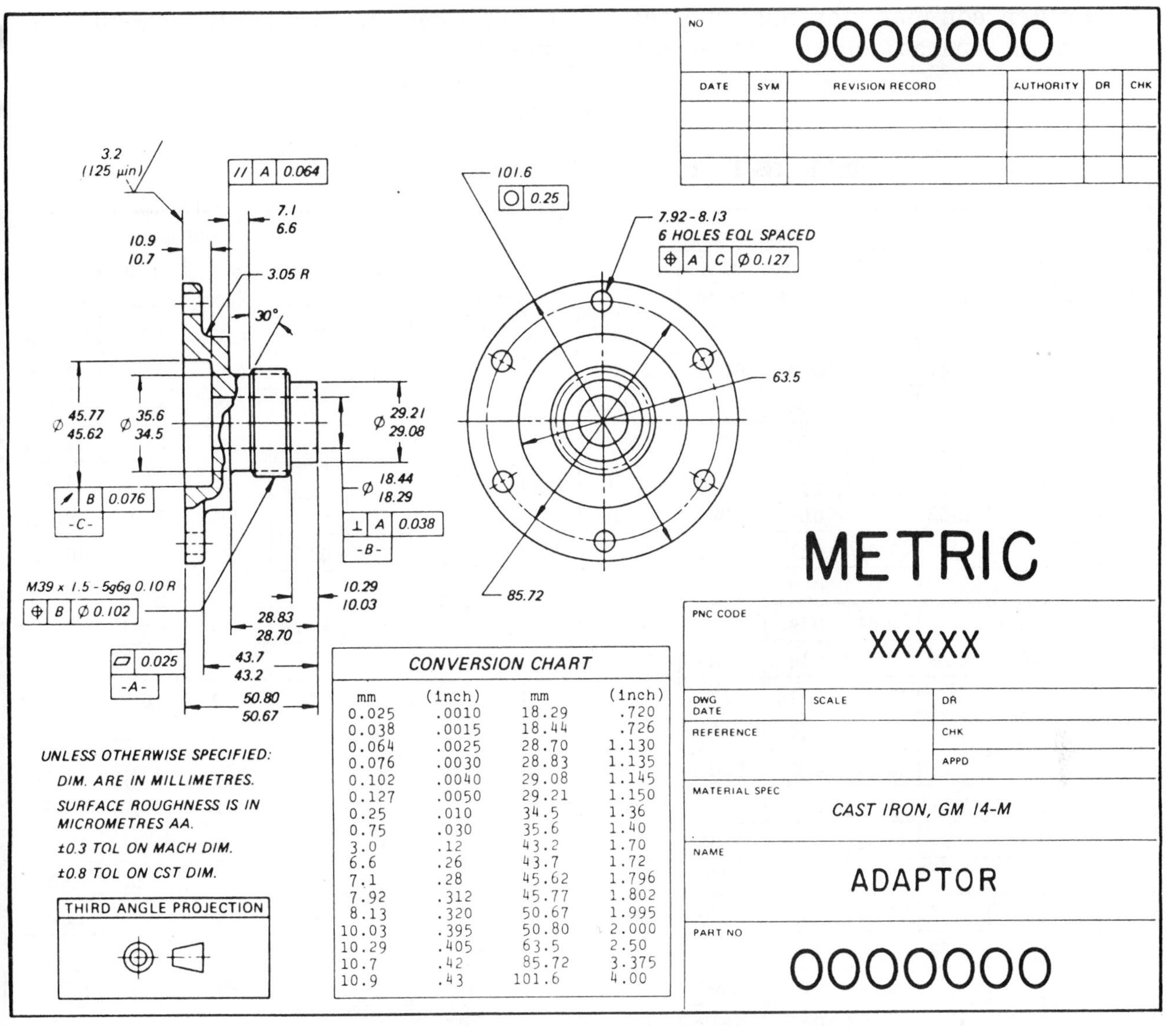

mm	(inch)	mm	(inch)
0.025	.0010	18.29	.720
0.038	.0015	18.44	.726
0.064	.0025	28.70	1.130
0.076	.0030	28.83	1.135
0.102	.0040	29.08	1.145
0.127	.0050	29.21	1.150
0.25	.010	34.5	1.36
0.75	.030	35.6	1.40
3.0	.12	43.2	1.70
6.6	.26	43.7	1.72
7.1	.28	45.62	1.796
7.92	.312	45.77	1.802
8.13	.320	50.67	1.995
10.03	.395	50.80	2.000
10.29	.405	63.5	2.50
10.7	.42	85.72	3.375
10.9	.43	101.6	4.00

Fig. 16–16 Metric print (*Courtesy of General Motors Corp.*)

A simple conversion rule to remember is that the metric dimension (millimeter) will read approximately 25 times larger than the equivalent inch dimension.

Metric Decimal Dimensions

In order to maintain consistency with *ASME, ANSI Y14.5M — 1982* standards, the following rules in metric decimal dimensions now apply:*

1. When the dimension is less than one millimeter, a zero precedes the decimal point (for example, 0.16).

2. When the dimension is a whole number, neither the zero nor decimal point is shown (for example, 24).

3. When the dimension exceeds a whole number by a decimal fraction of one millimeter, the last digit to the right of the decimal point is *not* followed by a zero (for example, 1.2).

*(*Courtesy of ASME, extracted from ANSI Y14.5M — 1982*)

4. Neither commas nor spaces shall be used to separate digits into groups in specifying millimeter dimensions on drawings.

Table 16-1 Inch—metric conversion table

Fractions					Decimals		mm	Fractions					Decimals		mm
4ths	8ths	16ths	32nds	64ths	to 2 places	to 3 place		4ths	8ths	16ths	32nds	64ths	to 2 places	to 3 places	
				1/64	0.02	0.016	0.3969					33/64	0.52	0.516	13.0968
			1/32		0.03	0.031	0.7937				17/32		0.53	0.531	13.4937
				3/64	0.05	0.047	1.1906					35/64	0.55	0.547	13.8906
		1/16			0.06	0.062	1.5875			9/16			0.56	0.562	14.2875
				5/64	0.08	0.078	1.9844					37/64	0.58	0.578	14.6844
			3/32		0.09	0.094	2.3812				19/32		0.59	0.594	15.0812
				7/64	0.11	0.109	2.7781					39/64	0.61	0.609	15.4781
	1/8				0.12	0.125	3.1750		5/8				0.62	0.625	15.8750
				9/64	0.14	0.141	3.5719					41/64	0.64	0.641	16.2719
			5/32		0.16	0.156	3.9687				21/32		0.66	0.656	16.6687
				11/64	0.17	0.172	4.3656					43/64	0.67	0.672	17.0656
		3/16			0.19	0.188	4.7625			11/16			0.69	0.688	17.4625
				13/64	0.20	0.203	5.1594					45/64	0.70	0.703	17.8594
			7/32		0.22	0.219	5.5562				23/32		0.72	0.719	18.2562
				15/64	0.23	0.234	5.9531					47/64	0.73	0.734	18.6532
1/4					0.25	0.250	6.3500	3/4					0.75	0.750	19.0500
				17/64	0.27	0.266	6.7469					49/64	0.77	0.766	19.4469
			9/32		0.28	0.281	7.1437				25/32		0.78	0.781	19.8433
				19/64	0.30	0.297	7.5406					51/64	0.80	0.797	20.2402
		5/16			0.31	0.312	7.9375			13/16			0.81	0.812	20.6375
				21/64	0.33	0.328	8.3344					53/64	0.83	0.828	21.0344
			11/32		0.34	0.344	8.7312				27/32		0.84	0.844	21.4312
				23/64	0.36	0.359	9.1281					55/64	0.86	0.859	21.8281
	3/8				0.38	0.375	9.5250		7/8				0.88	0.875	22.2250
				25/64	0.39	0.391	9.9219					57/64	0.89	0.891	22.6219
			13/32		0.41	0.406	10.3187				29/32		0.91	0.906	23.0187
				27/64	0.42	0.422	10.7156					59/64	0.92	9.922	23.4156
		7/16			0.44	0.438	11.1125			15/16			0.94	0.938	23.8125
				29/64	0.45	0.453	11.5094					61/64	0.95	0.953	24.2094
			15/32		0.47	0.469	11.9062				31/32		0.94	0.969	24.6062
				31/64	0.48	0.484	12.3031					63/64	0.98	0.984	25.0031
1/2					0.50	0.500	12.7000	1					1.00	1.000	25.4000

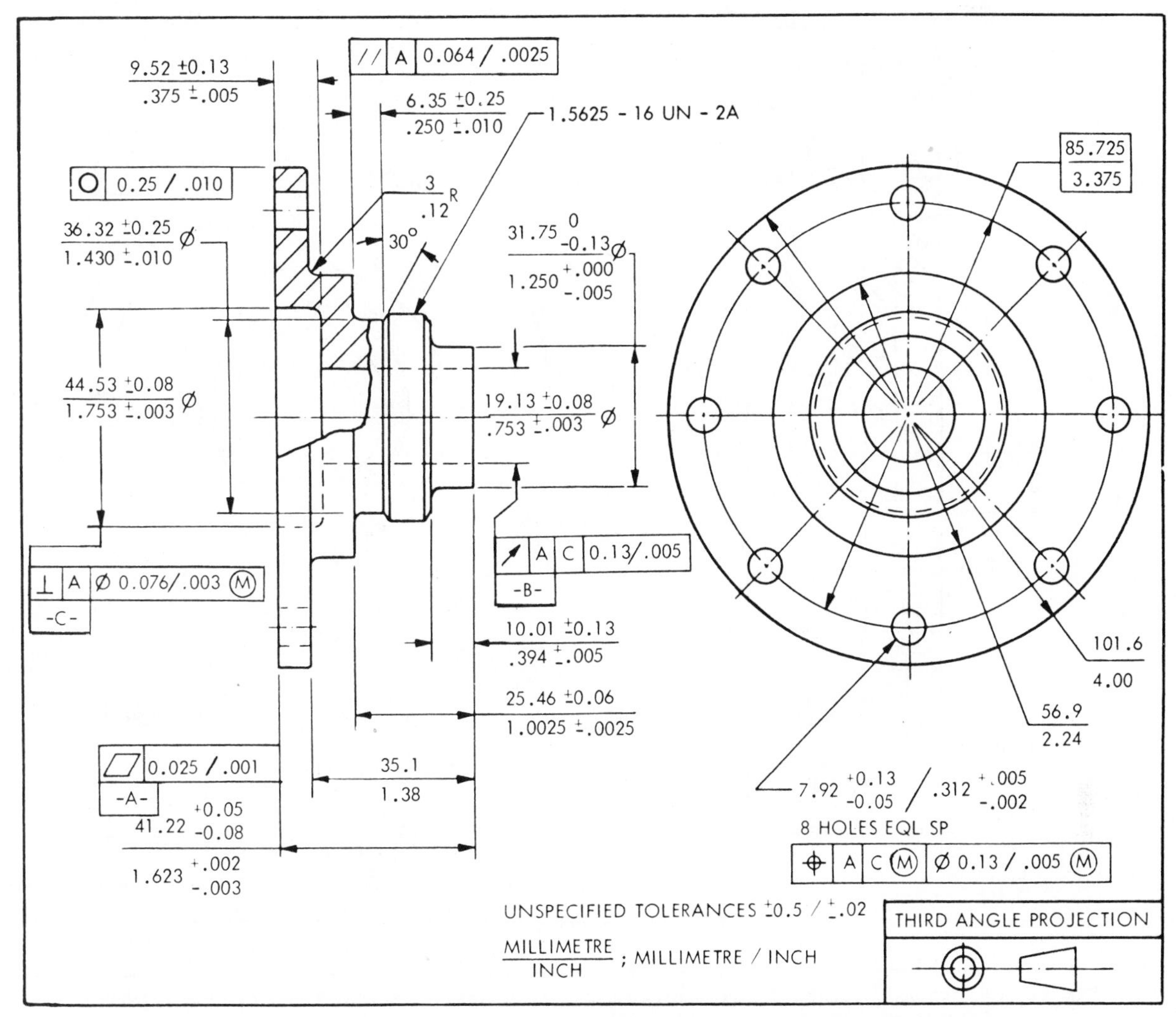

Fig. 16–17 Position method of dual dimensioning (*Courtesy of ASME, ANSI Y14.5M — 1982*)

DUAL DIMENSIONING—

Most smaller US companies are more conservative toward conversion to the metric system and are thus indicating both metric and inch dimensions on their prints. This dual dimensioning system seems to satisfy the immediate needs of these companies.

The American National Standards Institute, in *ANSI Y14.5—1973*, established standards for dual dimensioning. The two standards are the position method and the bracket method. This has been maintained in the 1982 revision of this standard.

Position Method

In the *position method* of dual dimensioning, the millimeter dimensions are located above the inch locations. The two dimensions are separated by a horizontal line. Any necessary tolerances are shown to the right of the dimensions. A key located near the title block specifies how the dimensions and tolerances are shown on the print. See Figure 16-17.

Bracket Method

In the *bracket method* of dual dimensioning, both metric and inch dimensions are shown on the print. However,

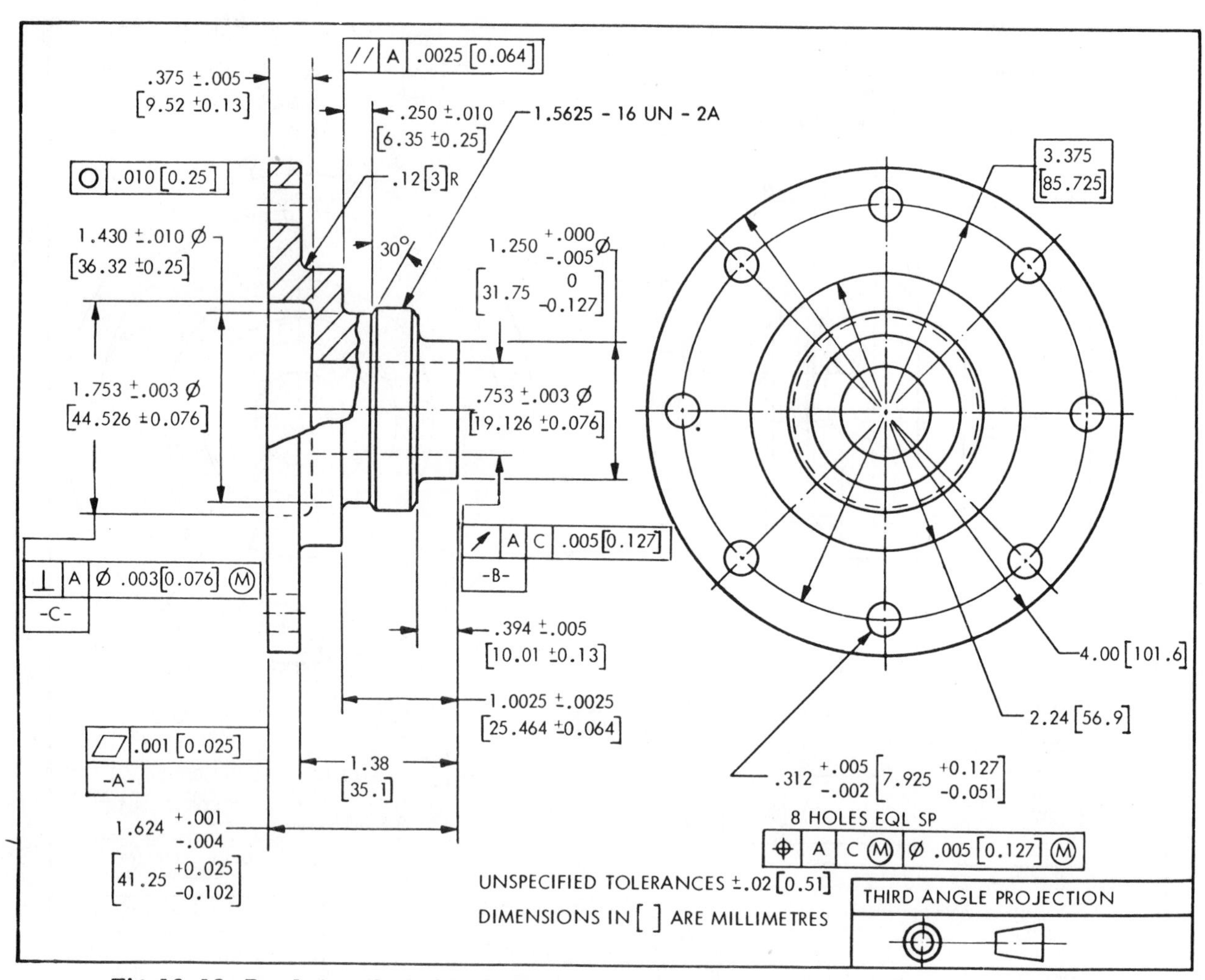

Fig. 16–18 Bracket method of dual dimensioning (*Courtesy of ASME, ANSI Y14.5M — 1982*)

the dimensions are not separated by a horizontal line. The location of these dimensions may vary relative to each other, but the metric dimension will always be enclosed by brackets. As in the position method, a key close to the title block specifies how the dimensions and tolerances are shown on the print. See Figure 16-18.

FIRST AND THIRD ANGLE PROJECTIONS

In the United States, prints are drawn in the *third angle projection*, Figure 16-19. This system allows the views on the prints to be in their normal positions. The top view is above the front view.

In Europe, the *first angle projection* is used, Figure 16-19. In this system, the top view is shown below the front view, and the left side view is to the right of the front view.

Even though the United States is leaning toward metric usage, there is no indication that the third angle projection system will be abandoned. To eliminate confusion on prints that may be used in both the United States and Europe, a notation placed near the title block tells which type of projection is used.

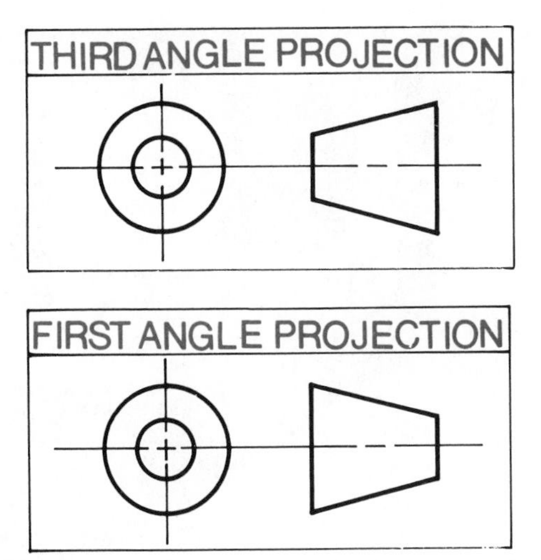

Fig. 16–19 First and third angle projections

PRODUCT LIABILITY NOTATIONS

In recent years, increasing numbers of lawsuits regarding product design have been filed against various companies. As a result, many companies have established stricter production controls to ensure that their products will be manufactured in close compliance with the design specifications.

Certain critical design and assembly operations are pointed out on the print as being "safety related." Currently there are no standard procedures for this special notation on prints. One company handles this problem by placing a large triangle above or near the title block. This triangle identifies the part or feature of the part as being safety related. The symbol alerts the methods or quality-control department that certain actions must be taken to satisfy the design requirements. Inspection reports must be maintained for at least 10 years, and all manufacturing deviations or changes must be approved by the chief engineer of the appropriate design department.

Sometimes a number is placed inside the triangle safety-related symbol. The number indicates how many safety-related items must be controlled on the print. A smaller triangle, with the appropriate number, is then placed close to the part feature that must be controlled, Figure 16-20.

Additional examples of safety-related notations are shown in Chapter 14.

COMPUTER-AIDED DRAFTING (CAD)

The use of *computer-aided drafting* (CAD) *systems* has increased at a fast rate since the early 1970s. CAD is no longer a specialized drafting technique used by one or two industries. CAD systems are now used by many different types of industries. Although the equipment (hardware) used in CAD has been available for many years, until recently it was too expensive for many potential users. However, as in personal computers, the advances in technology and market competition have reduced equipment costs. Now even small companies can afford CAD systems. This trend indicates an even greater use of CAD in the future.

The shop worker and others using industrial prints need not be CAD technicians or drafters. However, basic knowledge of CAD is helpful. The print reader should be able to distinguish between a hand-drawn print and a CAD print. Those

Fig. 16–20 Safety-related symbols

Fig. 16–21 Typical computer-aided drafting hardware (*Courtesy of Auto Trol Technology Corp.*)

who wish to further their training in the CAD drafting field can study textbooks devoted to the subject or take some of the many CAD courses now being offered in schools. The information in this section covers only the basics of computer aided drafting.

Computer aided drafting is a tool that can be used by the drafter; it is not intended to replace the drafter. CAD systems do not draw. The drafter uses CAD equipment to reduce the time spent in producing an industrial print.

A typical CAD hardware system is shown in Figure 16-21. The hardware used in a typical CAD system includes the following:

1. A processor that includes a terminal-style keyboard used to establish or maintain communications between the operator and computer. The system may also include a cathode ray tube (CRT) that displays the digital information that is stored in the computer. See Figure 16-22.

2. A graphic display (CRT) that displays graphic information pictorially, as would be seen on a drawing. See Figure 16-23.

Fig. 16–22 Central CAD processor (*Courtesy of PSI Corp.*)

Fig. 16–23 CAD graphics display (*Courtesy of PSI Corp.*)

3. A digitizer which converts graphic information (lines, dimensions, letters, etc.) into digital data for computer use. See Figure 16-24. The type of data (called *menu*) that needs to be digitized is illustrated in Figure 16-25.

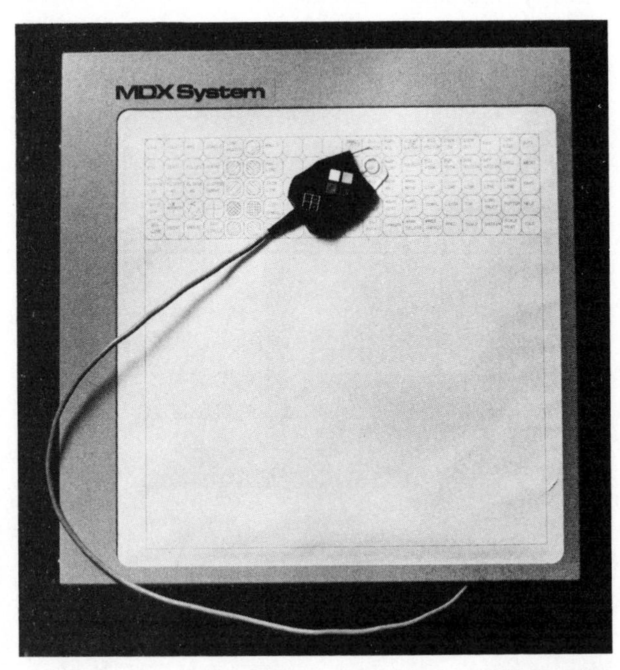

Fig. 16–24 CAD digitizer (*Courtesy of PSI Corp.*)

Fig. 16–25 CAD menu and cursor (*Courtesy of PSI Corp.*)

4. A plotter for converting the graphic data into a hard copy print. See Figure 16-26.

5. A printer which produces a hard-copy manuscript of the digital data that has been stored in the computer. See Figure 16-27.

It should be noted that the CAD equipment mentioned above does not make up a total CAD system. This equipment has been available for some time. However, the CAD software (instructions and programming) has been developed only recently.

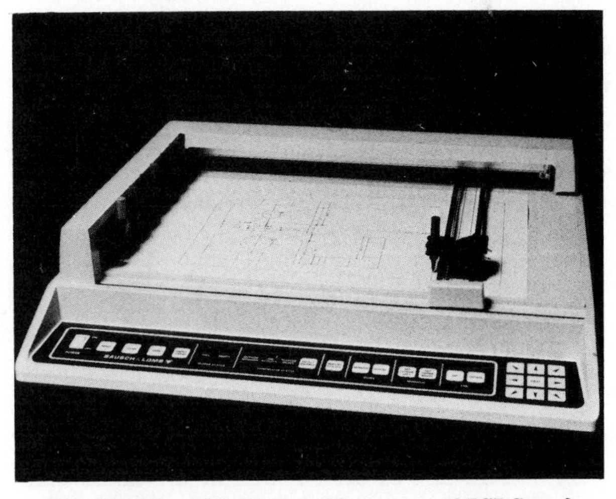

Fig. 16–26 CAD plotter (*Courtesy of PSI Corp.*)

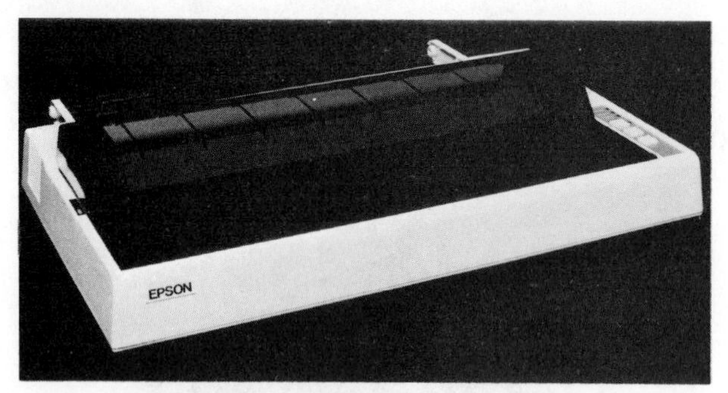

Fig. 16–27 CAD character printer (*Courtesy of PSI Corp.*)

A CAD version of the print in Figure 16-9 is shown in Figure 16-28. This print is included to illustrate the differences between hand-drawn prints and CAD-produced prints. The CAD printer produced the print in about five minutes. The printer made all numbers, letters, and linework, and the border lines and title blocks. Imperfections can be seen at line intersections, and uneven ink flow occurs at some numbers and letters. This is typical of CAD prints.

CAD Users

A *CAD operator* does not have to be a computer programmer to operate the equipment, if the software has already been developed. A CAD operator is trained to operate the CAD hardware, but may not have any training in drafting. A *CAD technician,* however, is trained to operate the CAD equipment but also has a strong background in the drafting field. Therefore, a CAD technician is more useful to a company than a CAD operator.

The Future of CAD

The CAD system is suitable for any type of drafting. CAD is also used for product design. The system is limited only by the development of the software programs.

The use of CAD reduces the time and cost of producing functional industrial prints. It also provides an excellent means of storing related data, such as numerical control information, needed to produce the parts on the prints. Print revisions can be made easily on the computer-stored data. Since the print data is stored on a small floppy disc, very little storage space is required with a CAD system. This greatly reduces the floor space normally needed for print storage cabinets. In addition, print copies can be enlarged or reduced easily.

The benefits of using the CAD system are only now being realized. The future of CAD is limited only by the resourcefulness of its users.

COMMONLY USED HIGH-TECHNOLOGY ABBREVIATIONS

The increased use of abbreviations in technology can be confusing. The abbreviations related to the new drafting technology are listed here.

ASCII—American Standards Communications Information Interface

CAD—Computer-Aided Drafting

CADD—Computer-Aided Design and Drafting

CAE—Computer-Aided Engineering

CAI—Computer-Aided Instruction

CAM—Computer-Aided Manufacturing

CAPP—Computer-Aided Process Planning

CMI—Computer-Managed Instruction

CNC—Computerized Numerical Control

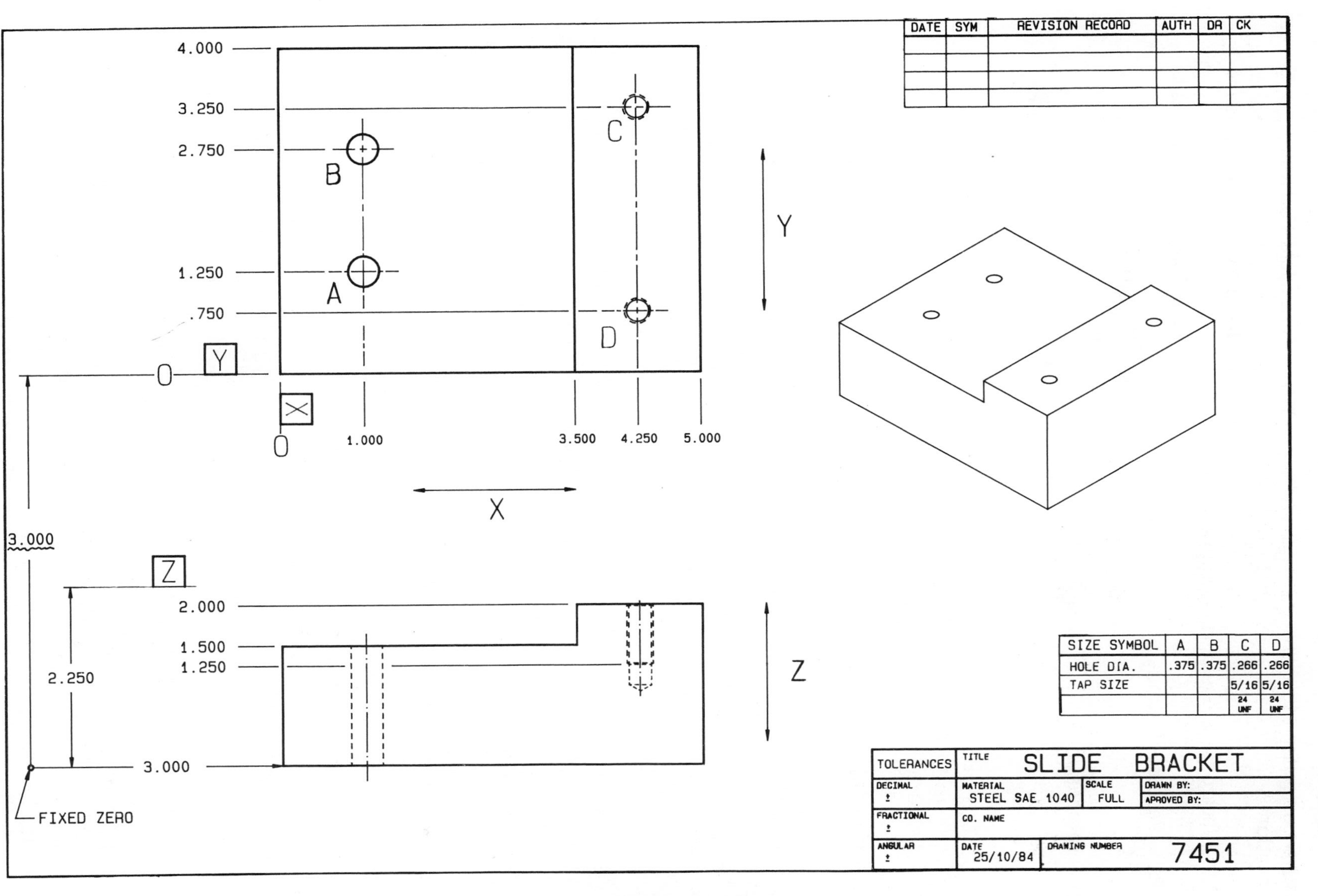

Fig. 16–28 CAD print

CATI—Computer-Aided Testing and Inspection

EIA—Electronic Industries Association

FAPT—Fannuc Automatic Programming Tooling

ISO—International Standards Organization

NC—Numerical Control

NCS—Numerical Code Standard

PRP—Principal Reference Plane

TCE—Tape Command Editor

REVIEW

This review is provided to serve as reinforcement study material. Fill in the appropriate word(s) to complete the sentences below.

1. Prints that display dimensions next to the part features, but no dimension lines or arrowheads, use ______________________ dimensioning.

2. Prints that have no dimensions at the part features, but whose dimensions are displayed in a separate table, use __________________ dimensioning.

3. The horizontal line on a print would be concerned with the ______________________ coordinate.

4. A vertical line as viewed in the top view would be concerned with the ______________________ coordinate.

5. A vertical line, as viewed from the front and side views, would be concerned with the ______________________ coordinate.

6. The *X* and *Y* coordinates would be plus (+) in quadrant ________ .

7. The *X* and *Y* coordinates would be minus (−) in quadrant ______ .

8. The starting point (0-0) can be at any position in (a) (an) ______________________ zero system.

9. (A) (An) ______________________ zero system usually uses a starting point at one corner of the machine table.

10. Before an N/C punched tape can be made, (a) (an) _____________ must be developed.

11. One inch equals approximately ___________________ millimeters.

12. In Europe, ______________________ angle projection is used.

13. When (a) (an) ______________________ is displayed on a print, it indicates that a portion of the part is safety related.

(Continued)

14. ________________________ is a tool that can be used by the drafter; it is not intended to replace the drafter.

15. A CAD ________________________ has been trained to operate the CAD equipment, and also has a strong background in the drafting field.

NAME ______________

DATE ______________

SCORE ______________

EXERCISE B16-1

Print #7451

Answer the following questions after studying Print #7451 found after this exercise.

1. Where is the fixed zero point located for numerical control machining of this part in the *X* direction, in the *Y* direction, and in the *Z* direction from the lower corner of the part?

1. *X* ______
Y ______
Z ______

2. Where is the location of the cutting tool for drilling hole *A* in the *X* direction and the *Y* direction from the fixed zero point?

2. *X* ______
Y ______

3. Hole *C* is how far behind hole *D*?

3. ______

4. Holes *C* and *D* are how far to the right of holes *A* and *B*?

4. ______

5. What is the overall size of this part?

5. Length ______
Width ______
Height ______

6. How deep is the tapped hole drilled?

6. ______

7. What is the clearance of the cutting tool over the left end of the part and over the right end?

7. Left ______
Right ______

8. How deep are holes *C* and *D* tapped?

8. ______

9. What is the size of the drill for the tapped hole in letter size and in inches?

9. Letter ______
Inches ______

10. This part is made of what material?

10. ______

11. What is the last machining operation?

11. ______

12. What is the thread size?

12. ______

NAME ______
DATE ______
SCORE ______

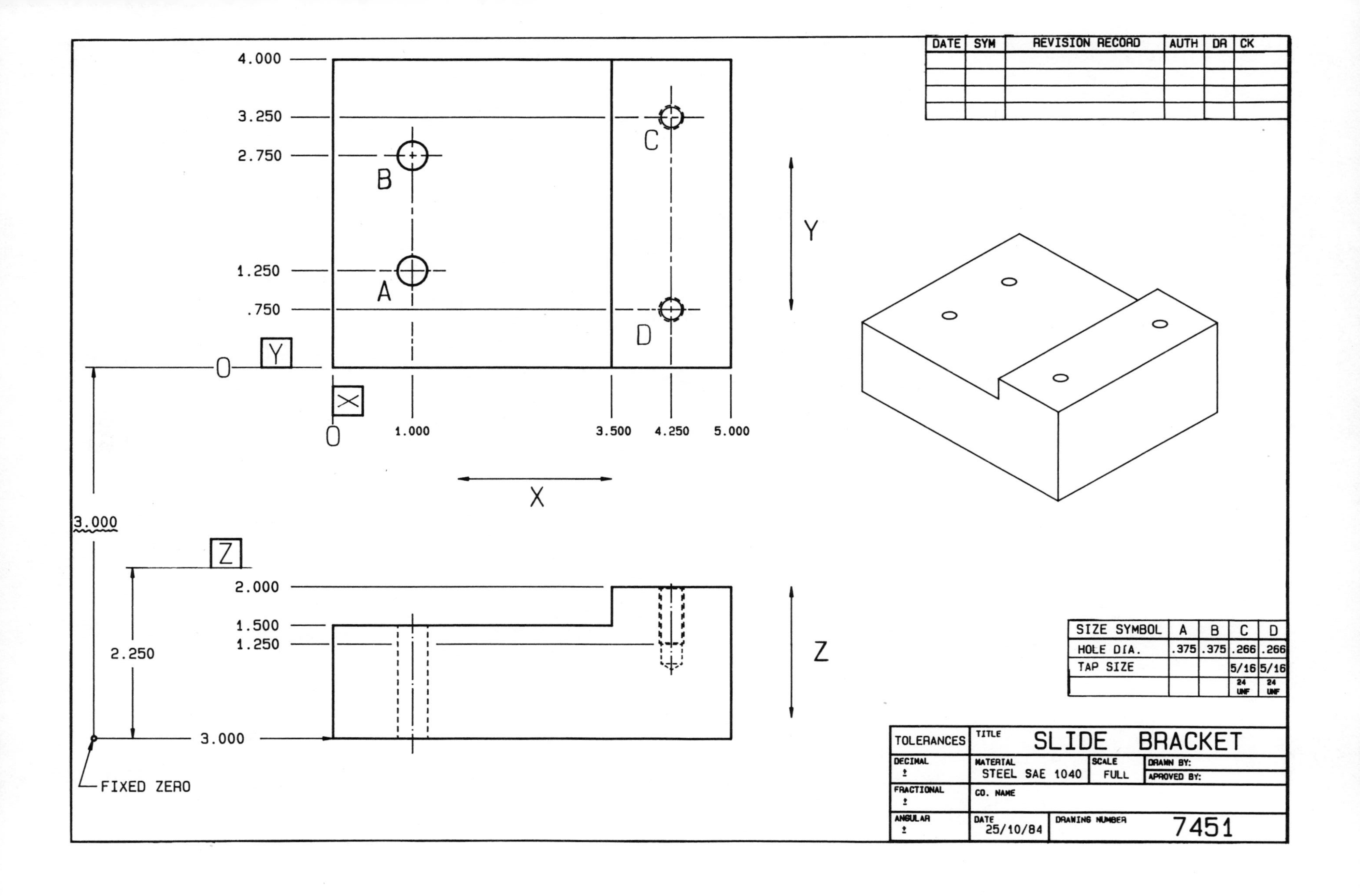
DATE
SYM
REVISION RECORD
AUTH
DR
CK
4.000
3.250
2.750
1.250
.750
0
Y
X
0
1.000
3.500
4.250
5.000
A
B
C
D
X
Y
Z
3.000
Z
2.000
1.500
1.250
2.250
3.000
FIXED ZERO
SIZE SYMBOL
A
B
C
D
HOLE DIA.
.375
.375
.266
.266
TAP SIZE
5/16
5/16
24 UNF
24 UNF
TOLERANCES
TITLE
SLIDE BRACKET
DECIMAL
±
MATERIAL
STEEL SAE 1040
SCALE
FULL
DRAWN BY:
APROVED BY:
FRACTIONAL
±
CO. NAME
ANGULAR
±
DATE
25/10/84
DRAWING NUMBER
7451

APPENDIX A

Tables

Table 1. Thread Elements and Tap Drill Sizes—Unified National Coarse Thread Series

Sizes	Threads Per Inch	DIAMETERS (Basic)				TAP DRILLS	
		Major Diameter —Inches	Pitch Diameter —Inches	Minor Diameter —Inches		Tap Drill To Produce Approx. 75% Full Thread	Decimal Equivalent of Tap Drill —Inches
				Ext. Thds.	Int. Thds.		
1	64	.073	.0629	.0538	.0561	No. 53	.0595
2	56	.086	.0744	.0641	.0667	No. 50	.0700
3	48	.099	.0855	.0734	.0764	No. 47	.0785
4	40	.112	.0958	.0813	.0849	No. 43	.0890
5	40	.125	.1088	.0943	.0979	No. 38	.1015
6	32	.138	.1177	.0997	.1042	No. 36	.1065
8	32	.164	.1437	.1257	.1302	No. 29	.1360
10	24	.190	.1629	.1389	.1449	No. 25	.1495
12	24	.216	.1889	.1649	.1709	No. 16	.1770
1/4	20	.2500	.2175	.1887	.1959	No. 7	.2010
5/16	18	.3125	.2764	.2443	.2524	Let. F	.2570
3/8	16	.3750	.3344	.2983	.3073	5/16	.3125
7/16	14	.4375	.3911	.3499	.3602	Let. U	.3680
1/2	13	.5000	.4500	.4056	.4167	27/64	.4219
9/16	12	.5625	.5084	.4603	.4723	31/64	.4844
5/8	11	.6250	.5660	.5135	.5266	17/32	.5312
3/4	10	.7500	.6850	.6273	.6417	21/32	.6562
7/8	9	.8750	.8028	.7387	.7547	49/64	.7656
1	8	1.0000	.9188	.8466	.8647	7/8	.8750
1 1/8	7	1.1250	1.0322	.9407	.9704	63/64	.9844
1 1/4	7	1.2500	1.1572	1.0747	1.0954	1 7/64	1.1093
1 1/2	6	1.5000	1.3917	1.2955	1.3196	1 11/32	1.4218
1 3/4	5	1.7500	1.6201	1.5046	1.5335	1 9/16	1.5625
2	4 1/2	2.0000	1.8557	1.7274	1.7594	1 25/32	1.7812
2 1/4	4 1/2	2.2500	2.1057	1.9774	2.0094	2 1/32	2.0312
2 1/2	4	2.5000	2.3376	2.1933	2.2294	2 1/4	2.2500
2 3/4	4	2.7500	2.5876	2.4433	2.4794	2 1/2	2.5000
3	4	3.0000	2.8376	2.6933	2.7294	2 3/4	2.7500

Table 2. Thread Elements and Tap Drill Sizes — Unified National Fine Thread Series

Sizes	Threads Per Inch	DIAMETERS (Basic) Major Diameter — Inches	Pitch Diameter — Inches	Minor Diameter — Inches Ext. Thds.	Minor Diameter — Inches Int. Thds.	TAP DRILLS Tap Drill To Produce Approx. 75% Full Thread	Decimal Equivalent of Tap Drill — Inches
0	80	.060	.0519	.0447	.0465	3/64	.0469
1	72	.073	.0640	.0560	.0580	No. 53	.0595
2	64	.086	.0759	.0668	.0691	No. 50	.0700
3	56	.099	.0874	.0771	.0797	No. 45	.0820
4	48	.112	.0985	.0864	.0894	No. 42	.0935
5	44	.125	.1102	.0971	.1004	No. 37	.1040
6	40	.138	.1218	.1073	.1109	No. 33	.1130
8	36	.164	.1460	.1299	.1339	No. 29	.1360
10	32	.190	.1697	.1517	.1562	No. 21	.1590
12	28	.216	.1928	.1722	.1773	No. 14	.1820
1/4	28	.2500	.2268	.2060	.2113	No. 3	.2130
5/16	24	.3125	.2854	.2614	.2674	Let. I	.2720
3/8	24	.3750	.3479	.3239	.3299	Let. Q	.3346
7/16	20	.4375	.4050	.3762	.3834	25/64	.3906
1/2	20	.5000	.4675	.4387	.4459	29/64	.4531
9/16	18	.5625	.5264	.4943	.5024	33/64	.5156
5/8	18	.6250	.5889	.5568	.5649	37/64	.5781
3/4	16	.7500	.7094	.6733	.6823	11/16	.6875
7/8	14	.8750	.8286	.7874	.7977	13/16	.8125
1	12	1.0000	.9549	.8978	.9098	59/64	.9219
1 1/8	12	1.1250	1.0709	1.0228	1.0348	1 3/64	1.0469
1 1/4	12	1.2500	1.1959	1.1478	1.1598	1 11/64	1.1719
1 1/2	12	1.5000	1.4459	1.3978	1.4098	1 27/64	1.4219

Table 3. Metric Tap Drill Sizes

Metric Tap Size	Recommended Metric Drill				Closest Recommended Inch Drill			
	Drill Size (mm)	Inch Equiv.	Probable Hole Size (Inches)	Probable Percent of Thread	Drill Size	Inch Equiv.	Probable Hole Size (Inches)	Probable Percent of Thread
M1.6 × 0.35	1.25	0.0492	0.0507	69	—	—	—	—
M1.8 × 0.35	1.45	0.0571	0.0586	69	—	—	—	—
M2 × 0.4	1.60	0.0630	0.0647	69	#52	0.0635	0.0652	66
M2.2 × 0.45	1.75	0.0689	0.0706	70	—	—	—	—
M2.5 × 0.45	2.05	0.0807	0.0826	69	#46	0.0810	0.0829	67
M3 × 0.5	2.50	0.0984	0.1007	68	#40	0.0980	0.1003	70
M3.5 × 0.6	2.90	0.1142	0.1168	68	#33	0.1130	0.1156	72
M4 × 0.7	3.30	0.1299	0.1328	69	#30	0.1285	0.1314	73
M4.5 × 0.75	3.70	0.1457	0.1489	74	#26	0.1470	0.1502	70
M5 × 0.8	4.20	0.1654	0.1686	69	#19	0.1660	0.1692	68
M6 × 1	5.00	0.1968	0.2006	70	#9	0.1960	0.1998	71
M7 × 1	6.00	0.2362	0.2400	70	15/64	0.2344	0.2382	73
M8 × 1.25	6.70	0.2638	0.2679	74	17/64	0.2656	0.2697	71
M8 × 1	7.00	0.2756	0.2797	69	J	0.2770	0.2811	66
M10 × 1.5	8.50	0.3346	0.3390	71	Q	0.3320	0.3364	75
M10 × 1.25	8.70	0.3425	0.3471	73	11/32	0.3438	0.3483	71
M12 × 1.75	10.20	0.4016	0.4063	74	Y	0.4040	0.4087	71
M12 × 1.25	10.80	0.4252	0.4299	67	27/64	0.4219	0.4266	72
M14 × 2	12.00	0.4724	0.4772	72	15/32	0.4688	0.4736	76
M14 × 1.5	12.50	0.4921	0.4969	71	—	—	—	—
M16 × 2	14.00	0.5512	0.5561	72	35/64	0.5469	0.5518	76
M16 × 1.5	14.50	0.5709	0.5758	71	—	—	—	—
M18 × 2.5	15.50	0.6102	0.6152	73	39/64	0.6094	0.6144	74
M18 × 1.5	16.50	0.6496	0.6546	70	—	—	—	—
M20 × 2.5	17.50	0.6890	0.6942	73	11/16	0.6875	0.6925	74
M20 × 1.5	18.50	0.7283	0.7335	70	—	—	—	—
M22 × 2.5	19.50	0.7677	0.7729	73	49/64	0.7656	0.7708	75
M22 × 1.5	20.50	0.8071	0.8123	70	—	—	—	—
M24 × 3	21.00	0.8268	0.8327	73	53/64	0.8281	0.8340	72
M24 × 2	22.00	0.8661	0.8720	71	—	—	—	—
M27 × 3	24.00	0.9449	0.9511	73	15/16	0.9375	0.9435	78
M27 × 2	25.00	0.9843	0.9913	70	63/64	0.9844	0.9914	70
M30 × 3.5	26.50	1.0433						
M30 × 2	28.00	1.1024						
M33 × 3.5	29.50	1.1614						
M33 × 2	31.00	1.2205						
M36 × 4	32.00	1.2598						
M36 × 3	33.00	1.2992						
M39 × 4	35.00	1.3780						
M39 × 3	36.00	1.4173						

FORMULA FOR METRIC TAP DRILL SIZE

$$\underset{\text{(mm)}}{\text{Basic Major Dia.}} - \frac{\text{\% Thread} \times \text{Pitch (mm)}}{76.980} = \text{DRILLED HOLE SIZE (mm)}$$

FORMULA FOR PERCENT OF THREAD

$$\frac{76.980}{\text{Pitch (mm)}} \times \left[\underset{\text{(mm)}}{\text{Basic Major Dia.}} - \underset{\text{(mm)}}{\text{Drilled Hole Size}} \right] = \text{Percent of Thread}$$

Table 4. Basic Elements of Metrics

Metric Unit Prefixes

Value	Multiples and Submultiples	Prefixes	Symbols
1000	10^3	kilo	k
100	10^2	hecto	h
10	10^1	deka	da
0.1	10^{-1}	deci	d
0.01	10^{-2}	centi	c
0.001	10^{-3}	milli	m

Linear Measurements

1000 meters (m) = 1 kilometer (km)
100 meters = 1 hectometer (hm)
10 meters = 1 dekameter (dam)
1/10 meter = 1 decimeter (dm)
1/100 meter = 1 centimeter (cm)
1/1000 meter = 1 millimeter (mm)

Metric Conversions

Metric values to English system

millimeters × 0.039 = inches (in.)
centimeters × 0.39 = inches
meters × 39.4 = inches
centimeters × 0.033 = feet (ft)
meters × 3.28 = feet
meters × 1.09 = yards (yd)
kilometers × 0.62 = miles (mi)

English system to metric values

inches × 25.4 = millimeters
inches × 2.5 = centimeters
inches × 0.025 = meters
feet × 30.5 = centimeters
feet × 0.305 = meters
yards × 0.91 = meters
miles × 1.6 = kilometers

Table 5. Decimal and Millimeter Equivalents

Fraction	DECIMALS	MILLIMETERS
1/64	0.015625	0.397
1/32	.03125	0.794
3/64	.046875	1.191
1/16	.0625	1.588
5/64	.078125	1.984
3/32	.09375	2.381
7/64	.109375	2.778
1/8	.1250	3.175
9/64	.140625	3.572
5/32	.15625	3.969
11/64	.171875	4.366
3/16	.1875	4.763
13/64	.203125	5.159
7/32	.21875	5.556
15/64	.234375	5.953
1/4	.2500	6.350
17/64	.265625	6.747
9/32	.28125	7.144
19/64	.296875	7.541
5/16	.3125	7.938
21/64	.328125	8.334
11/32	.34375	8.731
23/64	.359375	9.128
3/8	.3750	9.525
25/64	.390625	9.922
13/32	.40625	10.319
27/64	.421875	10.716
7/16	.4375	11.113
29/64	.453125	11.509
15/32	.46875	11.906
31/64	.484375	12.303
1/2	.5000	12.700

1 mm = .03937″

Fraction	DECIMALS	MILLIMETERS
33/64	0.515625	13.097
17/32	.53125	13.494
35/64	.546875	13.891
9/16	.5625	14.288
37/64	.578125	14.684
19/32	.59375	15.081
39/64	.609375	15.478
5/8	.6250	15.875
41/64	.640625	16.272
21/32	.65625	16.669
43/64	.671875	17.066
11/16	.6875	17.463
45/64	.703125	17.859
23/32	.71875	18.256
47/64	.734375	18.653
3/4	.7500	19.050
49/64	.765625	19.447
25/32	.78125	19.844
51/64	796875	20.241
13/16	.8125	20.638
53/64	.828125	21.034
27/32	.84375	21.431
55/64	.859375	21.828
7/8	.8750	22.225
57/64	.890625	22.622
29/32	.90625	23.019
59/64	.921875	23.416
15/16	.9375	23.813
61/64	.953125	24.209
31/32	.96875	24.606
63/64	.984375	25.003
1	1.000	25.400

.001″ = .0254 mm

MM	INCHES	MM	INCHES
.1	.0039	46	1.8110
.2	.0079	47	1.8504
.3	.0118	48	1.8898
.4	.0157	49	1.9291
.5	.0197	50	1.9685
.6	.0236	51	2.0079
.7	.0276	52	2.0472
.8	.0315	53	2.0866
.9	.0354	54	2.1260
1	.0394	55	2.1654
2	.0787	56	2.2047
3	.1181	57	2.2441
4	.1575	58	2.2835
5	.1969	59	2.3228
6	.2362	60	2.3622
7	.2756	61	2.4016
8	.3150	62	2.4409
9	.3543	63	2.4803
10	.3937	64	2.5197
11	.4331	65	2.5591
12	.4724	66	2.5984
13	.5118	67	2.6378
14	.5512	68	2.6772
15	.5906	69	2.7165
16	.6299	70	2.7559
17	.6693	71	2.7953
18	.7087	72	2.8346
19	.7480	73	2.8740
20	.7874	74	2.9134
21	.8268	75	2.9528
22	.8661	76	2.9921
23	.9055	77	3.0315
24	.9449	78	3.0709
25	.9843	79	3.1102
26	1.0236	80	3.1496
27	1.0630	81	3.1890
28	1.1024	82	3.2283
29	1.1417	83	3.2677
30	1.1811	84	3.3071
31	1.2205	85	3.3465
32	1.2598	86	3.3858
33	1.2992	87	3.4252
34	1.3386	88	3.4646
35	1.3780	89	3.5039
36	1.4173	90	3.5433
37	1.4567	91	3.5827
38	1.4961	92	3.6220
39	1.5354	93	3.6614
40	1.5748	94	3.7008
41	1.6142	95	3.7402
42	1.6535	96	3.7795
43	1.6929	97	3.8189
44	1.7323	98	3.8583
45	1.7717	99	3.8976
		100	3.9370

Table 6. American National Standards of Particular Interest to Designers, Architects, and Draftsmen

TITLE OF STANDARD	
Abbreviations	Y1.1-1972
American National Standard Drafting Practices	
Size and Format	Y14.1-1980
Line Conventions and Lettering	Y14.2M-1979
Multi and Sectional View Drawings	Y14.3-1975(R1980)
Pictorial Drawing	Y14.4-1957
Dimensioning and Tolerancing	Y14.5M-1982
Screw Threads	Y14.6-1978
Screw Threads (Metric Supplement)	Y14.6aM-1981
Gears and Splines	
Spur, Helical, and Racks	Y14.7.1-1971
Bevel and Hyphoid	Y14.7.2-1978
Forgings	Y14.9-1958
Springs	Y14.13M–1981
Electrical and Electronic Diagram	Y14.15-1966(R1973)
Interconnection Diagrams	Y14.15a-1971
Information Sheet	Y14.15b-1973
Fluid Power Diagrams	Y14.17-1966(R1980)
Digital Representation for Communication of Product Definition Data	Y14.26M-1981
Computer-Aided Preparation of Product Definition Data Dictionary of Terms	Y14.26.3-1975
Digital Representation of Physical Object Shapes	Y14 Report
Guideline–User Instructions	Y14 Report No. 2
Guideline–Design Requirements	Y14 Report No. 3
Ground Vehicle Drawing Practices	In Preparation
Chassis Frames	Y14.32.1-1974
Parts Lists, Data Lists, and Index Lists	Y14.34M-1982
Surface Texture Symbols	Y14.36-1978
Illustrations for Publication and Projection	Y15.1M-1979
Time Series Charts	Y15.2M-1979
Process Charts	Y15.3M-1979
Graphic Symbols for:	
Electrical and Electronics Diagrams	Y32.2-1975
Plumbing	Y32.4-1977
Use on Railroad Maps and Profiles	Y32.7-1972(R1979)
Fluid Power Diagrams	Y32.10-1967(R1974)
Process Flow Diagrams in Petroleum and Chemical Industries	Y32.11-1961
Mechanical and Acoustical Element as Used in Schematic Diagrams	Y32.18-1972(R1978)
Pipe Fittings, Valves, and Piping	Z32.2.3-1949(R1953)
Heating, Ventilating, and Air Conditioning	Z32.2.4-1949(R1953)
Heat Power Apparatus	Z32.2.6-1950(R1956)
Letter Symbols for:	
Glossary of Terms Concerning Letter Symbols	Y10.1-1972
Hydraulics	Y10.2-1958
Quantities Used in Mechanics for Solid Bodies	Y10.3-1968
Heat and Thermodynamics	Y10.4-1982
Quantities Used in Electrical Science and Electrical Engineering	Y10.5-1968
Aeronautical Sciences	Y10.7-1954
Structural Analysis	Y10.8-1962
Meteorology	Y10.10-1953(R1973)
Acoustics	Y10.11-1953(R1959)
Chemical Engineering	Y10.12-1955(R1973)
Rocket Propulsion	Y10.14-1959
Petroleum Reservoir Engineering and Electric Logging	Y10.15-1958(R1973)
Shell Theory	Y10.16-1964(R1973)
Guide for Selecting Greek Letters Used as Symbols for Engineering Mathematics	Y10.17-1961(R1973)
Illuminating Engineering	Y10.18-1967(R1977)
Mathematical Signs and Symbols for Use in Physical Sciences and Technology	Y10.20-1975

The Above Standards May Be Purchased From:

ANSI American National Standards Institute, Inc.
1430 Broadway
New York, New York 10018

The American Society of Mechanical Engineers
United Engineering Center
345 East 47th Street
New York, New York 10017

Table 7. COMMON DRAFTING ABBREVIATIONS

Aluminum	AL
Anneal	ANL
Assembly	ASSY
Auxiliary	AUX
Babbit	BAB
Base Line	BL
Between Centers	BC
Bevel	BEV
Bill of Material	B/M
Blueprint	BP
Bolt Circle	BC
Bracket	BRKT
Brass	BRS
Broach	BRO
Bronze	BRZ
Carburize	CARB
Case Harden	CH
Cast Iron	CI
Cast Steel	CS
Casting	CSTG
Center Line	CL or ℄
Chamfer	CHAM
Circle	CIR
Cold Drawn Steel	CDS
Cold Rolled Steel	CRS
Counterbore	CBORE or ⌴
Countersink	CSK or ∨
Cross Section	XSEC
Cyanide	CYN
Cylinder	CYL
Depth	DP or ↧
Detail	DET
Diameter (common)	DIA or Ø
(spherical)	SØ
Diametral Pitch	DP
Dimension	DIM
Drawing	DWG
Drop Forge	DF
Fillet	FIL
Finish	FIN
Finish All Over	FOA
Fixture	FIX
Flat Head	FH
Forged Steel	FST
Forging	FORG
Foundry	FDRY
Gage	GA
Grind	GRD
Harden	HDN
Head	HD
Heat Treat	HT TR
Hexagon	HEX
High-speed Steel	HSS
Inside Diameter	ID
Keyway	KWY
Left-hand Thread	LH THD
Machine	MACH
Material	MATL
Maximum	MAX
Micro (10^{-6})	μ or U
Millimeter	MM
Minimum	MIN
National Coarse (thread)	NC
National Fine (thread)	NF
National Special (thread)	NS
Number	NO. or #
Origin of Dimension	⊕
Outside Diameter	OD
Overall	OA
Pitch Diameter	PD
Punch	PCH
Radial	RAD
Radius (common)	R
(spherical)	SR
Ream	RM
Repetitive Feature	X
Right-hand Thread	RH THD
Rockwell Hardness	RH
Round	RD
Screw	SCR
Section	SECT
Spot Face	SF or ⌴
Square	SQ or □
Stainless Steel	SST
Standard	STD
Steel	STL
Stock	STK
Straight	STR
Taper (flat)	TPR or ▻
(conical)	TPR or ⊳
Tensile Strength	TS
Threads Per Inch	TPI
Tolerance	TOL
Tool Steel	TS
Total Indicator Reading	TIR
Typical	TYP
Unified National Coarse (thread)	UNC
Unified National Fine (thread)	UNF
Unified National Special (thread)	UNS
Width	W
Wrought Iron	WI

APPENDIX B

Selected Answers to Chapter Reviews

CHAPTER 1

2. black or dark; white
4. shape; additional production information; feature size; additional business information

CHAPTER 2

2. front
4. lines
6. length; height
8. height; width

CHAPTER 3

2. isometric; oblique
4. oval
6. round

CHAPTER 4

2. six
4. front
6. minimum

CHAPTER 5

2. cutting plane lines
4. hidden lines
6. phantom lines

CHAPTER 6

2. close
4. two
6. diameter
8. object

CHAPTER 7

2. part function; manufacturing method; interchangeability; combination of these
4. machining costs
6. one-millionth (or .000001)

CHAPTER 8

2. reduced
4. wavy line
6. tolerance dimension

CHAPTER 9

2. thread diameter; number of threads; thread type; thread series; class of fit
4. internal thread
6. millimeters
8. machine screws; bolts; studs; nuts

CHAPTER 10

2. spur
4. worm
6. diametral pitch; number of teeth; pressure angle
8. gear rack
10. press

CHAPTER 11

2. full section
4. crosshatched
6. opposite directions
8. offset section
10. cast-iron

CHAPTER 12

2. lower right
4. √
6. upper right
8. close to
10. microfilming

CHAPTER 13

2. iron; steel
4. .75; 1.70
6. percent
8. casting; wrought
10. Rockwell

CHAPTER 14

2. machine shop
4. assembly
6. double line; single line
8. electrical control
10. fillet; plug; spot; seam; groove

CHAPTER 15

2. consistency; easily drawn; reduces translation errors; uses less print space; conveys more information
4. three
6. first
8. Ø

CHAPTER 16

2. tabular
4. Y
6. I
8. floating
10. program document (sheet)
12. first
14. computer aided drafting

APPENDIX C

Selected Answers to Chapter Exercises

EXERCISE A2-2

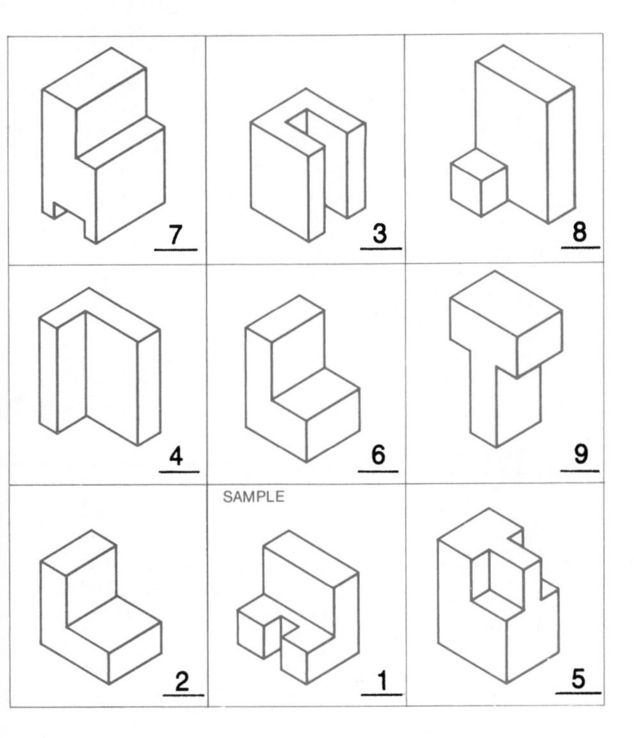

EXERCISE A3-2

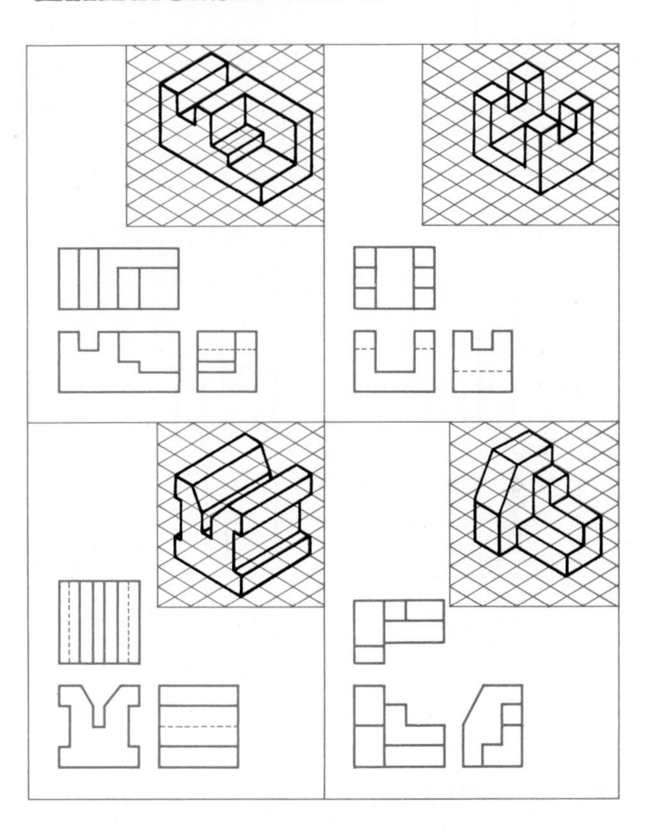

EXERCISE A4-2

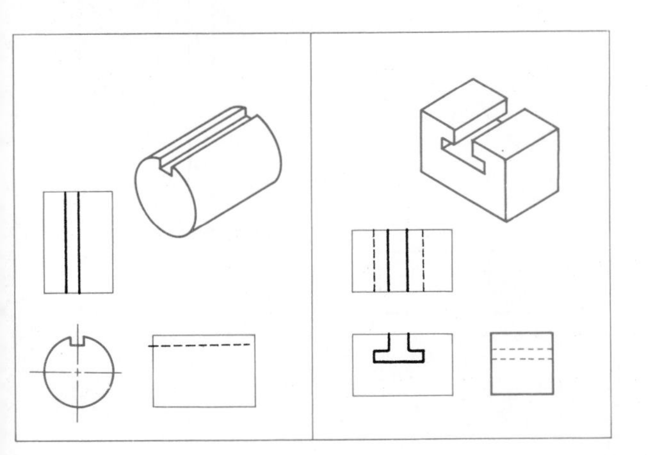

EXERCISE A4-4

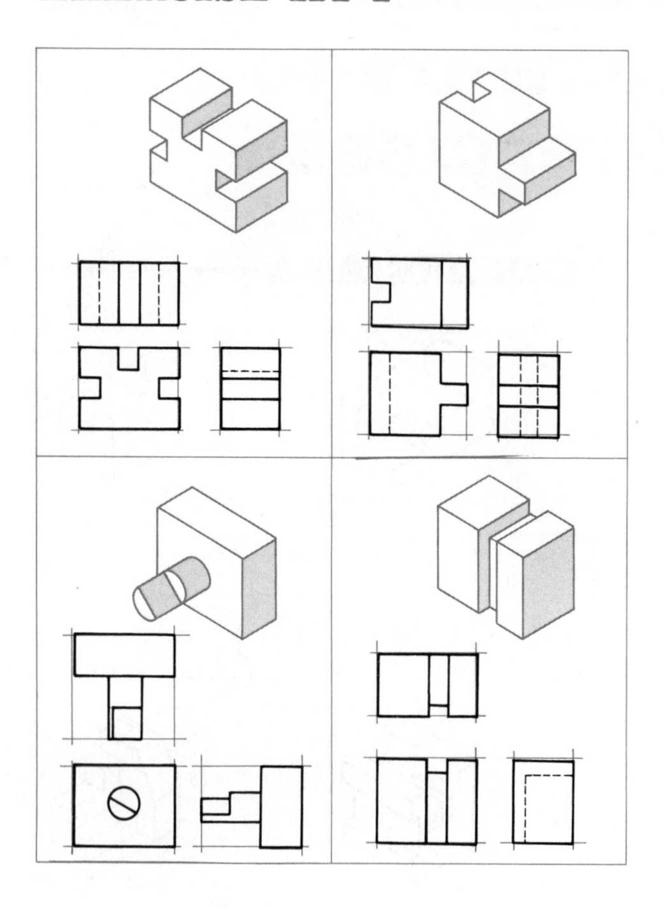

EXERCISE A4-6

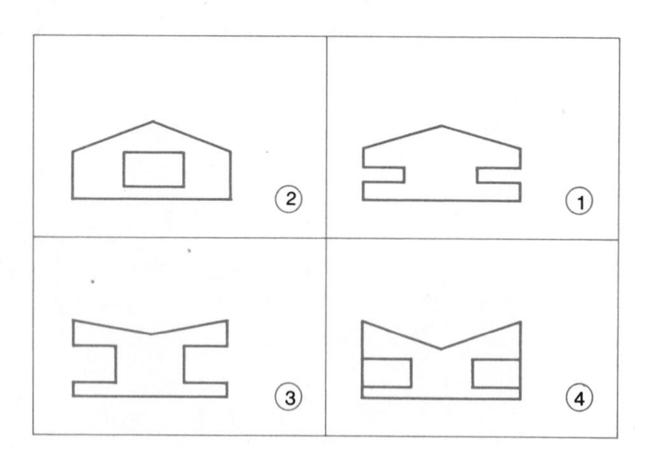

EXERCISE A4-8

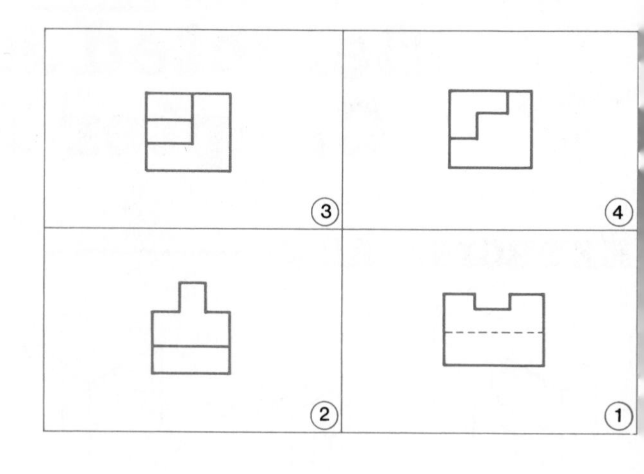

EXERCISE A5-2

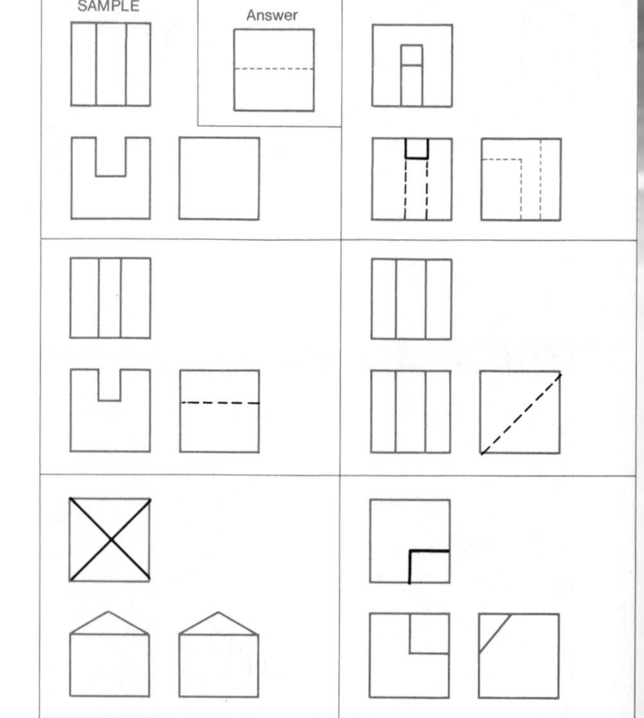

EXERCISE A5-4

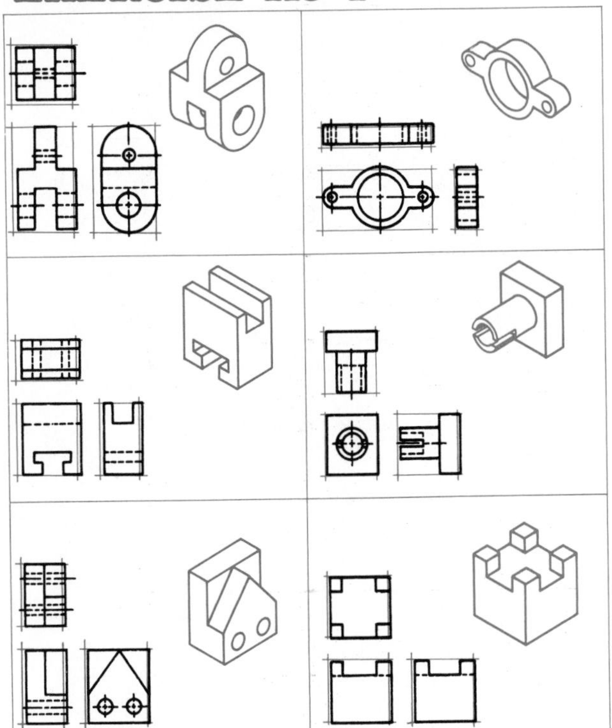

EXERCISE A11-2

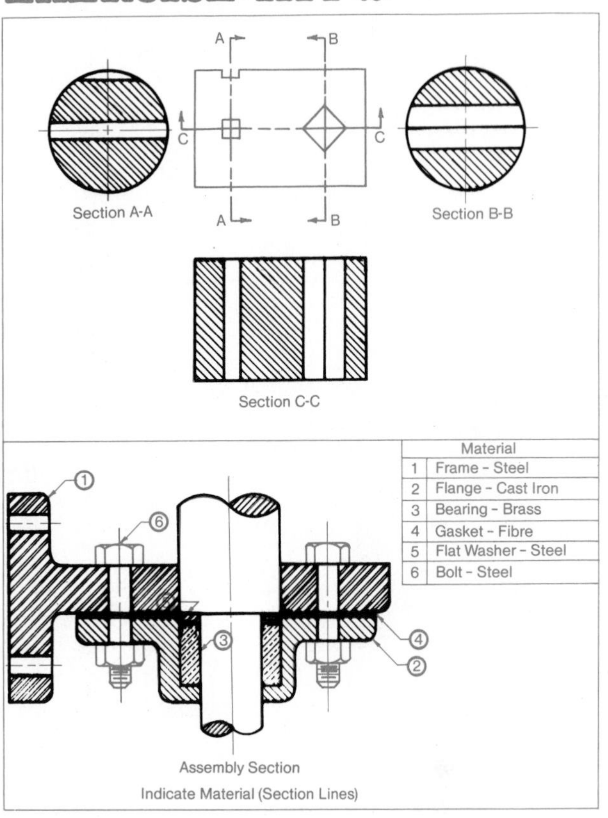

APPENDIX D

Advanced Exercises

The following exercises, series C and D, are provided to further test your knowledge of print reading. These print exercises are more difficult than the chapter exercises (series A and B). The series D exercises are more advanced than the series C exercises.

The answers can be determined by studying the information on the prints and by reviewing the material covered in the book. Some questions require the use of tables included in the chapters or the Appendix.

EXERCISE C-1

Print #DSB-31

QUESTIONS

1. What relation does this print have to Print #E-139 in Chapter 12?
2. How many points does this socket have?
3. What size is the drive?
4. This socket is made of what material?
5. What kind of finish does it have?
6. What hardness rating does this part have?
7. Prior to 5-21-76, what was the hardness rating?
8. What size and shape is the drive end of this part?
9. What is this part?
10. What is the purpose of the internal groove at the drive end (left)?
11. How is this part identified?
12. This socket will fit what size of spark plug?
13. What is the tolerance on the 3/8 drive?
14. What is the depth of the external groove?

ANSWERS

1. ____________________
2. ____________________
3. ____________________
4. ____________________
5. ____________________
6. ____________________
7. ____________________
8. ____________________
9. ____________________
10. ____________________
11. ____________________
12. Inch ____________________
    Millimeter ____________________
13. ____________________
14. ____________________

NAME ____________________

DATE ____________________

SCORE ____________________

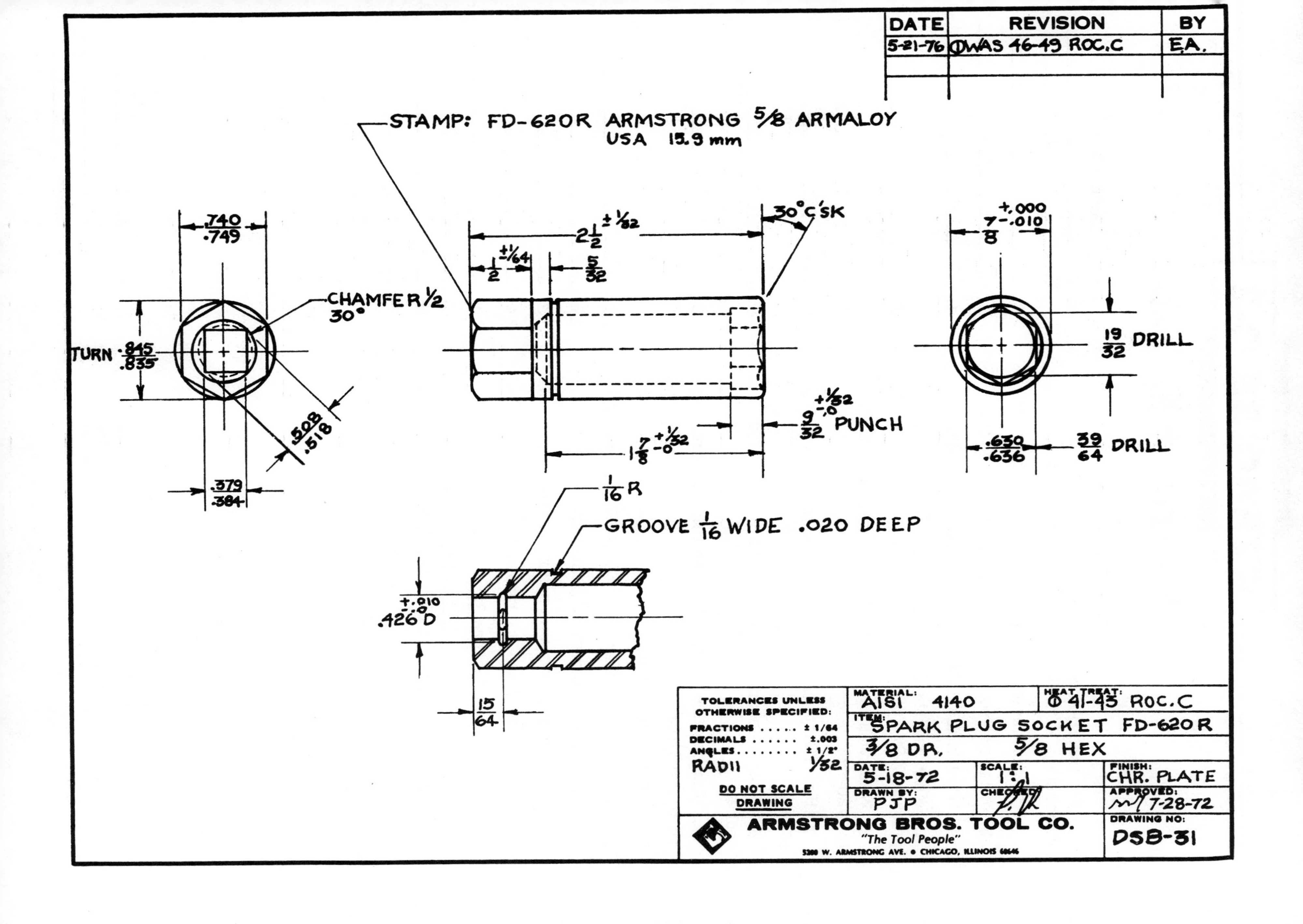

DATE
REVISION
BY
5-21-76
① WAS 46-49 ROC.C
EA.
STAMP: FD-620R ARMSTRONG 5/8 ARMALOY
USA 15.9 mm
.740
.749
CHAMFER 1/2
30°
TURN .845
.835
.508
.518
.379
.384
2 1/2 ± 1/32
1/2 ± 1/64
5/32
30° C'SK
9/32 +1/32 -0 PUNCH
1 7/8 +1/32 -0
+.000
7/8 -.010
19/32 DRILL
.630
.636
39/64 DRILL
1/16 R
GROOVE 1/16 WIDE .020 DEEP
.426 D +.010 -.0
15/64
TOLERANCES UNLESS OTHERWISE SPECIFIED:
FRACTIONS ...... ± 1/64
DECIMALS .... ±.003
ANGLES .... ± 1/2°
RADII 1/32
DO NOT SCALE DRAWING
MATERIAL: AISI 4140
HEAT TREAT: ① 41-45 ROC.C
ITEM: SPARK PLUG SOCKET FD-620R
3/8 DR. 5/8 HEX
DATE: 5-18-72
SCALE: 1:1
FINISH: CHR. PLATE
DRAWN BY: PJP
CHECKED:
APPROVED: 7-28-72
DRAWING NO: DSB-31
ARMSTRONG BROS. TOOL CO.
"The Tool People"
5200 W. ARMSTRONG AVE. • CHICAGO, ILLINOIS 60646

EXERCISE C-2

Print #E-125

QUESTIONS	ANSWERS
1. For the 16-mm size socket, indicate the dimensions *A*, *B*, *D*, and *L*.	1. *A.* ________ *B.* ________ *D.* ________ *L.* ________
2. Of what material are the metric sockets made?	2. ________
3. What size square drive will these sockets fit?	3. ________
4. What is the drill size for the 12-mm socket?	4. ________
5. What was the drill size for the 19-mm socket prior to 7-27-76?	5. ________
6. What was the length of the 22-mm socket prior to 1-20-77?	6. ________
7. How many sizes of metric sockets apply to this print?	7. ________
8. What hole size remains the same for all listed sockets?	8. ________
9. What feature of this part remains the same for all socket sizes?	9. ________
10. What scale is used?	10. ________
11. What is the tolerance for the .426$^{\phi}$ dimension?	11. ________
12. What heat treatment is indicated?	12. ________
13. On what date was this print drawn?	13. ________
14. On what date was it approved?	14. ________
15. What is the tolerance for all fractional dimensions?	15. ________
16. What type of section view is shown?	16. ________

NAME ________

DATE ________

SCORE ________

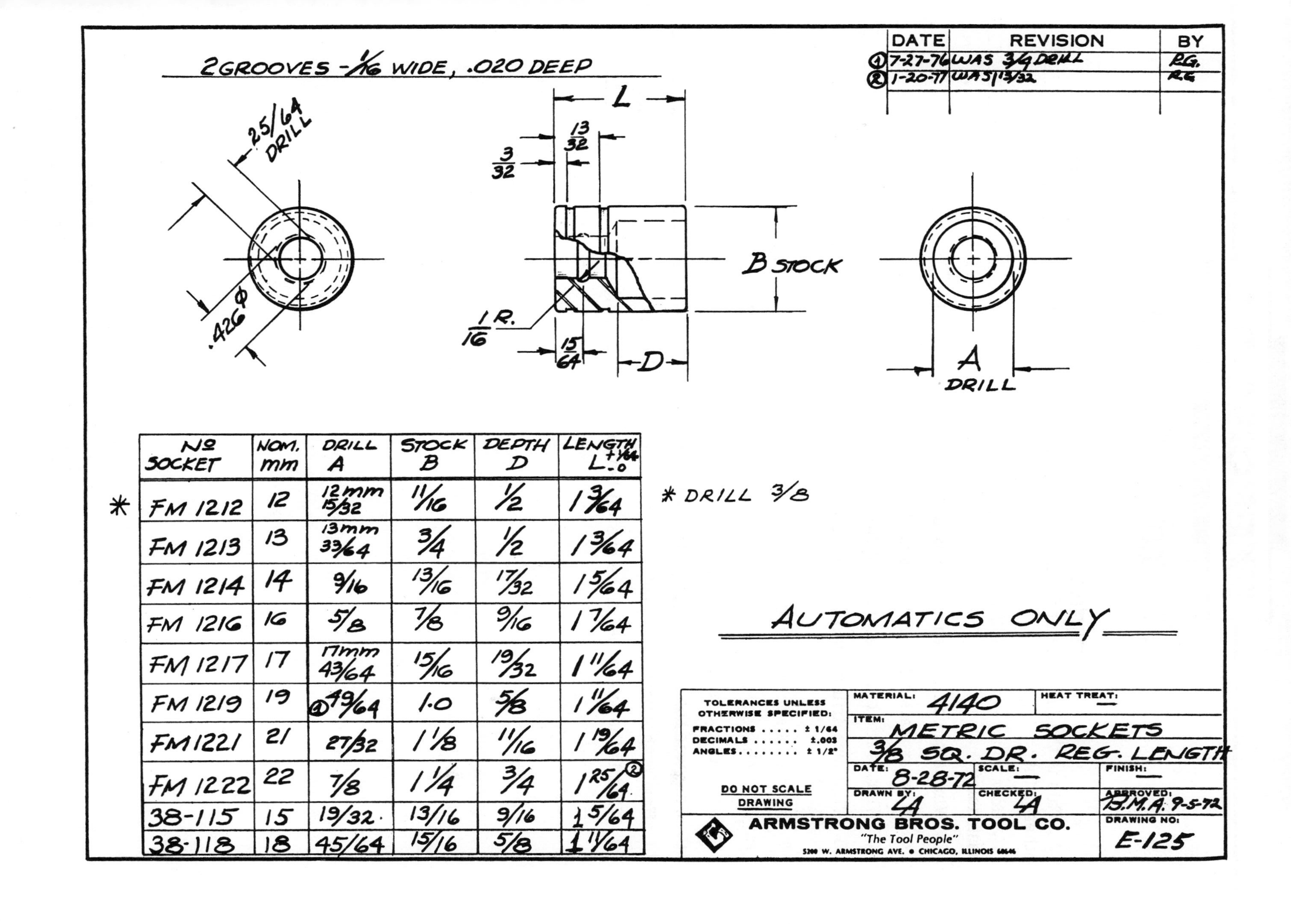

	№ SOCKET	NOM. mm	DRILL A	STOCK B	DEPTH D	LENGTH L +1/64 -0
*	FM 1212	12	12mm 15/32	11/16	1/2	1 3/64
	FM 1213	13	13mm 33/64	3/4	1/2	1 3/64
	FM 1214	14	9/16	13/16	17/32	1 5/64
	FM 1216	16	5/8	7/8	9/16	1 7/64
	FM 1217	17	17mm 43/64	15/16	19/32	1 11/64
	FM 1219	19	① 49/64	1.0	5/8	1 11/64
	FM 1221	21	27/32	1 1/8	11/16	1 19/64
	FM 1222	22	7/8	1 1/4	3/4	1 25/64 ②
	38-115	15	19/32	13/16	9/16	1 5/64
	38-118	18	45/64	15/16	5/8	1 11/64

EXERCISE C-3

Print #DSC-108

QUESTIONS	ANSWERS
1. This is the finish print for the part on what other print?	1. ______
2. How many parts are used on the finished assembly?	2. ______
3. What size ball is used on the external drive?	3. ______
4. Of what material is this part made?	4. ______
5. What is the hardness value of this part?	5. ______
6. What size drill is used for the hole which retains the ball?	6. ______
7. What is the tolerance for the internal drive hole?	7. ______
8. What shape are the drives?	8. ______
9. What kind of finish is used?	9. ______
10. What was the hardness prior to 5-21-76?	10. ______
11. How is this part identified?	11. ______
12. What is the tolerance on the overall length?	12. ______

NAME ______

DATE ______

SCORE ______

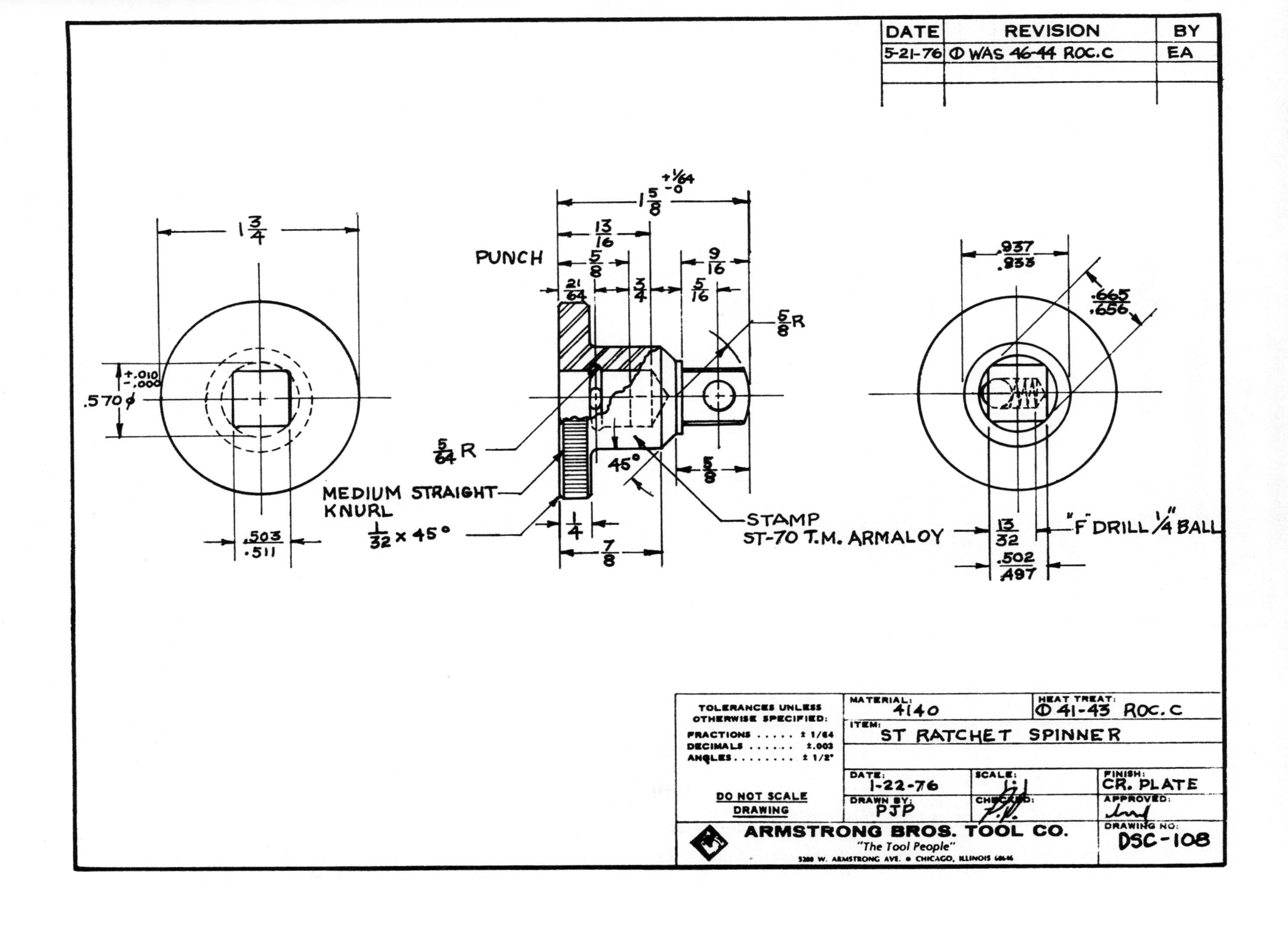
DATE
REVISION
BY
5-21-76
① WAS 46-44 ROC.C
EA
PUNCH
1 5/8 +1/64 -0
13/16
5/8
9/16
21/64
3/4
5/16
5/8 R
1 3/4
.570 ϕ +.010 -.000
5/64 R
45°
5/8
MEDIUM STRAIGHT KNURL
1/32 x 45°
1/4
7/8
.503 .511
STAMP ST-70 T.M. ARMALOY
.937 .933
.665 .656
13/32
"F" DRILL 1/4" BALL
.502 .497
TOLERANCES UNLESS OTHERWISE SPECIFIED:
FRACTIONS ± 1/64
DECIMALS ±.003
ANGLES ± 1/2°
DO NOT SCALE DRAWING
MATERIAL: 4140
HEAT TREAT: ① 41-43 ROC. C
ITEM: ST RATCHET SPINNER
DATE: 1-22-76
SCALE: 1:1
FINISH: CR. PLATE
DRAWN BY: PJP
CHECKED:
APPROVED:
ARMSTRONG BROS. TOOL CO.
"The Tool People"
5200 W. ARMSTRONG AVE. • CHICAGO, ILLINOIS 60646
DRAWING NO: DSC-108

EXERCISE C-4

Print #E-152

QUESTIONS	ANSWERS
1. On what date was this print drawn?	1. ______
2. Who drew this print?	2. ______
3. Who checked and approved this print?	3. ______
4. Of what material is this part made?	4. ______
5. What kind and size of knurl is used?	5. ______
6. What is the tolerance of the small outside diameter?	6. ______
7. What is this part?	7. ______
8. What scale is used?	8. ______
9. What are the size and angle of the chamfers?	9. ______
10. What is the overall length of this part?	10. ______
11. What is the overall diameter of this part?	11. ______

NAME ______

DATE ______

SCORE ______

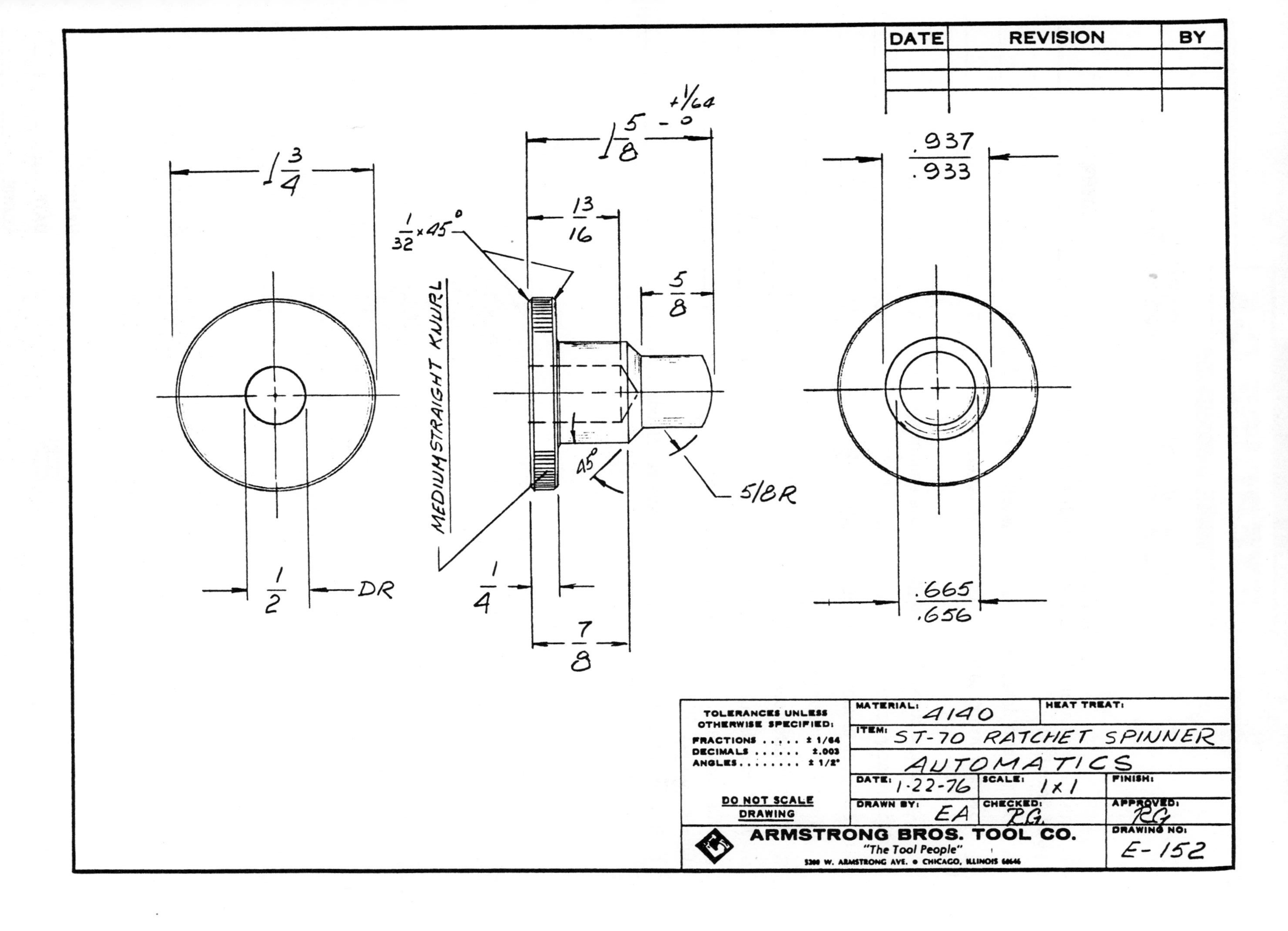
DATE
REVISION
BY
1 3/4
1/2
DR
1 5/8 +1/64 -0
13/16
1/32 x 45°
MEDIUM STRAIGHT KNURL
5/8
45°
5/8R
1/4
7/8
.937
.933
.665
.656
TOLERANCES UNLESS OTHERWISE SPECIFIED:
FRACTIONS ..... ± 1/64
DECIMALS ...... ±.003
ANGLES .... ± 1/2°
DO NOT SCALE DRAWING
MATERIAL: 4140
HEAT TREAT:
ITEM: ST-70 RATCHET SPINNER
AUTOMATICS
DATE: 1-22-76
SCALE: 1x1
FINISH:
DRAWN BY: EA
CHECKED: RG.
APPROVED: RG
ARMSTRONG BROS. TOOL CO.
"The Tool People"
5200 W. ARMSTRONG AVE. • CHICAGO, ILLINOIS 60646
DRAWING NO: E-152

EXERCISE C-5

Print #JM 6938

QUESTIONS	ANSWERS
1. The bill of material indicates that this is what type of drawing?	1. ______
2. How far can detail #1 slide into detail #4?	2. ______
3. How can you determine what material each part is made of?	3. ______
4. What is the maximum length of this assembly?	4. ______
5. How many parts are in this assembly?	5. ______
6. What part is most likely purchased?	6. ______
7. Of what material is detail #3 made?	7. ______
8. What is this assembly?	8. ______
9. What scale is used on this print?	9. ______
10. This assembly is what detail number on subassembly CT 48813?	10. ______
11. How many of these subassemblies are required?	11. ______

NAME ______

DATE ______

SCORE ______

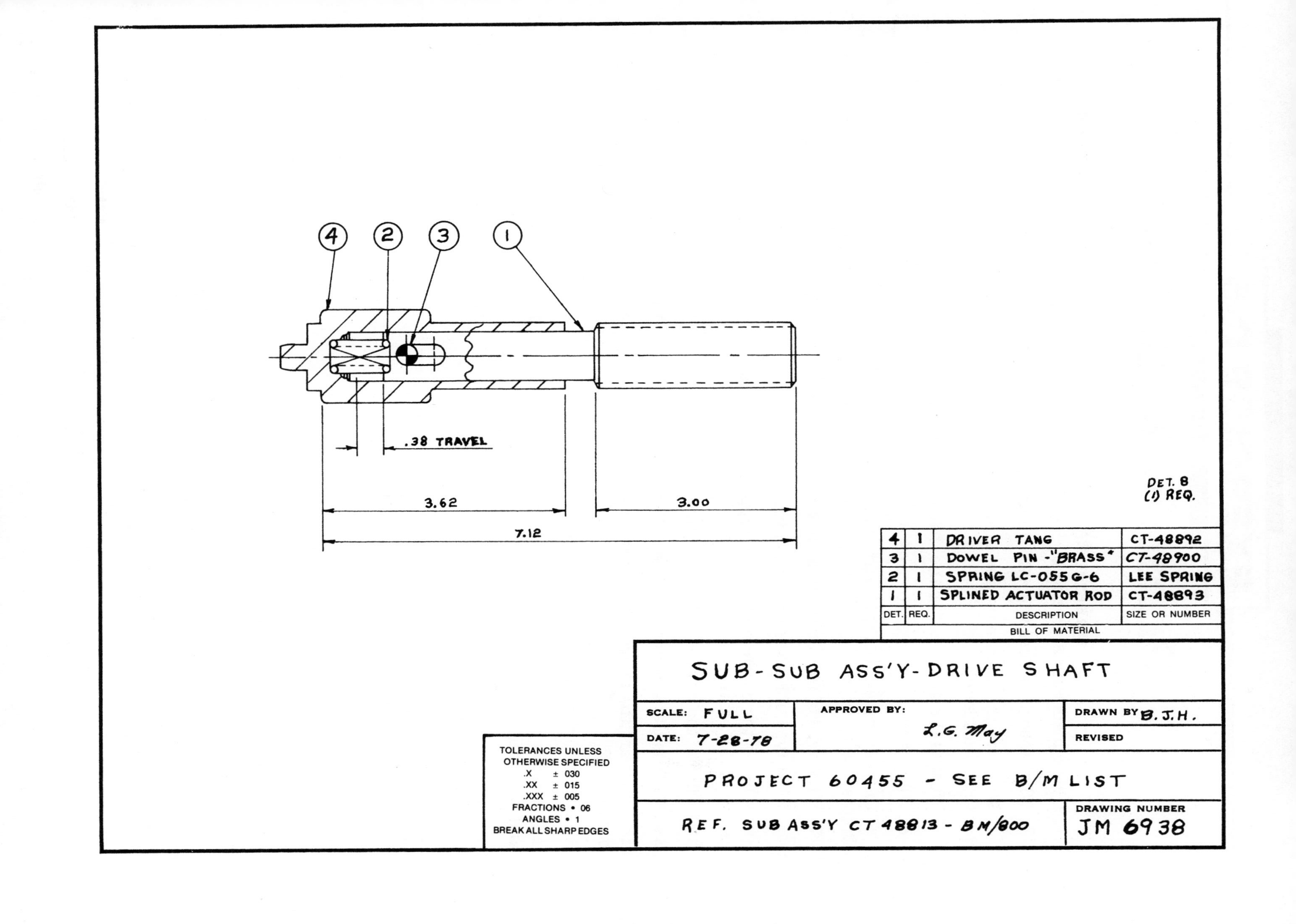

4
2
3
1
.38 TRAVEL
3.62
3.00
7.12
DET. 8
(1) REQ.
4 | 1 | DRIVER TANG | CT-48892
3 | 1 | DOWEL PIN -"BRASS" | CT-48900
2 | 1 | SPRING LC-055G-6 | LEE SPRING
1 | 1 | SPLINED ACTUATOR ROD | CT-48893
DET. | REQ. | DESCRIPTION | SIZE OR NUMBER
BILL OF MATERIAL
SUB-SUB ASS'Y-DRIVE SHAFT
SCALE: FULL
DATE: 7-28-78
APPROVED BY:
L.G. May
DRAWN BY B.J.H.
REVISED
PROJECT 60455 - SEE B/M LIST
REF. SUB ASS'Y CT 48813 - BM/800
DRAWING NUMBER
JM 6938
TOLERANCES UNLESS
OTHERWISE SPECIFIED
.X ± 030
.XX ± 015
.XXX ± 005
FRACTIONS • 06
ANGLES • 1
BREAK ALL SHARP EDGES

EXERCISE C-6

Print #JM 6938-7

QUESTIONS

1. Give dimensions *A*, *B*, and *C*.
2. What tolerance does the .625 dimension have?
3. What are the maximum and minimum sizes for the .625 dimension?
4. What is the drill size for the three tapped holes?
5. What is the stock size for this part?
6. Of what material is this part made? (See Chapter 13.)
7. What type of section view is shown?
8. What type of heat treatment is indicated?
9. What is the depth of the large tapped hole at the center of the part?
10. What is the hardness range of this part?
11. What is the tolerance for the 2.50 dimension?
12. What are the micro-finishes on the front and back sides?
13. What surface is the main datum?
14. What is the parallel tolerance between the front and back surfaces?

ANSWERS

1. *A* _____ *B* _____ *C* _____
2. ____________________
3. ____________________
4. ____________________
5. ____________________
6. ____________________
7. ____________________
8. ____________________
9. ____________________
10. ____________________
11. ____________________
12. ____________________
13. ____________________
14. ____________________

NAME ____________________

DATE ____________________

SCORE ____________________

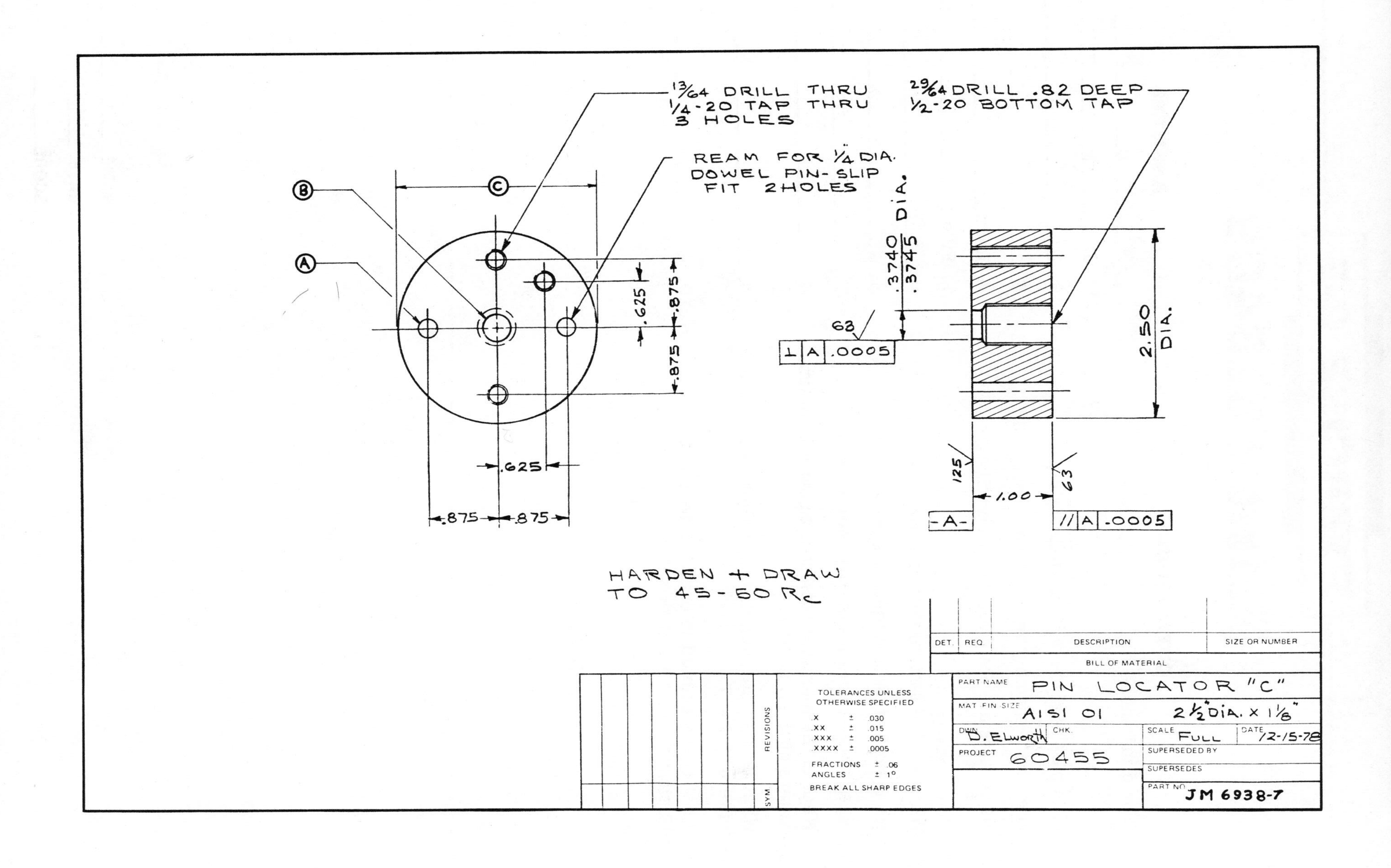
13/64 DRILL THRU
1/4-20 TAP THRU
3 HOLES
29/64 DRILL .82 DEEP
1/2-20 BOTTOM TAP
REAM FOR 1/4" DIA.
DOWEL PIN-SLIP
FIT 2 HOLES
.3740
.3745 DIA.
68
⊥ A .0005
2.50 DIA.
125
63
1.00
-A-
// A .0005
.625
.875
.875
.625
.875
.875
A
B
C
HARDEN + DRAW
TO 45-50 Rc
DET.
REQ.
DESCRIPTION
SIZE OR NUMBER
BILL OF MATERIAL
PART NAME PIN LOCATOR "C"
MAT. FIN. SIZE AISI 01 2 1/2" DIA. x 1 1/8"
DWN. D. ELWORTH
CHK.
SCALE FULL
DATE 12-15-78
PROJECT 60455
SUPERSEDED BY
SUPERSEDES
PART NO. JM 6938-7
TOLERANCES UNLESS OTHERWISE SPECIFIED
.X ± .030
.XX ± .015
.XXX ± .005
.XXXX ± .0005
FRACTIONS ± .06
ANGLES ± 1°
BREAK ALL SHARP EDGES
REVISIONS
SYM

EXERCISE C-7

Print #315409

QUESTIONS	ANSWERS
1. Give dimensions *A*, *B*, *C*, *D*, *E*, and *F*.	1. *A*____ *B*____ *C*____ *D*____ *E*____ *F*____
2. This part is made of what material?	2. ____
3. For what assembly was this part designed?	3. ____
4. What sizes are the small radii and fillets?	4. Radii ____ Fillets ____
5. What kind of section view is B-B?	5. ____
6. What kind of section view is A-A?	6. ____
7. What is the tolerance on the .890 hole?	7. ____
8. What is the maximum distance allowed between the centers of the .710 holes?	8. ____
9. What is the minimum distance allowed between the centers of the .710 holes?	9. ____
10. What is the amount of draft allowed on the cored holes?	10. ____

NAME ____

DATE ____

SCORE ____

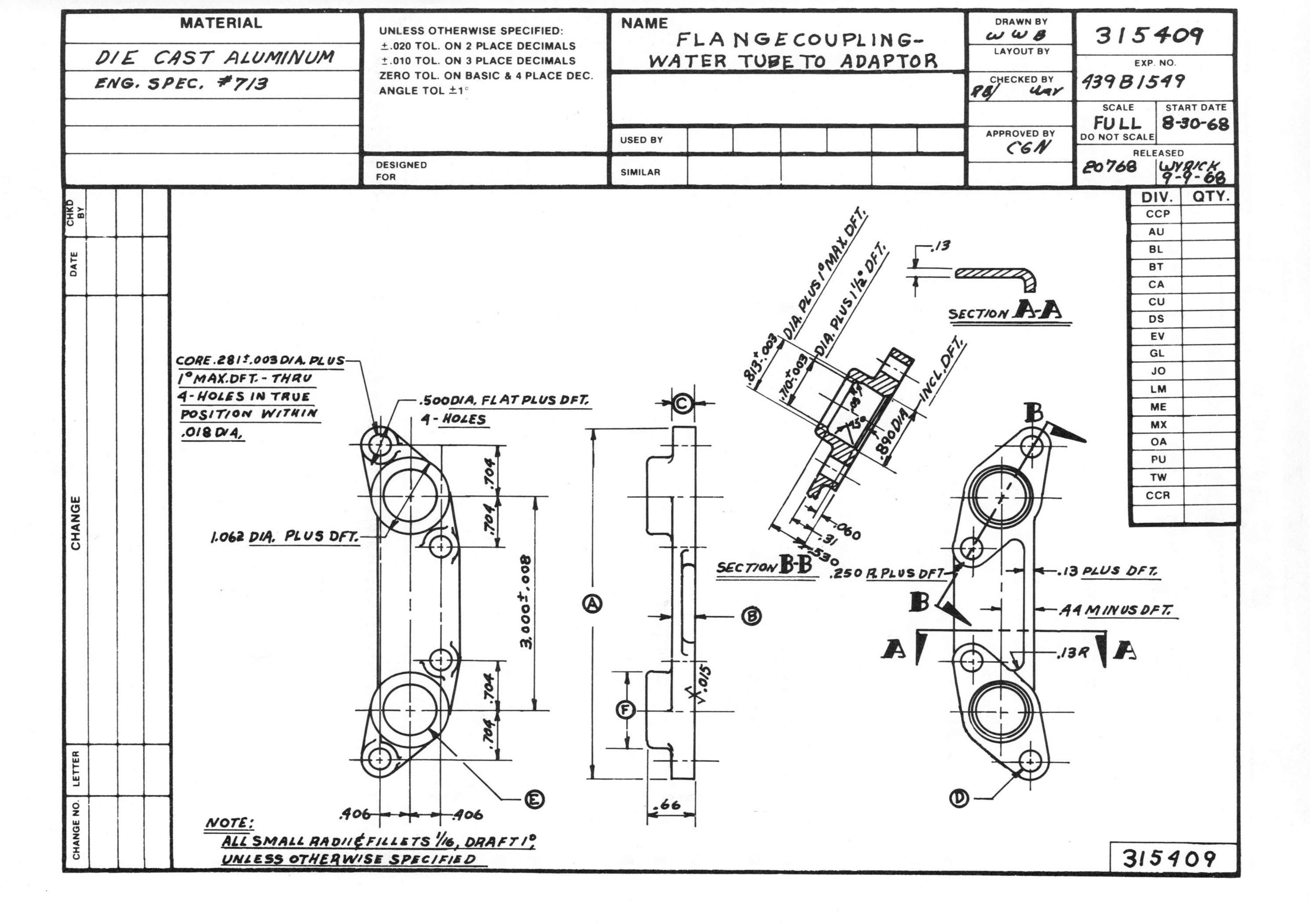

MATERIAL
DIE CAST ALUMINUM
ENG. SPEC. #713
UNLESS OTHERWISE SPECIFIED:
±.020 TOL. ON 2 PLACE DECIMALS
±.010 TOL. ON 3 PLACE DECIMALS
ZERO TOL. ON BASIC & 4 PLACE DEC.
ANGLE TOL ±1°
DESIGNED FOR
NAME
FLANGE COUPLING-
WATER TUBE TO ADAPTOR
USED BY
SIMILAR
DRAWN BY
WWB
LAYOUT BY
CHECKED BY
APPROVED BY
CGN
315409
EXP. NO.
439B1549
SCALE
FULL
DO NOT SCALE
START DATE
8-30-68
RELEASED
20768
WYRICK 9-9-68
DIV.
QTY.
CCP
AU
BL
BT
CA
CU
DS
EV
GL
JO
LM
ME
MX
OA
PU
TW
CCR
CHKD BY
DATE
CHANGE
LETTER
CHANGE NO.
CORE .281±.003 DIA. PLUS 1° MAX. DFT. - THRU 4-HOLES IN TRUE POSITION WITHIN .018 DIA.
.500 DIA. FLAT PLUS DFT. 4-HOLES
1.062 DIA. PLUS DFT.
.704
.704
.704
.704
3.000±.008
.406
.406
E
A
B
C
F
.015
.66
.813±.003 DIA. PLUS 1° MAX. DFT.
.710±.003 DIA. PLUS 1½° DFT.
.890 DIA INCL. DFT.
15°
.060
.31
.530
SECTION B-B
.13
SECTION A-A
.250 R. PLUS DFT.
.13 PLUS DFT.
.44 MINUS DFT.
.13R
A
A
B
B
D
NOTE:
ALL SMALL RADII & FILLETS 1/16, DRAFT 1°
UNLESS OTHERWISE SPECIFIED
315409

EXERCISE C-8

Print #312105

QUESTIONS

1. Of what material is this part made?
2. What is the hardness range of this part?
3. What scales are used?
4. What size cutter is used to machine the keyway?
5. This shaft has how many spline teeth?
6. What is the total length of this shaft?
7. What was the dimension indicated by B_1 prior to 3-10-65?
8. Who drew this print?
9. What size and angle chamfer is required at both ends of the shaft?
10. What is the width of the keyway?
11. What is the full length of the splines?
12. What is the outside diameter of this shaft?
13. What runout is required at the center of the shaft?
14. What is the minor diameter of the splines?
15. What fillet length is required at the minor diameter of the splines?
16. How are dimensions determined for the gages which are used to inspect the tolerance dimensions of this part?

ANSWERS

1. ____________________
2. ____________________
3. ____________________
4. ____________________
5. ____________________
6. ____________________
7. ____________________
8. ____________________
9. ____________________
10. ____________________
11. ____________________
12. ____________________
13. ____________________
14. ____________________
15. ____________________
16. ____________________

____

NAME ____________________

DATE ____________________

SCORE ____________________

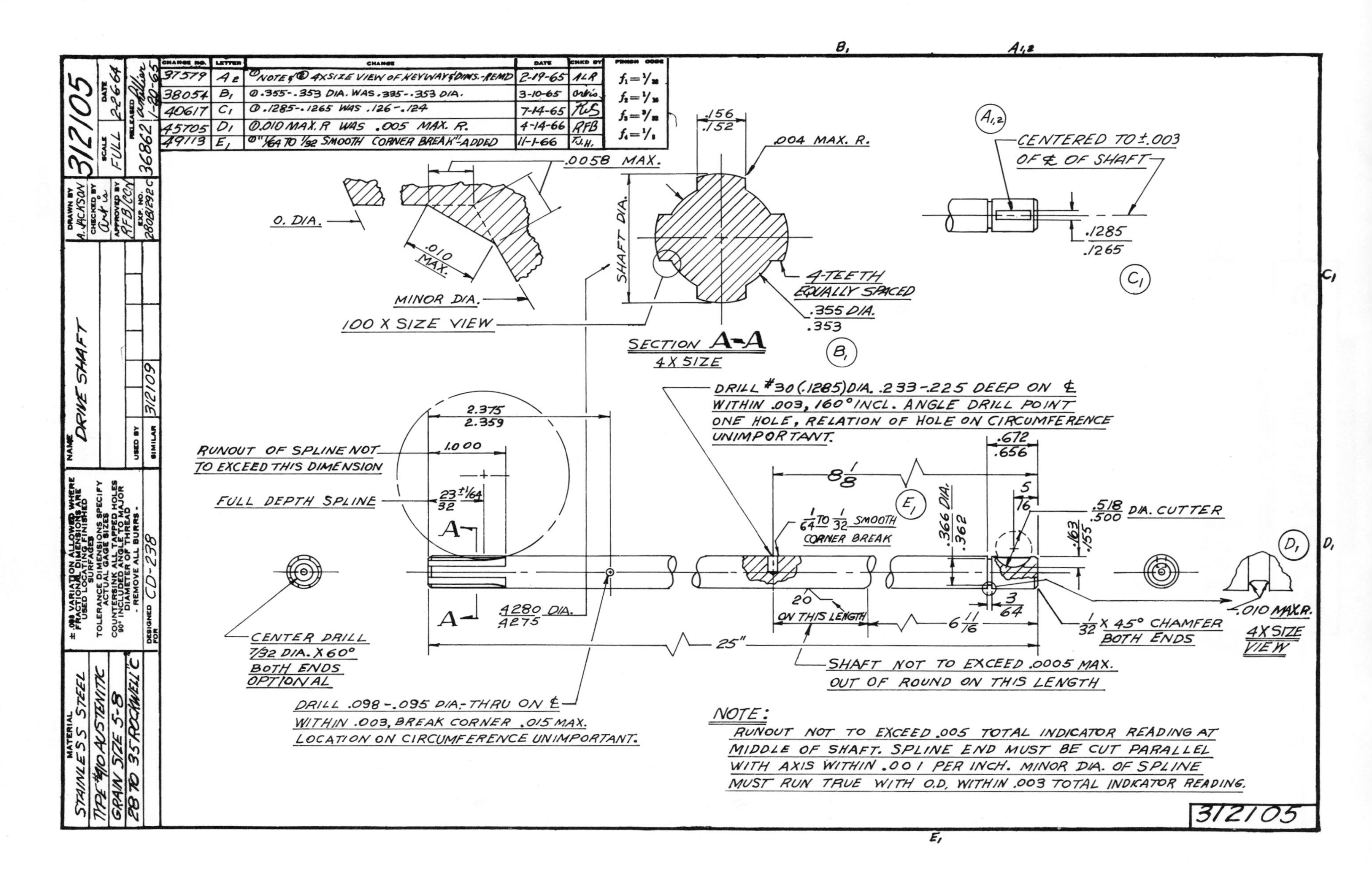

CHANGE NO.	LETTER	CHANGE	DATE	CHKD BY
37579	A2	① NOTE & ② 4X SIZE VIEW OF KEYWAY & DIMS.-REMD	2-19-65	ALR
38054	B1	① .355-.353 DIA. WAS .335-.353 DIA.	3-10-65	
40617	C1	① .1285-.1265 WAS .126-.124	7-14-65	RCS
45705	D1	① .010 MAX. R WAS .005 MAX. R.	4-14-66	RFB
49113	E1	① "1/64 TO 1/32 SMOOTH CORNER BREAK" ADDED	11-1-66	T.J.H.

FINISH CODE
$f_1 = 1/32$
$f_2 = 1/16$
$f_3 = 3/32$
$f_4 = 1/8$
.0058 MAX.
O. DIA.
.010 MAX.
MINOR DIA.
100 X SIZE VIEW
SHAFT DIA.
.156
.152
.004 MAX. R.
4-TEETH EQUALLY SPACED
.355 DIA.
.353
SECTION A-A
4 X SIZE
CENTERED TO ±.003 OF ℄ OF SHAFT
.1285
.1265
DRILL #30 (.1285) DIA. .233-.225 DEEP ON ℄ WITHIN .003, 160° INCL. ANGLE DRILL POINT ONE HOLE, RELATION OF HOLE ON CIRCUMFERENCE UNIMPORTANT.
2.375
2.359
1.000
RUNOUT OF SPLINE NOT TO EXCEED THIS DIMENSION
FULL DEPTH SPLINE
23/32 ±1/64
A
8 1/8
.672
.656
5/16
.518
.500 DIA. CUTTER
1/64 TO 1/32 SMOOTH CORNER BREAK
.366 DIA.
.362
.163
.155
20°
ON THIS LENGTH
6 11/16
3/64
1/32 X 45° CHAMFER BOTH ENDS
.010 MAX. R.
4X SIZE VIEW
.4280 DIA.
.4275
25"
CENTER DRILL 7/32 DIA. X 60° BOTH ENDS OPTIONAL
SHAFT NOT TO EXCEED .0005 MAX. OUT OF ROUND ON THIS LENGTH
DRILL .098-.095 DIA.-THRU ON ℄ WITHIN .003, BREAK CORNER .015 MAX. LOCATION ON CIRCUMFERENCE UNIMPORTANT.
NOTE:
RUNOUT NOT TO EXCEED .005 TOTAL INDICATOR READING AT MIDDLE OF SHAFT. SPLINE END MUST BE CUT PARALLEL WITH AXIS WITHIN .001 PER INCH. MINOR DIA. OF SPLINE MUST RUN TRUE WITH O.D. WITHIN .003 TOTAL INDICATOR READING.
312105
MATERIAL
STAINLESS STEEL
TYPE #410 AUSTENITIC
GRAIN SIZE 5-8
28 TO 35 ROCKWELL "C"
± .03 VARIATION ALLOWED WHERE FRACTIONAL DIMENSIONS ARE USED LOCATING FINISHED SURFACES
TOLERANCE DIMENSIONS SPECIFY ACTUAL GAGE SIZES
COUNTERSINK ALL TAPPED HOLES 90° INCLUDED ANGLE TO MAJOR DIAMETER OF THREAD
- REMOVE ALL BURRS -
DESIGNED FOR CD-238
NAME DRIVE SHAFT
USED BY
SIMILAR 312109
DRAWN BY A. JACKSON
CHECKED BY
APPROVED BY RFB/CCN
EXP. NO. 280B1292C
SCALE FULL
DATE 2-26-64
RELEASED 36862 1-20-65
312105

EXERCISE C-9

Print #324359

QUESTIONS	ANSWERS
1. Give dimensions *A*, *B*, and *C*.	1. *A*______ *B*______ *C*______
2. Of what material is this part made?	2. __________
3. What was .219 ±.002 [↗ | X(S) | .003] prior to 8-2-77?	3. __________
4. What size thread is used?	4. __________
5. What does P.D. .1688-.1658 mean?	5. __________
6. What is the tolerance for the .0750 dimension?	6. __________
7. What is the tolerance for the .395 dimension?	7. __________
8. What is the minimum and maximum allowable diameter of the knurled end?	8. __________
9. What is this part?	9. __________
10. For what unit is this part designed?	10. __________
11. On what assembly does this part fit?	11. __________
12. Explain [↗ | X(S) | .003] [–Y–] CIR.	12. __________

NAME __________

DATE __________

SCORE __________

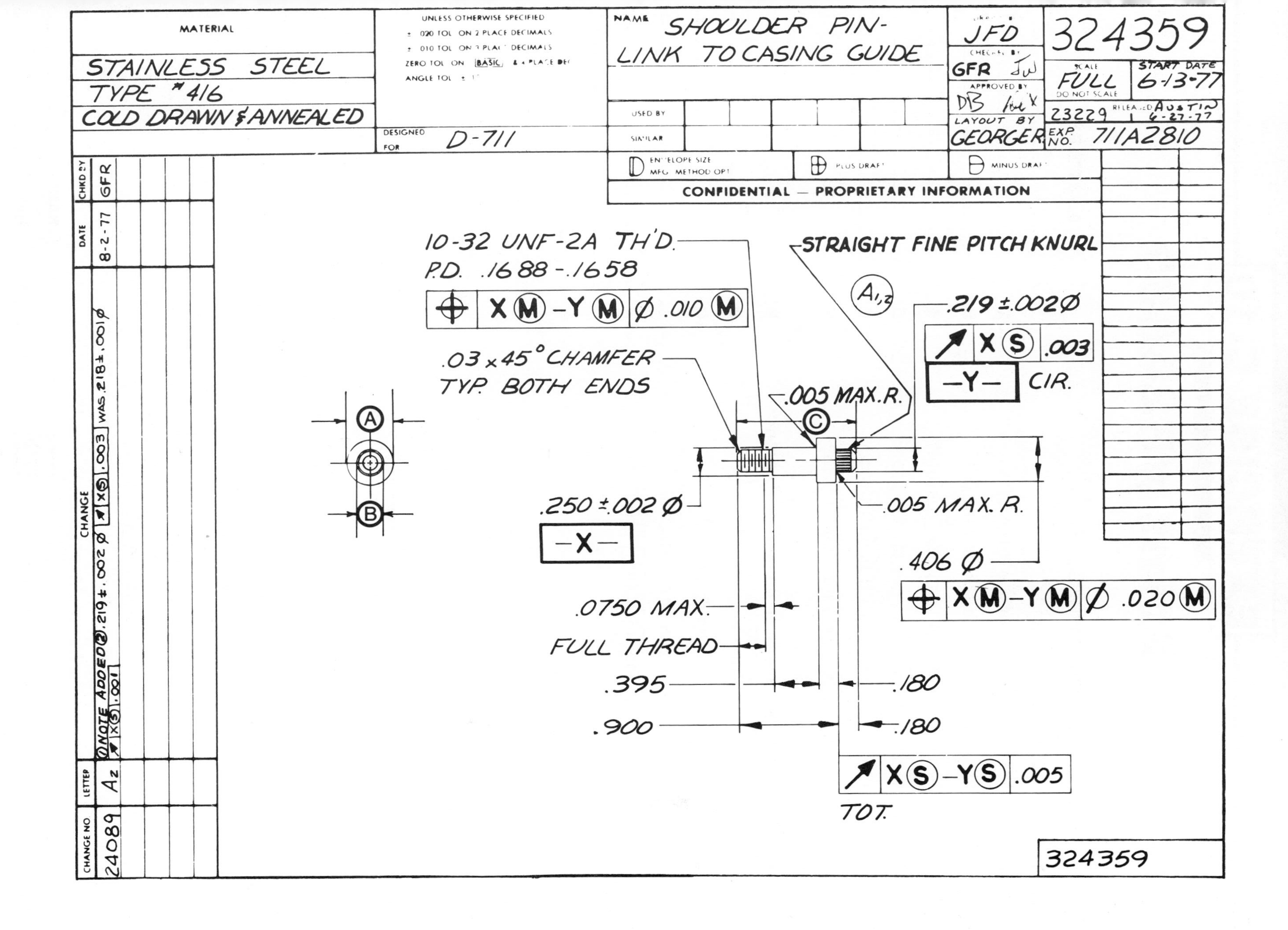
MATERIAL
STAINLESS STEEL
TYPE #416
COLD DRAWN & ANNEALED
UNLESS OTHERWISE SPECIFIED
± .020 TOL ON 2 PLACE DECIMALS
± .010 TOL ON 3 PLACE DECIMALS
ZERO TOL ON BASIC & 4 PLACE DEC
ANGLE TOL
DESIGNED FOR D-711
NAME SHOULDER PIN-
LINK TO CASING GUIDE
USED BY
SIMILAR
JFD
CHECKED BY GFR
APPROVED BY DB
LAYOUT BY GEORGER
324359
SCALE FULL
DO NOT SCALE
START DATE 6-13-77
23229
RELEASED AUSTIN 6-27-77
EXP NO. 711A2810
ENVELOPE SIZE MFG METHOD OPT
PLUS DRAFT
MINUS DRAFT
CONFIDENTIAL — PROPRIETARY INFORMATION
10-32 UNF-2A TH'D.
P.D. .1688-.1658
X M -Y M Ø .010 M
STRAIGHT FINE PITCH KNURL
A1,2
.219 ±.002Ø
X S .003
-Y-
CIR.
.03 x 45° CHAMFER
TYP. BOTH ENDS
.005 MAX. R.
C
A
B
.250 ±.002 Ø
-X-
.005 MAX. R.
.406 Ø
X M -Y M Ø .020 M
.0750 MAX.
FULL THREAD
.395
.900
.180
.180
X S -Y S .005
TOT.
324359
CHANGE NO.
LETTER
CHANGE
DATE
CHKD BY
24089
A2
① NOTE ADDED ② .219±.002 Ø X S .003 WAS .218±.001 Ø X S .001
8-2-77
GFR

EXERCISE C-10

Print #324687

QUESTIONS

1. Give dimensions *A*, *B*, and *C*.
2. What is the dimension of the large I.D.?
3. What is the dimension of the small I.D.?
4. What is the diameter of the [–X–] datum?
5. What is the maximum length of this part?
6. Into what other part does this bushing fit?
7. What was the .546 ±.001 dimension prior to 2-2-78?
8. How many changes were made on this print?
9. What kind of section view is A-A?
10. What is the angle of the chamfers?
11. For what unit was this part designed?
12. Give the readout for [⊕ | XⓂ | Ø.001Ⓜ].

ANSWERS

1. *A*______ *B*______ *C*______
2. ____________
3. ____________
4. ____________
5. ____________
6. ____________
7. ____________
8. ____________
9. ____________
10. ____________
11. ____________
12. ____________

NAME____________

DATE____________

SCORE____________

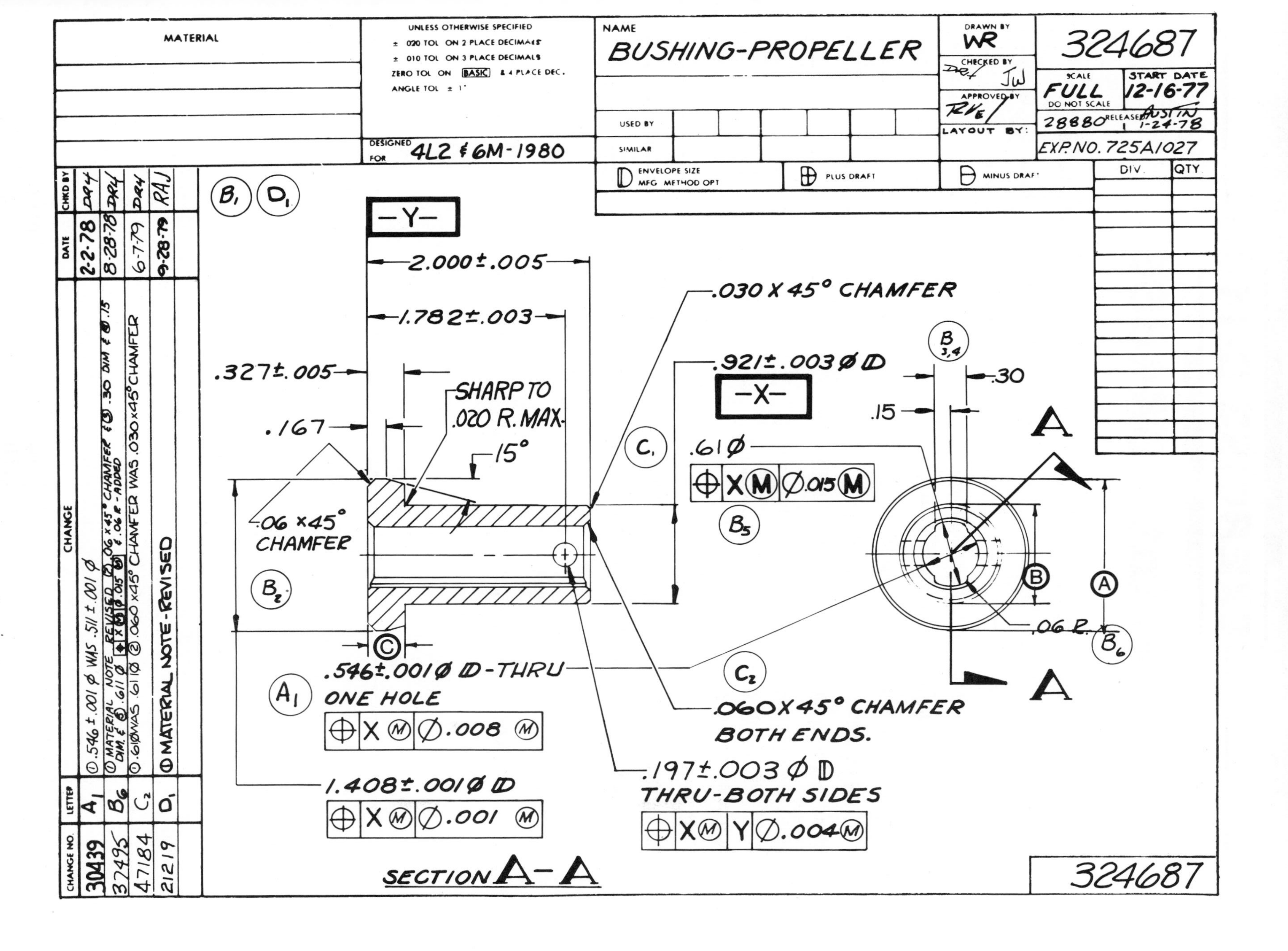
MATERIAL
UNLESS OTHERWISE SPECIFIED
± .020 TOL ON 2 PLACE DECIMALS
± .010 TOL ON 3 PLACE DECIMALS
ZERO TOL ON BASIC & 4 PLACE DEC.
ANGLE TOL ± 1°
DESIGNED FOR 4L2 & 6M-1980
NAME
BUSHING-PROPELLER
USED BY
SIMILAR
DRAWN BY WR
CHECKED BY JW
APPROVED BY
LAYOUT BY:
324687
SCALE FULL
DO NOT SCALE
START DATE 12-16-77
28880
RELEASED 1-24-78
EXP. NO. 725A1027
ENVELOPE SIZE MFG METHOD OPT
PLUS DRAFT
MINUS DRAFT
DIV.
QTY
-Y-
2.000±.005
1.782±.003
.327±.005
.167
SHARP TO .020 R. MAX.
15°
.06 x 45° CHAMFER
.030 X 45° CHAMFER
.921±.003 Ø D
-X-
.61 Ø
X M Ø.015 M
.15
.30
.06 R.
.546±.001 Ø D-THRU ONE HOLE
X M Ø.008 M
1.408±.001 Ø D
X M Ø.001 M
.060 X 45° CHAMFER BOTH ENDS.
.197±.003 Ø D THRU-BOTH SIDES
X M Y Ø.004 M
SECTION A-A
A
A
CHANGE NO.
LETTER
CHANGE
DATE
CHKD BY
30439 A1 ① .546±.001 Ø WAS .511±.001 Ø 2-2-78
37495 B6 ① MATERIAL NOTE REVISED ② .06 x 45° CHAMFER & ③ .30 DIM & ④ .15 DIM. & ⑤ .61 Ø X M Ø.015 M & .06 R - ADDED 8-28-78
47184 C2 ① .61 Ø WAS .611 Ø ② .060 x 45° CHAMFER WAS .030 x 45° CHAMFER 6-7-79
21219 D1 ① MATERIAL NOTE-REVISED 9-28-79 RAJ

EXERCISE C-11

Print #324856

QUESTIONS	ANSWERS
1. Give dimensions *A*, *B*, and *C*.	**1.** *A*______ *B*______ *C*______
2. What is this part?	**2.** ______
3. What other parts are similar to this part?	**3.** ______
4. How many basic dimensions are shown?	**4.** ______
5. At what scale was this print drawn?	**5.** ______
6. Of what material is this part made?	**6.** ______
7. What is the tolerance of the [1.720] dimension?	**7.** ______
8. What is the tolerance of the 1.840 dimension?	**8.** ______
9. What was the diameter of this part before it is machined?	**9.** ______
10. How many .052 holes are drilled?	**10.** ______
11. What feature of this part is datum [–W–] ?	**11.** ______
12. What is the wall thickness at the right end of this part?	**12.** ______
13. What is the difference between the .052 holes on the right end and those on the left end?	**13.** ______
14. How many changes were made on this print?	**14.** ______

NAME______

DATE______

SCORE______

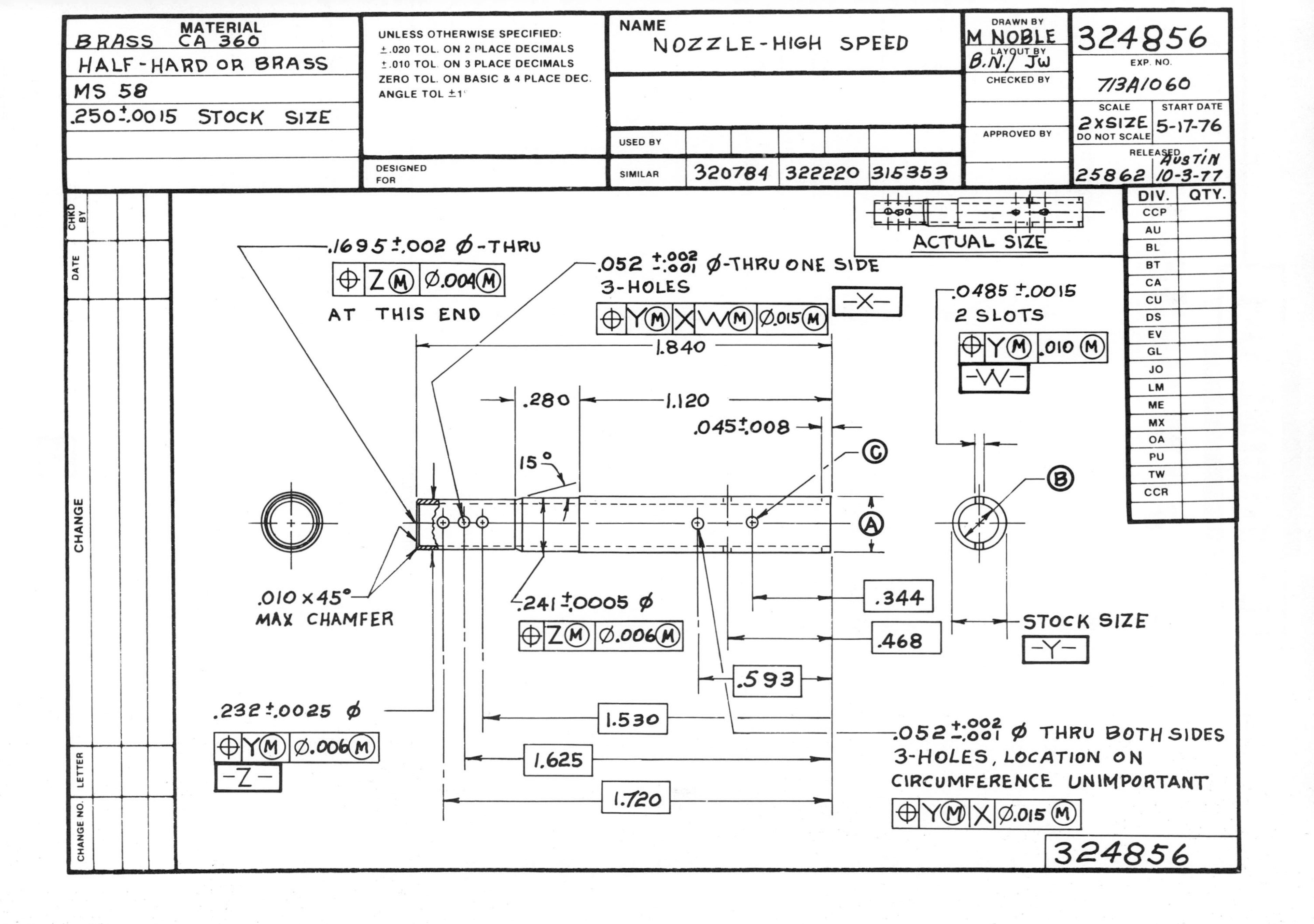
MATERIAL
BRASS CA 360
HALF-HARD OR BRASS
MS 58
.250±.0015 STOCK SIZE
UNLESS OTHERWISE SPECIFIED:
±.020 TOL. ON 2 PLACE DECIMALS
±.010 TOL. ON 3 PLACE DECIMALS
ZERO TOL. ON BASIC & 4 PLACE DEC.
ANGLE TOL ±1°
DESIGNED FOR
NAME
NOZZLE-HIGH SPEED
USED BY
SIMILAR
320784
322220
315353
DRAWN BY
M NOBLE
LAYOUT BY
B.N./ JW
CHECKED BY
APPROVED BY
324856
EXP. NO.
713A1060
SCALE
2XSIZE
DO NOT SCALE
START DATE
5-17-76
RELEASED
AUSTIN
25862
10-3-77
DIV.
QTY.
CCP
AU
BL
BT
CA
CU
DS
EV
GL
JO
LM
ME
MX
OA
PU
TW
CCR
ACTUAL SIZE
.1695±.002 Ø-THRU
Z Ø.004
AT THIS END
.052 +.002 -.001 Ø-THRU ONE SIDE
3-HOLES
Y X W Ø.015
-X-
.0485±.0015
2 SLOTS
Y .010
-W-
1.840
.280
1.120
.045±.008
15°
C
B
A
.010 x 45°
MAX CHAMFER
.241±.0005 Ø
Z Ø.006
.344
.468
.593
STOCK SIZE
-Y-
.232±.0025 Ø
Y Ø.006
-Z-
1.530
1.625
1.720
.052 +.002 -.001 Ø THRU BOTH SIDES
3-HOLES, LOCATION ON
CIRCUMFERENCE UNIMPORTANT
Y X Ø.015
324856
CHKD BY
DATE
CHANGE
LETTER
CHANGE NO.

EXERCISE D-1

Print #JM 6938-1

QUESTIONS	ANSWERS
1. What is this part?	**1.** ____________
2. On what assembly does this part fit?	**2.** ____________
3. Of what material is this part made?	**3.** ____________
4. What is the stock diameter and length of this part?	**4.** ____________
5. How many of these parts are required?	**5.** ____________
6. What is the depth of the .391 drilled hole?	**6.** ____________
7. Give dimensions *A*, *B*, *C*, and *D*. Indicate the correct tolerances.	**7.** *A*______ *B*______ *C*______ *D*______
8. What does the 63√ mean?	**8.** ____________
9. What is the tolerance for the 6.31 dimension? (Refer to Exercise C-5.)	**9.** ____________
10. What is the drill size for the tapped hole?	**10.** ____________
11. Give the read out of [◎ | B | .001].	**11.** ____________ ____________
12. What is the depth of the SAE 6B spline?	**12.** ____________
13. What is the tolerance of the small end of the shaft?	**13.** ____________
14. What is the number of threads per inch of the internal threads?	**14.** ____________
15. What is the width tolerance of the oval?	**15.** ____________
16. What is the width of each spline?	**16.** ____________
17. What are the size and angle of the chamfers?	**17.** ____________
18. How would you define datum [–A–]?	**18.** ____________ ____________ ____________

NAME ____________

DATE ____________

SCORE ____________

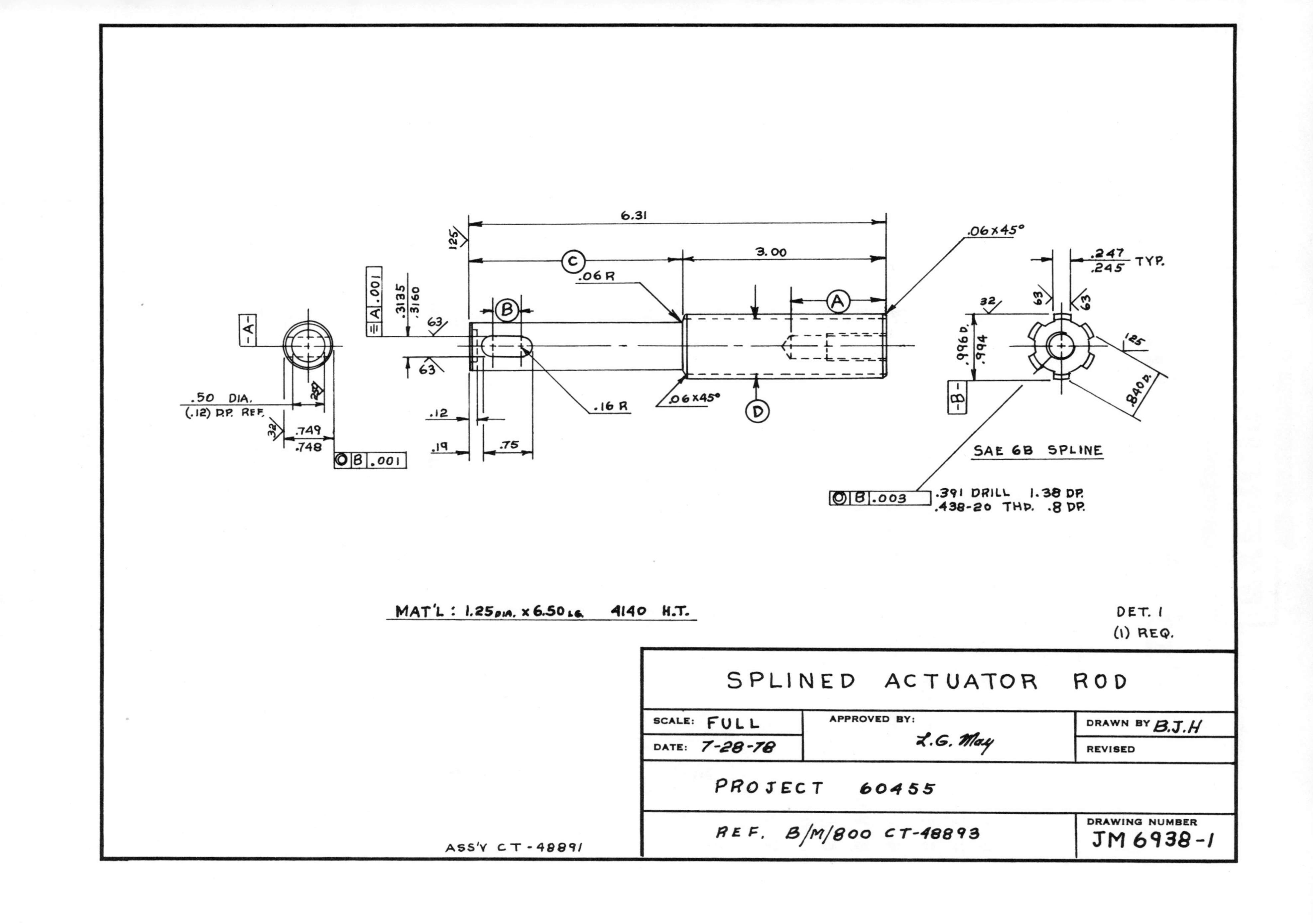
6.31
3.00
.06 R
.06x45°
.247
.245 TYP.
.996 D.
.994
.840 D.
-A-
-B-
.50 DIA.
(.12) DP. REF.
.749
.748
.3135
.3160
A .001
B .001
.16 R
.12
.19
.75
SAE 6B SPLINE
B .003
.391 DRILL 1.38 DP.
.438-20 THD. .8 DP.
MAT'L : 1.25 DIA. x 6.50 LG. 4140 H.T.
DET. 1
(1) REQ.
SPLINED ACTUATOR ROD
SCALE: FULL
DATE: 7-28-78
APPROVED BY:
L.G. May
DRAWN BY B.J.H
REVISED
PROJECT 60455
REF. B/M/800 CT-48893
DRAWING NUMBER
JM 6938-1
ASS'Y CT-48891

EXERCISE D-2

Print #JM 6938-4

QUESTIONS	ANSWERS
1. Give dimensions *A*, *B*, *C*, and *D*.	**1.** *A*______ *B*______ *C*______ *D*______
2. What is the stock diameter and length of this part?	**2.** ______
3. What is the overall finished size of this part?	**3.** ______
4. Of what material is this part made?	**4.** ______
5. What is the finish hardness range?	**5.** ______
6. What material was this part made of prior to 9-12-79?	**6.** ______
7. What is the diameter tolerance at the right end of this part?	**7.** ______
8. What is the depth of the .500 drilled hole?	**8.** ______
9. What is the depth of the .500 F.B. drilled hole?	**9.** ______
10. What is the depth of .734 drilled hole?	**10.** ______
11. What is the depth of the $\frac{.751}{.753}$ reamed hole?	**11.** ______
12. What is the finish on the $\frac{1.000}{.997}$ diameter?	**12.** ______
13. Give the read-out of ◎ | A | .002.	**13.** ______ ______
14. This part is a component of what assembly and is also a part of what project?	**14.** Assembly______ Project______
15. What are the minimum and maximum sizes of the .375 diameter projection? (Refer to Exercise C-5.)	**15.** Minimum ______ Maximum ______
16. What is the tolerance of the 1.250 dimension?	**16.** ______
17. What size of drill is used prior to reaming the $\frac{.3123}{.3119}$ hole?	**17.** ______

NAME______

DATE______

SCORE______

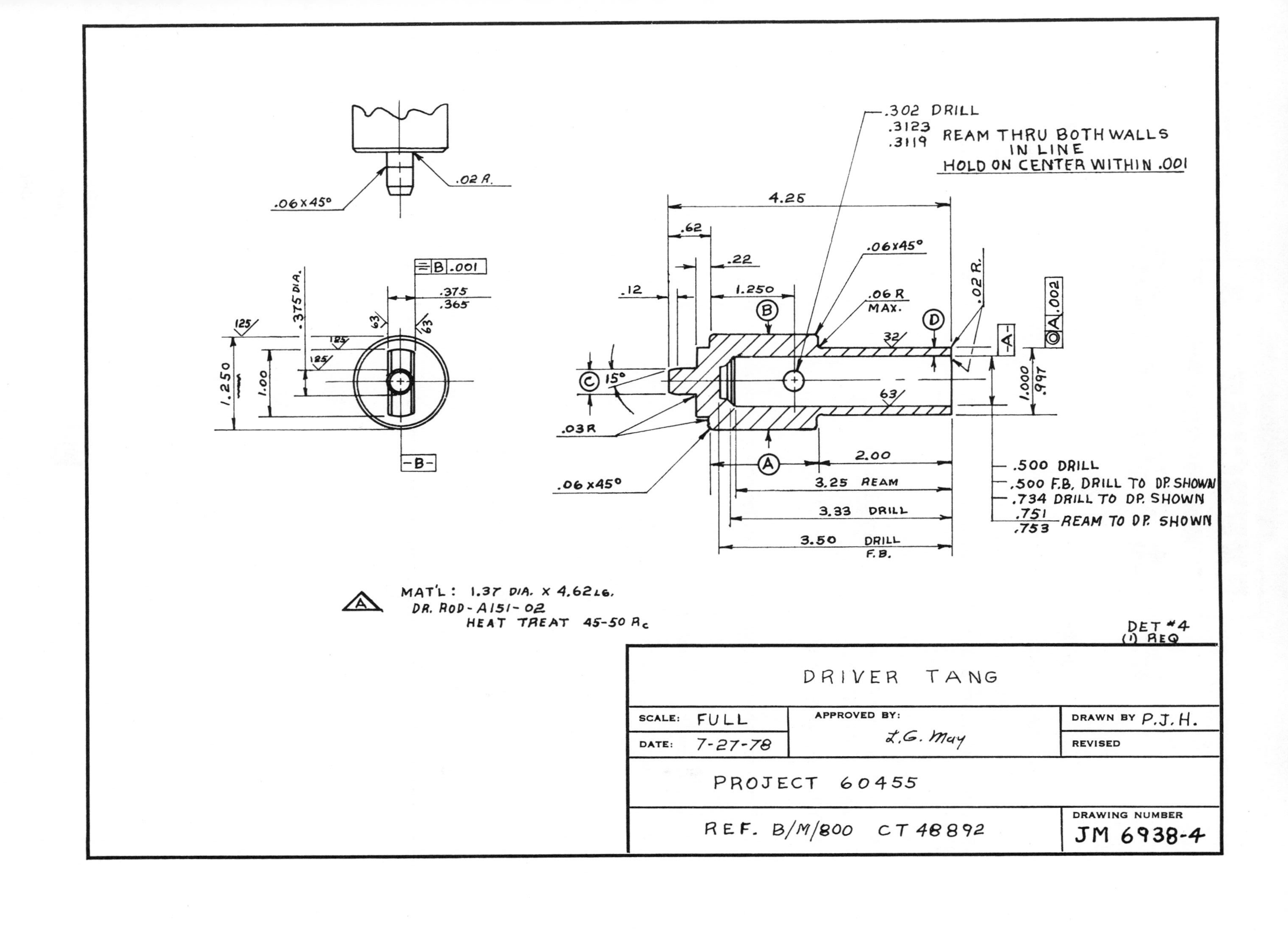

.06 X 45°
.02 R.
.302 DRILL
.3123
.3119 REAM THRU BOTH WALLS IN LINE
HOLD ON CENTER WITHIN .001
4.25
.62
.22
.12
1.250
.06 X 45°
.06 R MAX.
.02 R.
= B .001
.375
.365
.375 DIA.
125
63
1.250
1.00
-B-
-A-
◎ A .002
1.000
.997
32
63
15°
.03 R
.06 X 45°
2.00
3.25 REAM
3.33 DRILL
3.50 DRILL F.B.
.500 DRILL
.500 F.B. DRILL TO DP. SHOWN
.734 DRILL TO DP. SHOWN
.751
.753 REAM TO DP. SHOWN
MAT'L: 1.37 DIA. X 4.62 LG.
DR. ROD - A151 - 02
HEAT TREAT 45-50 Rc
DET #4
(1) REQ
DRIVER TANG
SCALE: FULL
DATE: 7-27-78
APPROVED BY: L.G. May
DRAWN BY P.J.H.
REVISED
PROJECT 60455
REF. B/M/800 CT 48892
DRAWING NUMBER
JM 6938-4

EXERCISE D-3

Print #JM 6938-6

QUESTIONS

1. What is the function of this part?
2. Of what materials is this part made?
3. Give dimensions *A*, *B*, *C*, *D*, *E*, and *F*.
4. What micro-finish is specified?
5. What is the meaning of the wavy lines below dimensions .31, .37 and .937?
6. What is the stock size of this part?
7. What are the two indicated chamfers?
8. On what assembly does this part fit?
9. What are the indicated datums?
10. Give the read-out of | // | B | .005 |.
11. Give the read-out of –B– | ⊥ | A | .003 |.
12. What tolerance is allowed for the 1.25 D. hole?
13. What tolerance is allowed for the $\frac{1.499}{1.503}$ bore?
14. What does the wavy line under this dimension mean?

ANSWERS

1. ____________
2. ____________
3. *A*______ *B*______ *C*______
   *D*______ *E*______ *F*______
4. ____________
5. ____________
6. Diameter ____________
   Thickness ____________
7. ____________
8. ____________
9. ____________
10. ____________
    ____
11. ____________
    ____
12. ____________
13. ____________
14. ____________

NAME ____________
DATE ____________
SCORE ____________

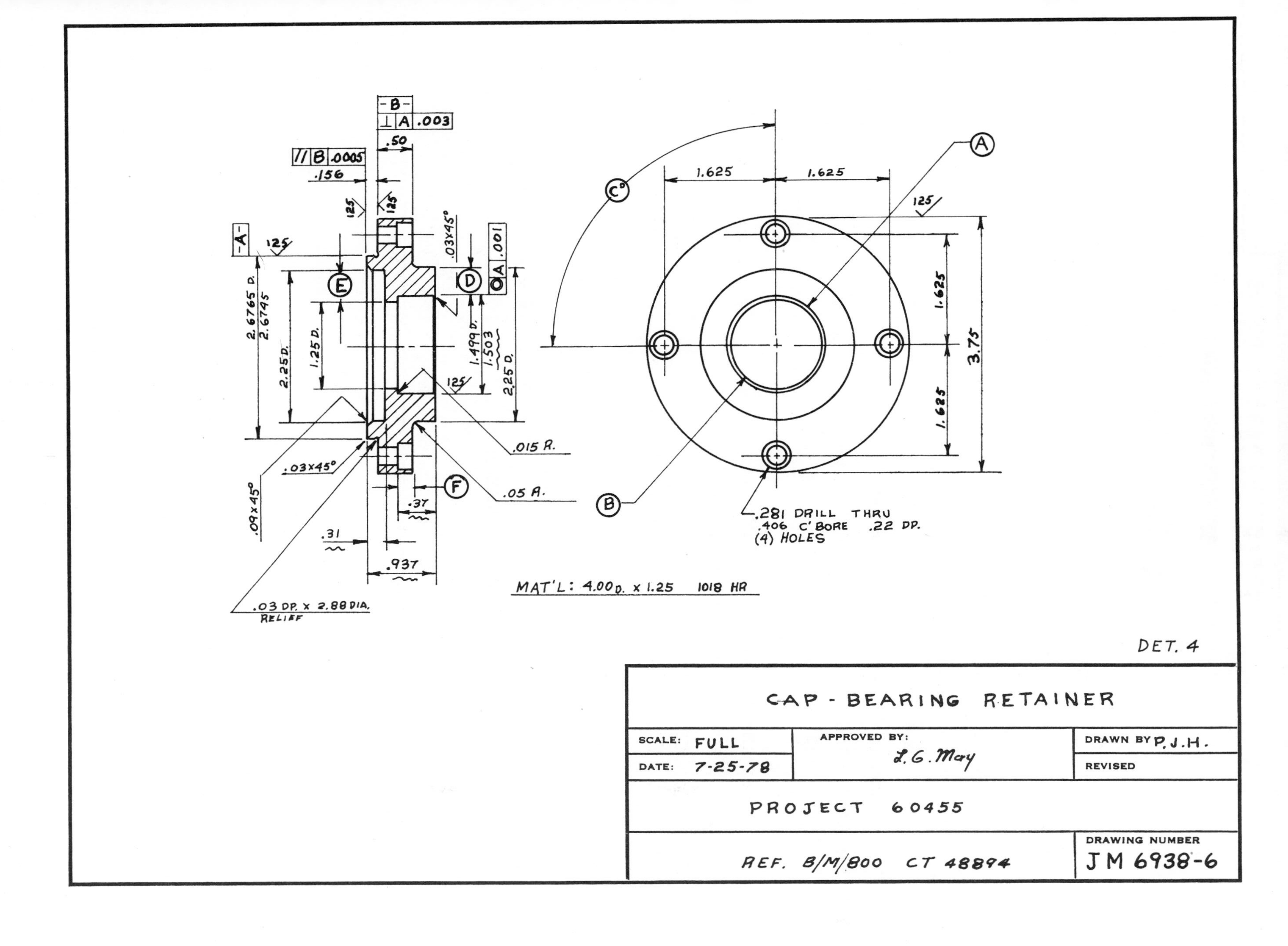

-B-
⊥ A .003
.50
// B .0005
.156
125
-A-
2.6765 D.
2.6745
2.25 D.
1.25 D.
E
D
.03X45°
O A .001
1.499 D.
1.503
2.25 D.
.015 R.
.05 R.
.03X45°
.09X45°
F
.37
.31
.937
.03 DP. X 2.88 DIA.
RELIEF
C°
1.625
1.625
1.625
1.625
3.75
A
B
.281 DRILL THRU
.406 C'BORE .22 DP.
(4) HOLES
MAT'L: 4.00 D. X 1.25 1018 HR
DET. 4
CAP - BEARING RETAINER
SCALE: FULL
DATE: 7-25-78
APPROVED BY:
L.G. May
DRAWN BY P.J.H.
REVISED
PROJECT 60455
REF. B/M/800 CT 48894
DRAWING NUMBER
JM 6938-6

EXERCISE D-4

Print #A63105

QUESTIONS	ANSWERS
1. Of what material is this gear made?	**1.** ____________
2. What kind of gear is shown?	**2.** ____________
3. What is the diametral pitch?	**3.** ____________
4. How many teeth are there on this gear?	**4.** ____________
5. What is the pitch diameter?	**5.** ____________
6. What is the surface hardness range?	**6.** ____________
7. What is the tolerance for the .620 dimension?	**7.** ____________
8. How many changes were made on this print since 1-8-71?	**8.** ____________
9. What would be the maximum squareness allowance between the gear face and tooth edge?	**9.** ____________
10. What is the case hardness thickness?	**10.** ____________
11. What is the purpose of the grooves on the faces of the gear?	**11.** ____________
12. What type of section view is shown?	**12.** ____________
13. Who drew this print?	**13.** ____________
14. What is the gear tooth inspection pin diameter?	**14.** ____________
15. What is the finish on the face of the gear?	**15.** ____________
16. What is the tooth side chamfer?	**16.** ____________
17. What are the thickness and angle of the chamfer on the shaft hole?	**17.** ____________

NAME ____________

DATE ____________

SCORE ____________

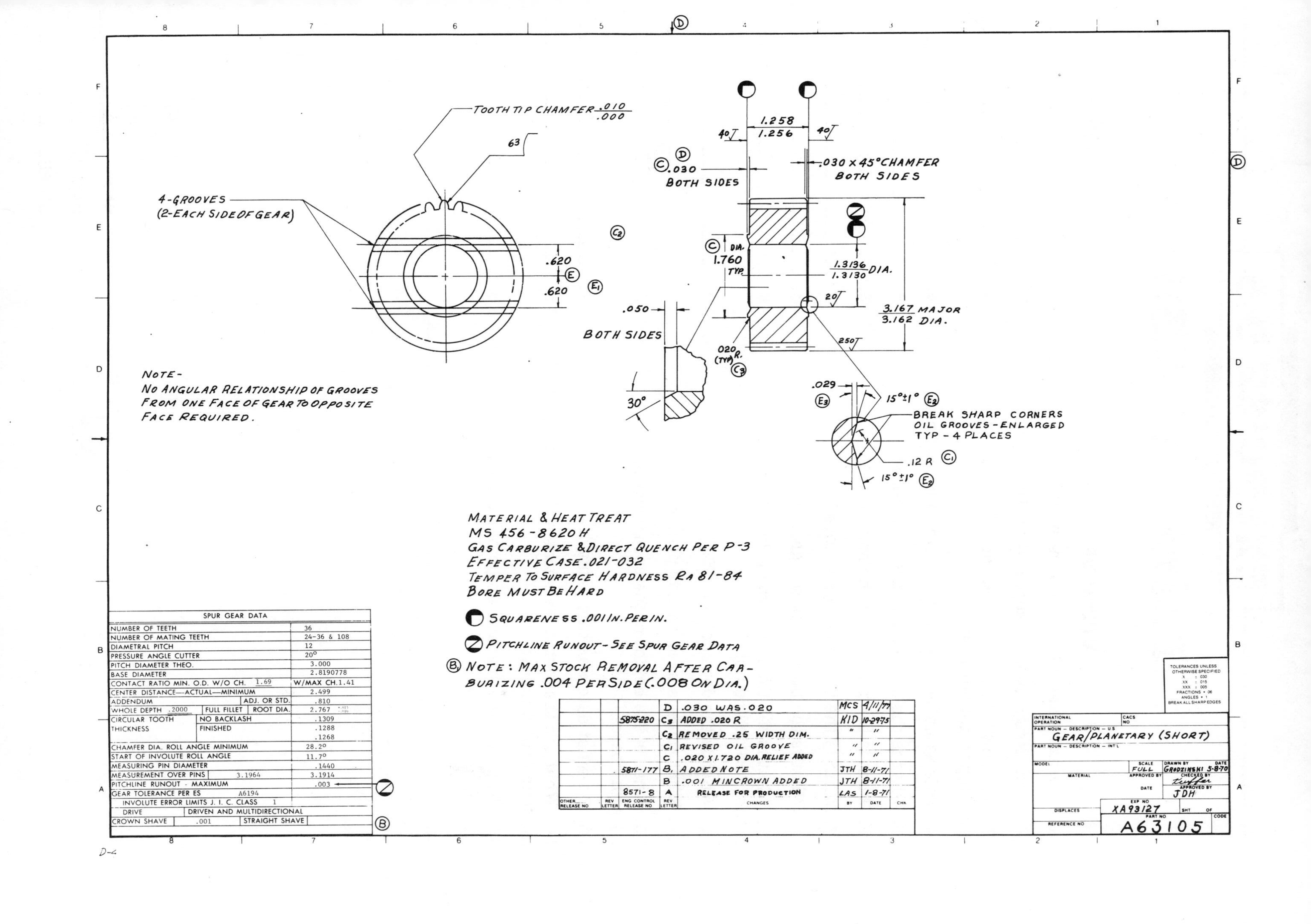
TOOTH TIP CHAMFER .010 / .000
63
4-GROOVES
(2-EACH SIDE OF GEAR)
.620
.620
E
E1
C2
NOTE-
NO ANGULAR RELATIONSHIP OF GROOVES
FROM ONE FACE OF GEAR TO OPPOSITE
FACE REQUIRED.
1.258 / 1.256
40
40
C D .030
BOTH SIDES
.030 X 45° CHAMFER
BOTH SIDES
C DIA.
1.760
TYP.
1.3136 / 1.3130 DIA.
20
3.167 / 3.162 MAJOR DIA.
250
.050
BOTH SIDES
30°
.020 R. (TYP) C3
.029
E3
15°±1° E2
BREAK SHARP CORNERS
OIL GROOVES-ENLARGED
TYP - 4 PLACES
.12 R C1
15°±1° E2
MATERIAL & HEAT TREAT
MS 456-8620H
GAS CARBURIZE & DIRECT QUENCH PER P-3
EFFECTIVE CASE .021-.032
TEMPER TO SURFACE HARDNESS Ra 81-84
BORE MUST BE HARD
SQUARENESS .001 IN. PER IN.
PITCHLINE RUNOUT - SEE SPUR GEAR DATA
B NOTE: MAX STOCK REMOVAL AFTER CARBURIZING .004 PER SIDE (.008 ON DIA.)
SPUR GEAR DATA
NUMBER OF TEETH 36
NUMBER OF MATING TEETH 24-36 & 108
DIAMETRAL PITCH 12
PRESSURE ANGLE CUTTER 20°
PITCH DIAMETER THEO. 3.000
BASE DIAMETER 2.8190778
CONTACT RATIO MIN. O.D. W/O CH. 1.69 W/MAX CH. 1.41
CENTER DISTANCE—ACTUAL—MINIMUM 2.499
ADDENDUM ADJ. OR STD. .810
WHOLE DEPTH .2000 FULL FILLET ROOT DIA. 2.767
CIRCULAR TOOTH THICKNESS NO BACKLASH .1309
FINISHED .1288 .1268
CHAMFER DIA. ROLL ANGLE MINIMUM 28.2°
START OF INVOLUTE ROLL ANGLE 11.7°
MEASURING PIN DIAMETER .1440
MEASUREMENT OVER PINS 3.1964 3.1914
PITCHLINE RUNOUT - MAXIMUM .003
GEAR TOLERANCE PER ES A6194
INVOLUTE ERROR LIMITS J. I. C. CLASS 1
DRIVE DRIVEN AND MULTIDIRECTIONAL
CROWN SHAVE .001 STRAIGHT SHAVE
B
D .030 WAS .020 MCS 4/11/77
5875220 C3 ADDED .020 R KID 10-29-75
C2 REMOVED .25 WIDTH DIM. " "
C1 REVISED OIL GROOVE " "
C .020 X 1.720 DIA. RELIEF ADDED " "
5871-177 B1 ADDED NOTE JTH 8-11-71
B .001 MIN CROWN ADDED JTH 8-11-71
8571-8 A RELEASE FOR PRODUCTION LAS 1-8-71
OTHER RELEASE NO
REV LETTER
ENG CONTROL RELEASE NO
REV LETTER
CHANGES
BY
DATE
CHK
TOLERANCES UNLESS OTHERWISE SPECIFIED
X ± .030
XX ± .015
XXX ± .005
FRACTIONS ± .06
ANGLES ± 1
BREAK ALL SHARP EDGES
INTERNATIONAL OPERATION
CACS NO
PART NOUN - DESCRIPTION - U.S.
GEAR/PLANETARY (SHORT)
PART NOUN - DESCRIPTION - INT'L
MODEL
MATERIAL
SCALE FULL
DRAWN BY GRODZINSKI
DATE 5-8-70
APPROVED BY
CHECKED BY
DATE
APPROVED BY JDH
DISPLACES
EXP NO XA93127
SHT
OF
REFERENCE NO
PART NO A63105
CODE
D-4

EXERCISE D-5

Print #12-2394B

QUESTIONS	ANSWERS
1. Of what material is this part made?	1. ______
2. What rough stock is supplied for this part?	2. ______
3. As shown on the print, what seems to be the most important surface?	3. ______
4. This same surface is machined to what finish?	4. ______
5. What is the diameter of the four holes?	5. ______
6. What is the purpose of the spot facing?	6. ______
7. What is the finish on the groove?	7. ______
8. What is the purpose of the groove?	8. ______
9. What is the finished size of the enlargement of the .750 drilled hole?	9. ______
10. What is the depth of the .750 drilled hole?	10. ______
11. What scale is used on this print?	11. ______
12. What is the height of the spot-faced surfaces from the base?	12. ______
13. On what date was this print drawn?	13. ______
14. What are the distances between the centers of the four drilled holes?	14. ______
15. What are the size and angle of the three chamfers?	15. ______

NAME ______

DATE ______

SCORE ______

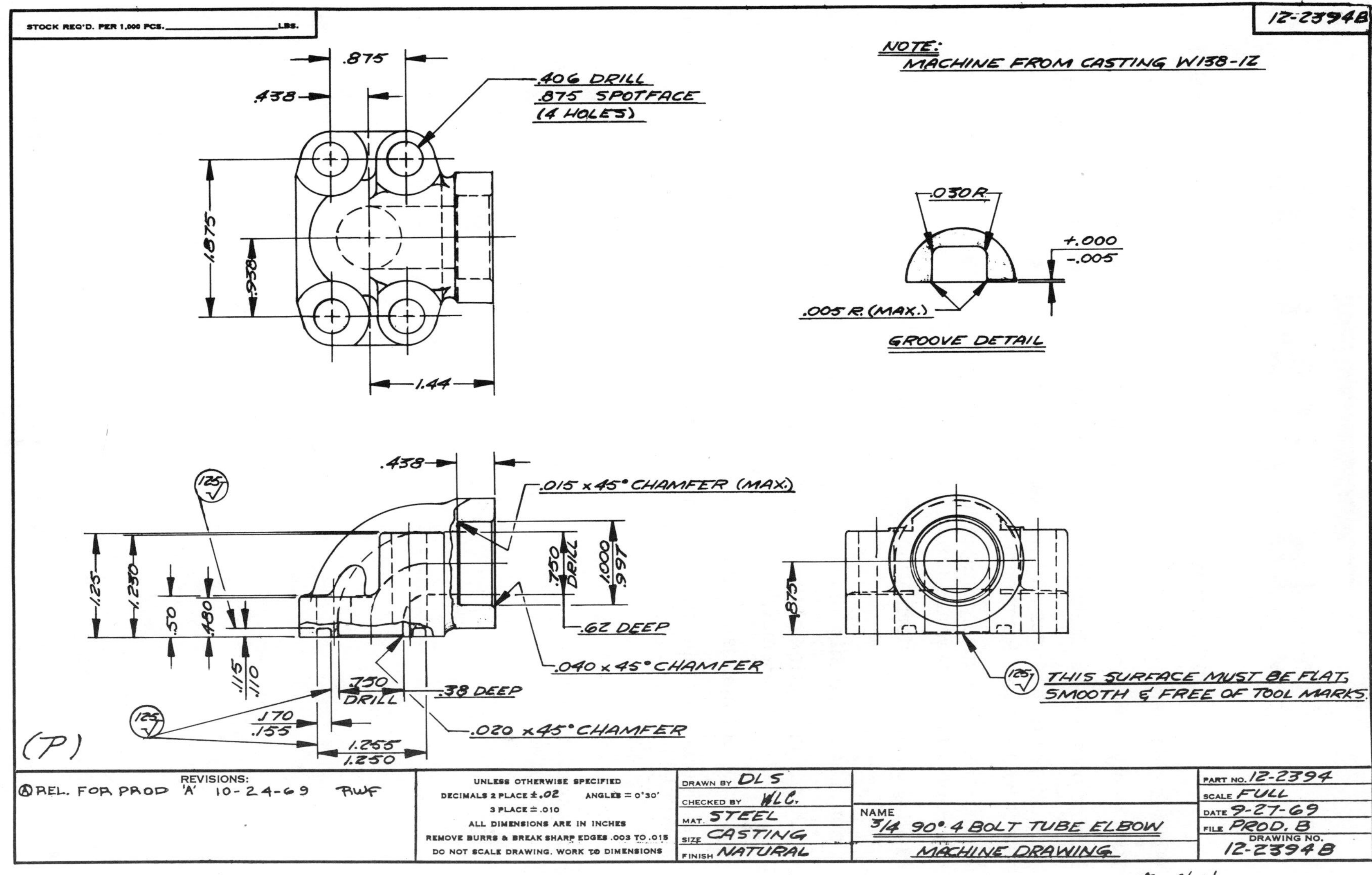
STOCK REQ'D. PER 1,000 PCS. LBS.
12-2394B
NOTE:
MACHINE FROM CASTING W138-12
.875
.438
.406 DRILL
.875 SPOTFACE
(4 HOLES)
1.875
.938
1.44
.030 R.
+.000
-.005
.005 R. (MAX.)
GROOVE DETAIL
.438
.015 x 45° CHAMFER (MAX.)
1.25
1.230
.50
.480
.750 DRILL
1.000
.997
.62 DEEP
.115
.110
.040 x 45° CHAMFER
.750 DRILL
.38 DEEP
.170
.155
.020 x 45° CHAMFER
1.255
1.250
.875
THIS SURFACE MUST BE FLAT, SMOOTH & FREE OF TOOL MARKS.
(P)
REVISIONS:
Ⓐ REL. FOR PROD 'A' 10-24-69 RWF
UNLESS OTHERWISE SPECIFIED
DECIMALS 2 PLACE ±.02 ANGLES = 0°30'
3 PLACE ±.010
ALL DIMENSIONS ARE IN INCHES
REMOVE BURRS & BREAK SHARP EDGES .003 TO .015
DO NOT SCALE DRAWING. WORK TO DIMENSIONS
DRAWN BY DLS
CHECKED BY WLC.
MAT. STEEL
SIZE CASTING
FINISH NATURAL
NAME
3/4 90° 4 BOLT TUBE ELBOW
MACHINE DRAWING
PART NO. 12-2394
SCALE FULL
DATE 9-27-69
FILE PROD. B
DRAWING NO.
12-2394B
3/26/79

EXERCISE D-6

Print #14-110C

QUESTIONS

1. What is this assembly?
2. How many parts are used on this assembly?
3. How is part #1592 attached to part #14-102?
4. Were there any changes on this assembly from the time it was released for production?
5. Is there any close machining on this assembly? If there is, on what part would it be located?
6. Is all the required information given for all the parts?
7. If not, how would this information be obtained?
8. Would the machining of the individual parts be done before or after assembly?
9. What type of finish is indicated on part #1586?
10. What special assembly instructions are given for part #1586?

ANSWERS

1. ______________________
2. ______________________
3. ______________________
4. ______________________
5. ______________________
6. ______________________
7. ______________________
8. ______________________
9. ______________________
10. ______________________

NAME ______________________

DATE ______________________

SCORE ______________________

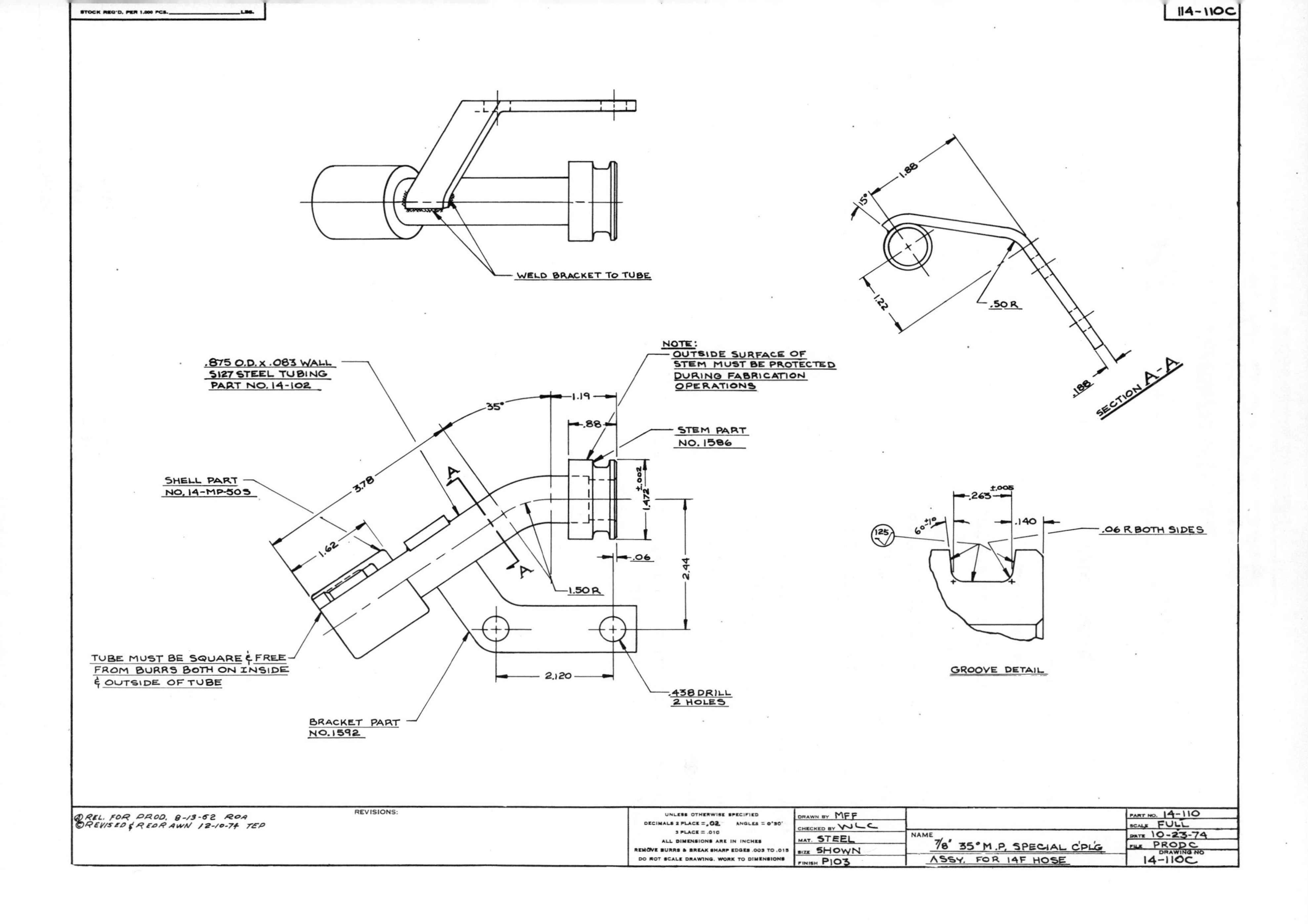

STOCK REQ'D. PER 1,000 PCS. LBS.
114-110C
WELD BRACKET TO TUBE
1.88
15°
1.22
.50 R
.188
SECTION A-A
.875 O.D. x .083 WALL
5127 STEEL TUBING
PART NO. 14-102
NOTE:
OUTSIDE SURFACE OF
STEM MUST BE PROTECTED
DURING FABRICATION
OPERATIONS
35°
1.19
.88
STEM PART
NO. 1586
SHELL PART
NO. 14-MP-505
3.78
1.62
1.472 ±.002
.06
2.44
1.50 R
TUBE MUST BE SQUARE & FREE
FROM BURRS BOTH ON INSIDE
& OUTSIDE OF TUBE
2.120
.438 DRILL
2 HOLES
BRACKET PART
NO. 1592
.263 ±.005
.140
60° ±1°
125
.06 R BOTH SIDES
GROOVE DETAIL
B REL. FOR PROD. 8-13-62 ROA
C REVISED & REDRAWN 12-10-74 TEP
REVISIONS:
UNLESS OTHERWISE SPECIFIED
DECIMALS 2 PLACE = .02 ANGLES = 0°30'
3 PLACE = .010
ALL DIMENSIONS ARE IN INCHES
REMOVE BURRS & BREAK SHARP EDGES .003 TO .015
DO NOT SCALE DRAWING. WORK TO DIMENSIONS
DRAWN BY MFF
CHECKED BY WLC
MAT. STEEL
SIZE SHOWN
FINISH P103
NAME
7/8" 35° M.P. SPECIAL CPLG
ASSY. FOR 14F HOSE
PART NO. 14-110
SCALE FULL
DATE 10-23-74
FILE PRODC
DRAWING NO
14-110C

EXERCISE D-7

Print #16-1379C

QUESTIONS	ANSWERS
1. What is this part?	1. ______
2. What is this part used for?	2. ______
3. How many section views are shown?	3. ______
4. What scale is used?	4. ______
5. Of what material is this part made?	5. ______
6. What type of finish is used?	6. ______
7. What size are the unspecified radii?	7. ______
8. What is the diameter of the four holes?	8. ______
9. What is the overall length of this part?	9. ______
10. What is the overall width of this part?	10. ______
11. What is the tolerance on all two-place dimensions?	11. ______
12. Were any changes made on this print since it was released for production?	12. ______

NAME ______

DATE ______

SCORE ______

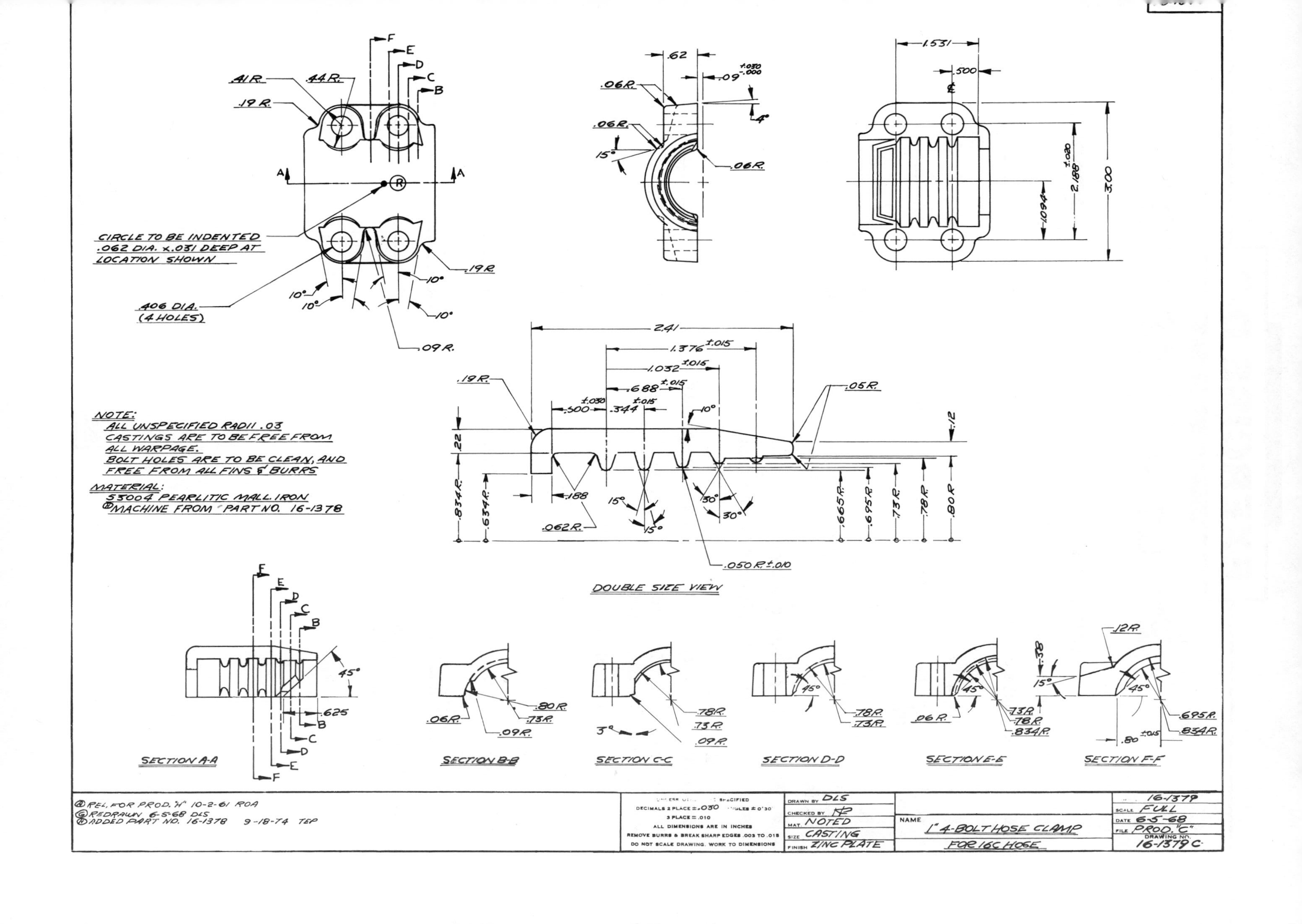

CIRCLE TO BE INDENTED .062 DIA. x .031 DEEP AT LOCATION SHOWN
.406 DIA. (4 HOLES)
NOTE:
ALL UNSPECIFIED RADII .03
CASTINGS ARE TO BE FREE FROM ALL WARPAGE.
BOLT HOLES ARE TO BE CLEAN, AND FREE FROM ALL FINS & BURRS
MATERIAL:
53004 PEARLITIC MALL. IRON
MACHINE FROM PART NO. 16-1378
DOUBLE SIZE VIEW
SECTION A-A
SECTION B-B
SECTION C-C
SECTION D-D
SECTION E-E
SECTION F-F
REL. FOR PROD. "H" 10-2-61 ROA
REDRAWN 6-5-68 DLS
ADDED PART NO. 16-1378 9-18-74 TEP
DECIMALS 2 PLACE ±.030 ANGLES ± 0°30'
3 PLACE ±.010
ALL DIMENSIONS ARE IN INCHES
REMOVE BURRS & BREAK SHARP EDGES .003 TO .015
DO NOT SCALE DRAWING. WORK TO DIMENSIONS
DRAWN BY DLS
MAT. NOTED
SIZE CASTING
FINISH ZINC PLATE
NAME 1" 4-BOLT HOSE CLAMP FOR 16C HOSE
16-1379
SCALE FULL
DATE 6-5-68
FILE PROD. "C"
DRAWING NO. 16-1379 C

EXERCISE D-8

Print #A148270

QUESTIONS	ANSWERS
1. What two parts are used in this assembly?	1. ______
2. Of what materials are these parts made? Explain.	2. ______
3. What scale is used?	3. ______
4. How many changes were made to this print since 1-22-76?	4. ______
5. What type and size of weld is used?	5. ______
6. What is the thickness range of the case hardness?	6. ______
7. What is the surface hardness?	7. ______
8. Is the entire part case hardened? Explain.	8. ______
9. What is the tolerance for the 1.25 dimension?	9. ______
10. On the right side view, what is the purpose of the two boxed-in numbers?	10. ______
11. What angle projection is used?	11. ______
12. What is the purpose of the arrowheads at the center of the vertical and horizontal border lines?	12. ______
13. What is the purpose of the circled letters along the border lines?	13. ______
14. What is this part?	14. ______
15. How many units of this assembly are required for each finished product?	15. ______

NAME ______

DATE ______

SCORE ______

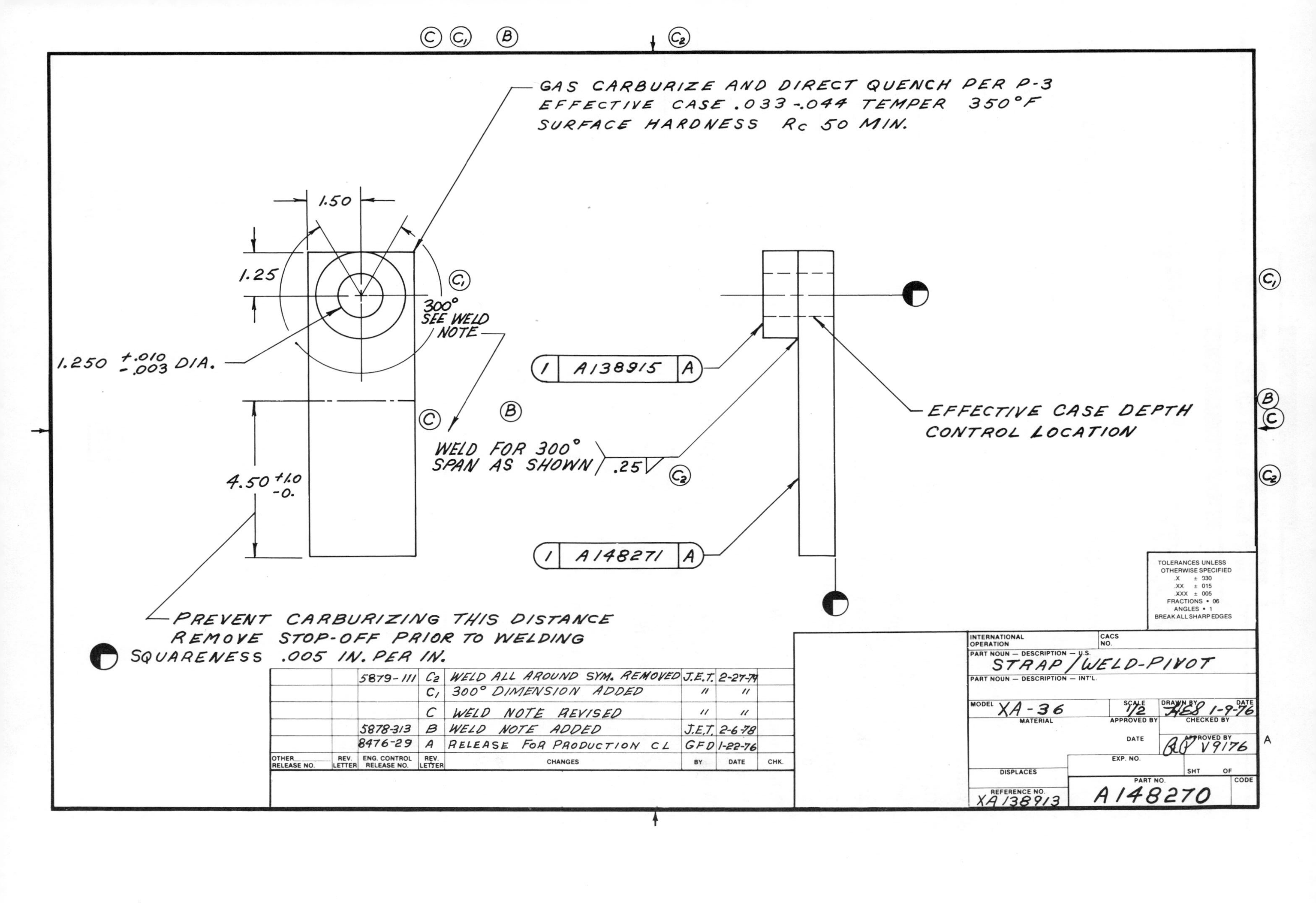
GAS CARBURIZE AND DIRECT QUENCH PER P-3
EFFECTIVE CASE .033-.044 TEMPER 350°F
SURFACE HARDNESS Rc 50 MIN.
1.50
1.25
300° SEE WELD NOTE
1.250 +.010 -.003 DIA.
4.50 +1.0 -0.
WELD FOR 300° SPAN AS SHOWN
.25
A138915
A148271
EFFECTIVE CASE DEPTH CONTROL LOCATION
PREVENT CARBURIZING THIS DISTANCE
REMOVE STOP-OFF PRIOR TO WELDING
SQUARENESS .005 IN. PER IN.
5879-111 C2 WELD ALL AROUND SYM. REMOVED J.E.T. 2-27-79
C1 300° DIMENSION ADDED
C WELD NOTE REVISED
5878-313 B WELD NOTE ADDED J.E.T. 2-6-78
8476-29 A RELEASE FOR PRODUCTION CL GFD 1-22-76
TOLERANCES UNLESS OTHERWISE SPECIFIED
STRAP/WELD-PIVOT
MODEL XA-36
SCALE 1/2
DRAWN BY
DATE 1-9-76
REFERENCE NO. XA138913
PART NO. A148270

EXERCISE D-9

Print #317856

QUESTIONS	ANSWERS
1. Give dimensions *A*, *B*, *C*, *D*, *E*, and *F*. Indicate the correct tolerances.	1. *A*______ *B*______ *C*______ *D*______ *E*______ *F*______
2. What is this part?	2. ____________
3. What scales are used?	3. ____________
4. Of what material is this part made?	4. ____________
5. What size rod is used for this part?	5. ____________
6. What size drill is used for the threaded hole?	6. ____________
7. To what finish is the center .3755 hole machined?	7. ____________
8. What does .0100 T.I.R. mean?	8. ____________
9. How many section views are shown?	9. ____________
10. What was the .0100 T.I.R. prior to 8-15-72?	10. ____________
11. What is the tolerance for the .580 dimension?	11. ____________
12. On what assembly is this part used?	12. ____________
13. What is the hardness range of this part?	13. ____________
14. What other part is similar to this part?	14. ____________

NAME____________

DATE____________

SCORE____________

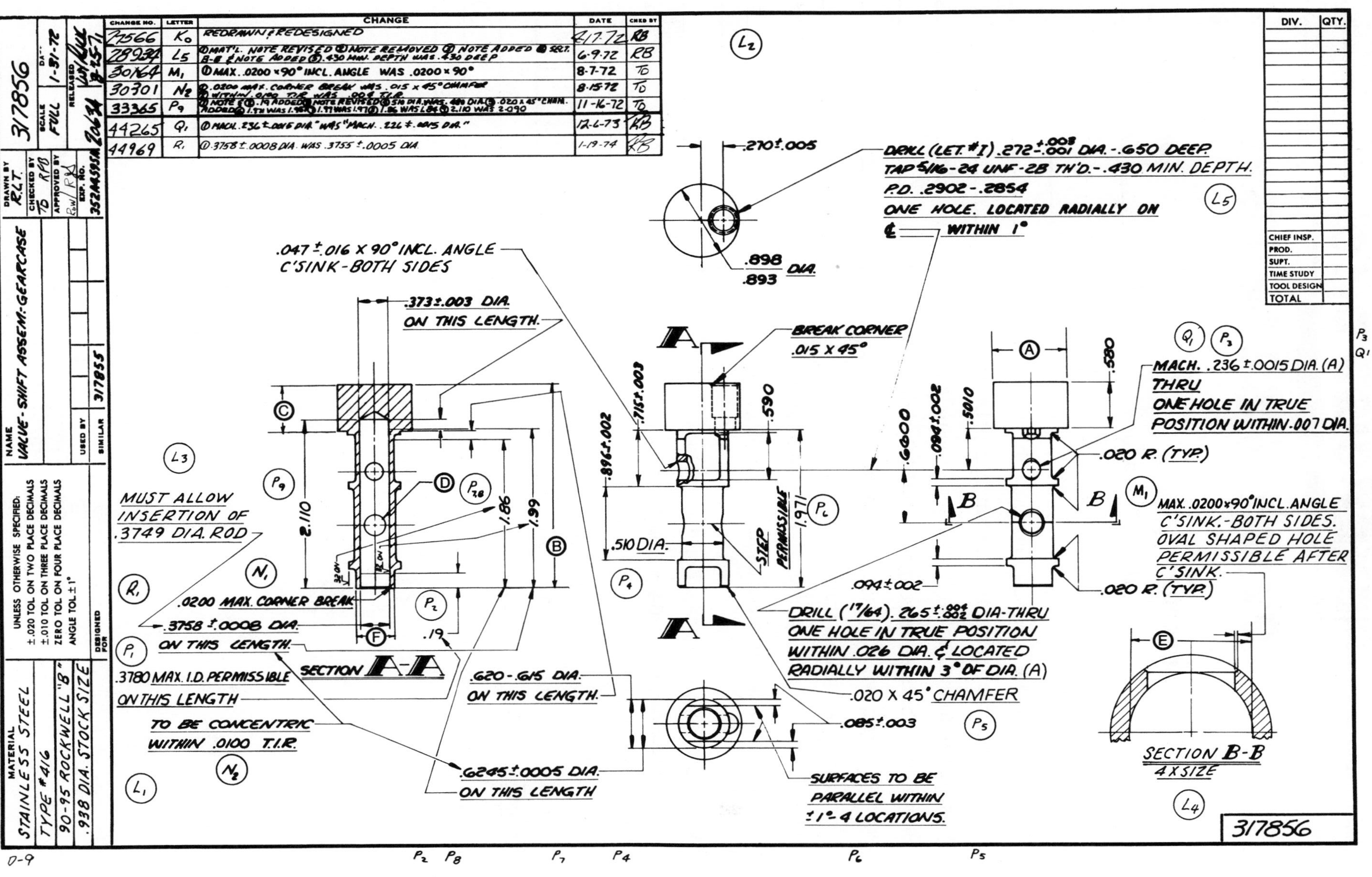

CHANGE NO. | LETTER | CHANGE | DATE | CHKD BY
27566 K0 REDRAWN & REDESIGNED 4-17-72 RB
30164 M1 MAX. .0200 x 90° INCL. ANGLE WAS .0200 x 90° 8-7-72 TO
33365 P9 11-16-72 TO
44265 Q1 MACH. .236 ±.0015 DIA. WAS "MACH. .226 ±.0015 DIA." 12-6-73
44969 R1 .3758 ±.0008 DIA. WAS .3755 ±.0005 DIA. 1-19-74
DRILL (LET. #I) .272 +.003 -.001 DIA. - .650 DEEP.
TAP 5/16-24 UNF-2B TH'D. - .430 MIN. DEPTH.
P.D. .2902 - .2854
ONE HOLE. LOCATED RADIALLY ON ℄ WITHIN 1°
.270 ±.005
.898 .893 DIA.
.047 ±.016 X 90° INCL. ANGLE C'SINK-BOTH SIDES
.373 ±.003 DIA. ON THIS LENGTH.
BREAK CORNER .015 X 45°
MACH. .236 ±.0015 DIA. (A) THRU
ONE HOLE IN TRUE POSITION WITHIN .007 DIA.
.020 R. (TYP.)
MAX. .0200 x 90° INCL. ANGLE C'SINK.-BOTH SIDES.
OVAL SHAPED HOLE PERMISSIBLE AFTER C'SINK.
.020 R. (TYP.)
MUST ALLOW INSERTION OF .3749 DIA. ROD
.0200 MAX. CORNER BREAK
.3758 ±.0008 DIA. ON THIS LENGTH.
.3780 MAX. I.D. PERMISSIBLE ON THIS LENGTH
TO BE CONCENTRIC WITHIN .0100 T.I.R.
SECTION A-A
.620 - .615 DIA. ON THIS LENGTH.
.6245 ±.0005 DIA. ON THIS LENGTH
2.110
1.86
1.99
.19
.896 ±.002
.715 ±.003
.590
.510 DIA.
STEP PERMISSIBLE
1.971
.6600
.094 ±.002
.5010
.580
.094 ±.002
DRILL (17/64) .265 +.004 -.002 DIA-THRU
ONE HOLE IN TRUE POSITION WITHIN .026 DIA. ℄ LOCATED RADIALLY WITHIN 3° OF DIA. (A)
.020 X 45° CHAMFER
.085 ±.003
SURFACES TO BE PARALLEL WITHIN ±1° - 4 LOCATIONS.
SECTION B-B 4 X SIZE
DIV. QTY.
CHIEF INSP. PROD. SUPT. TIME STUDY TOOL DESIGN TOTAL
MATERIAL STAINLESS STEEL TYPE #416 90-95 ROCKWELL "B" .938 DIA. STOCK SIZE
UNLESS OTHERWISE SPECIFIED: ±.020 TOL ON TWO PLACE DECIMALS ±.010 TOL ON THREE PLACE DECIMALS ZERO TOL ON FOUR PLACE DECIMALS ANGLE TOL ±1°
NAME VALVE-SHIFT ASSEM.-GEARCASE
DRAWN BY R.L.T.
CHECKED BY TO RPB
SCALE FULL
DATE 1-31-72
SIMILAR 317855
317856

EXERCISE D-10

Print #324673

QUESTIONS

1. Give dimensions *A*, *B*, *C*, *D*, *E*, *F*, *G*, and *H*.
2. To what assembly does this part attach?
3. What is the start date for drawing this print?
4. What is the tolerance on four-place dimensions?
5. How many changes were made on this print?
6. For what unit was this part designed?
7. What scale is used on this print?
8. How are the .215, .254, and .425 holes made?
9. How many basic dimensions are shown? List them.
10. What datums are indicated on this print?
11. Give the read-out of: .254 ±.002Ø
| ⌖ | Z | Y Ⓜ | Ø .004 Ⓜ | .
12. Give the read-out of: 425 ±.002 Ø
| ⊥ | Z | Ø.004 Ⓜ | .

ANSWERS

1. *A*______ *B*______ *C*______
*D*______ *E*______ *F*______
*G*______ *H*______
2. __________
3. __________
4. __________
5. __________
6. __________
7. __________
8. __________
9. __________
10. __________
11. __________
12. __________

NAME__________

DATE__________

SCORE__________

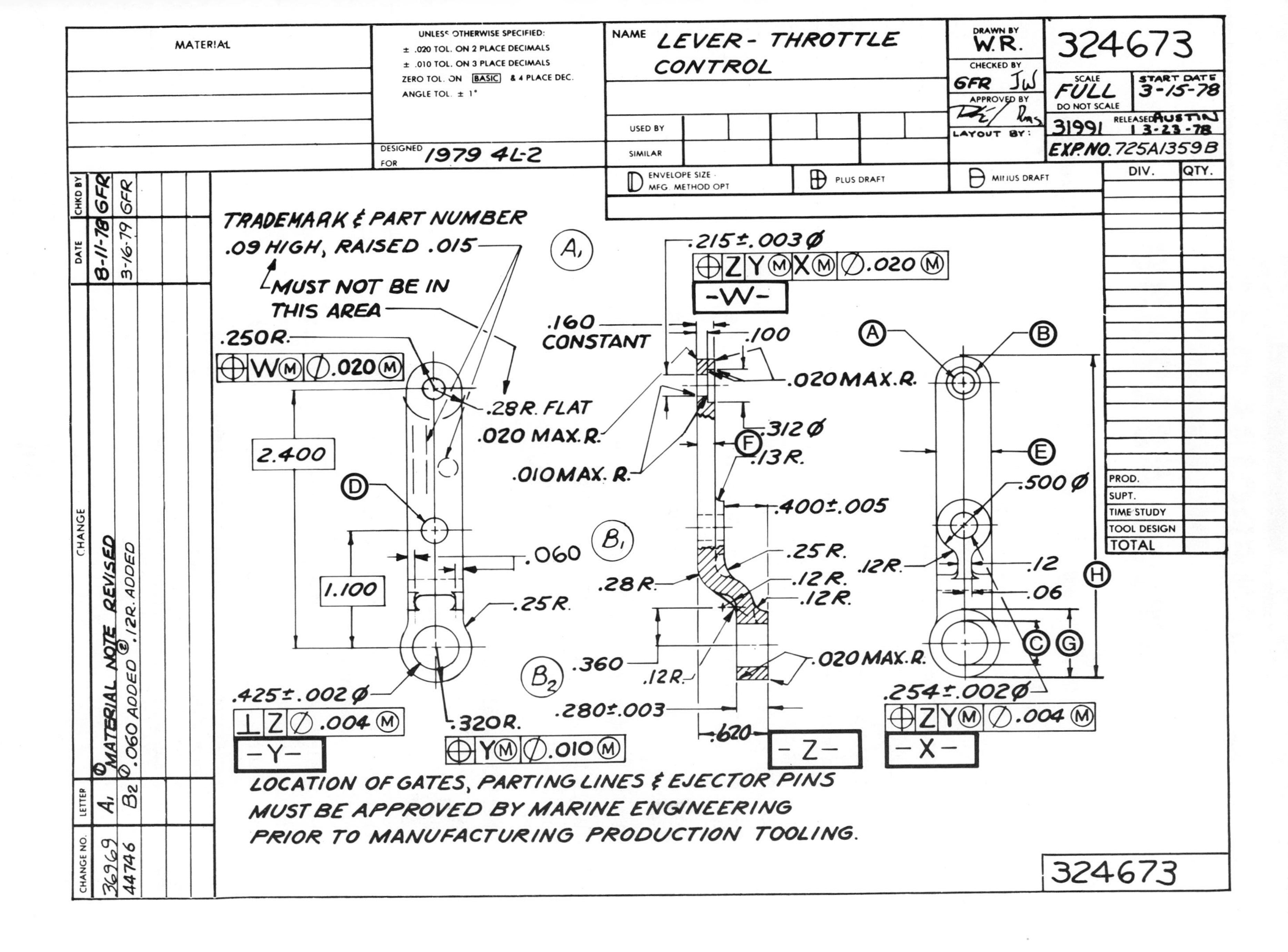

MATERIAL
UNLESS OTHERWISE SPECIFIED:
± .020 TOL. ON 2 PLACE DECIMALS
± .010 TOL. ON 3 PLACE DECIMALS
ZERO TOL. ON BASIC & 4 PLACE DEC.
ANGLE TOL. ± 1°
DESIGNED FOR 1979 4L-2
NAME LEVER - THROTTLE CONTROL
USED BY
SIMILAR
ENVELOPE SIZE - MFG. METHOD OPT
PLUS DRAFT
MINUS DRAFT
DRAWN BY W.R.
CHECKED BY GFR JW
APPROVED BY
LAYOUT BY:
324673
SCALE FULL
DO NOT SCALE
START DATE 3-15-78
31991
RELEASED AUSTIN 13-23-78
EXP. NO. 725A1359B
DIV.
QTY.
PROD.
SUPT.
TIME STUDY
TOOL DESIGN
TOTAL
TRADEMARK & PART NUMBER
.09 HIGH, RAISED .015
MUST NOT BE IN THIS AREA
A1
B1
B2
.250R.
W M .020 M
.28R. FLAT
.020 MAX. R.
.010 MAX. R.
2.400
D
1.100
.060
.25R.
.425±.002 Ø
Z .004 M
-Y-
.320R.
Y M .010 M
.215±.003 Ø
Z Y M X M .020 M
-W-
.160 CONSTANT
.100
.020 MAX. R.
.312 Ø
F
.13R.
.400±.005
.25R.
.28R.
.12R.
.12R.
.360
.12R.
.280±.003
.620
.020 MAX. R.
-Z-
A
B
E
.500 Ø
.12R.
.12
.06
H
C
G
.254±.002 Ø
Z Y M .004 M
-X-
LOCATION OF GATES, PARTING LINES & EJECTOR PINS
MUST BE APPROVED BY MARINE ENGINEERING
PRIOR TO MANUFACTURING PRODUCTION TOOLING.
CHANGE NO.
LETTER
CHANGE
DATE
CHKD BY
36969
A1
① MATERIAL NOTE REVISED
8-11-78
GFR
44746
B2
② .060 ADDED ② .12R. ADDED
3-16-79
GFR
324673

EXERCISE D-11

Print #317839

QUESTIONS	ANSWERS
1. What is this part?	1. ______
2. Of what material is this part made?	2. ______
3. To what assembly does this part attach?	3. ______
4. How many section views are shown?	4. ______
5. What kind of section view is C-C?	5. ______
6. What scales are used?	6. ______
7. Is this the original drawing? Explain.	7. ______ ______
8. How many changes were made on this print?	8. ______
9. What view is the one that locates the B-B section?	9. ______
10. What are three *X* points on the front view used for?	10. ______
11. What is the size of all small fillets and radii?	11. ______
12. What is the depth of the .735 ±.001 DIA. hole?	12. ______
13. How are the four holes around the outer edge formed?	13. ______
14. What is the diameter of the small hole at the center of the .735 ±.001 DIA. hole?	14. ______
15. How is this hole (answer to question #14) produced?	15. ______
16. What are the *E* letters and numbers along the right vertical border line used for?	16. ______
17. What is the tolerance for the gasket face surfaces?	17. ______
18. What is the typical draft used for this part?	18. ______
19. What is required for the unmachined surfaces around the four .296 ±.003 DIA. holes?	19. ______
20. What is the true position tolerance of these same four holes?	20. ______

NAME ______

DATE ______

SCORE ______